Smart Civil and Structural Engineering

Woodhead Publishing Series in Civil and Structural Engineering

Smart Civil and Structural Engineering

Integrating Innovative Technologies for Civil and Structural Excellence

Shahryar Habibi
University of Ferrara, Ferrara, Italy
Free University of Bozen-Bolzano, Bolzano, Italy

ELSEVIER
WP
WOODHEAD PUBLISHING
An imprint of Elsevier

Woodhead Publishing is an imprint of Elsevier
50 Hampshire Street, 5th Floor, Cambridge, MA 02139, United States
125 London Wall, London EC2Y 5AS, United Kingdom

For accessibility purposes, images in electronic versions of this book are accompanied by alt text descriptions independently provided by Elsevier, generated with the aid of AI. For more information, see https://www.elsevier.com/about/accessibility or Radarweg 29, 1043 NX Amsterdam, Netherlands.

Books and Journals published by Elsevier comply with applicable product safety requirements. For any product safety concerns or queries, please contact our authorised representative, Elsevier B.V., at productsafety@elsevier.com.

Publisher's note: Elsevier takes a neutral position with respect to territorial disputes or jurisdictional claims in its published content, including in maps and institutional affiliations.

Notices

Knowledge and best practice in this field are constantly changing. As new research and experience broaden our understanding, changes in research methods, professional practices, or medical treatment may become necessary.

Practitioners and researchers must always rely on their own experience and knowledge in evaluating and using any information, methods, compounds, or experiments described herein. In using such information or methods they should be mindful of their own safety and the safety of others, including parties for whom they have a professional responsibility.

To the fullest extent of the law, neither the Publisher nor the authors, contributors, or editors, assume any liability for any injury and/or damage to persons or property as a matter of products liability, negligence or otherwise, or from any use or operation of any methods, products, instructions, or ideas contained in the material herein.

ISBN: 978-0-443-30223-7 (print)
ISBN: 978-0-443-30224-4 (online)

Publisher: Matthew Deans
Acquisitions Editor: Glyn Jones
Editorial Project Manager: Riya Das
Production Project Manager: Venkatnadhan Rajagopalan
Cover Designer: Arumugam Kothandan

Typeset by MPS Limited, Chennai, India

Dedication

In loving memory of my father, **Parviz**,
whose wisdom, perseverance, and belief in the value of education
continue to inspire me.

And to my mother, **Shamsi**,
for her enduring support, encouragement, and strength
throughout my life.

—**Shahryar Habibi**

Contents

Preface

Architecture has always reflected humanity's aspirations, technological capabilities, and relationship with the environment. Throughout history, buildings have evolved from simple shelters into increasingly sophisticated systems that embody cultural values, engineering achievements, and scientific progress. Today, however, the built environment faces unprecedented challenges. Climate change, rapid urbanization, resource scarcity, technological disruption, and changing societal expectations demand new approaches to how buildings are conceived, designed, constructed, and operated. In response to these challenges, architecture is undergoing a profound transformation—one that extends beyond aesthetics and functionality toward intelligence, adaptability, resilience, and sustainability.

This book was conceived from the recognition that the future of architecture lies in the convergence of design creativity, structural engineering, computational technologies, and smart systems. As advances in material science, sensing technologies, artificial intelligence, digital twins, and responsive building systems continue to accelerate, the traditional boundaries between architecture and engineering are becoming increasingly blurred. Buildings are no longer passive objects occupying space; they are evolving into dynamic, responsive environments capable of sensing, learning, adapting, and interacting with both their occupants and the surrounding environment.

The integration of smart structures into architectural design represents one of the most significant developments in contemporary practice. Smart materials, adaptive façades, kinetic systems, tensegrity structures, embedded sensors, and intelligent control systems are fundamentally redefining how buildings perform and respond throughout their lifecycle. These innovations offer opportunities to improve energy efficiency, enhance occupant comfort, increase resilience, optimize resource utilization, and create more meaningful interactions between people and the built environment.

The purpose of this book is to provide a comprehensive exploration of this rapidly evolving field. It seeks to bridge the disciplines of architecture, structural engineering, and emerging technologies by presenting both theoretical foundations and practical applications. The chapters examine the principles that underpin smart structures, their architectural and structural implications, and their growing role in shaping future built environments. Through discussions of adaptive systems, kinetic architecture, tensegrity structures, computational design, sustainability, artificial intelligence, and digital technologies, this book presents a holistic perspective on the opportunities and challenges associated with intelligent architecture.

A central theme throughout this book is the importance of interdisciplinary collaboration. The complexity of contemporary architectural projects increasingly requires architects, engineers, computational designers, material scientists, and technology specialists to work together within integrated frameworks. Innovation often emerges at the intersection of disciplines, where creative thinking and technical expertise converge to address complex design challenges. This collaborative approach is essential for realizing buildings that are not only structurally efficient and environmentally responsible but also adaptable to the changing needs of society.

Sustainability serves as another foundational pillar of this work. While sustainability has traditionally focused on reducing environmental impact, this book explores a broader vision in which buildings actively contribute to ecological regeneration, resource optimization, and long-term resilience. Smart structures enable continuous performance monitoring, adaptive environmental control, and intelligent resource management, positioning architecture as an active participant in addressing global environmental challenges.

This book also recognizes that technological advancement must be accompanied by critical reflection. As buildings become increasingly connected, data-driven, and automated, important questions arise regarding privacy, cybersecurity, ethics, governance, and human-centered design. The responsible development and deployment of smart technologies will play a decisive role in ensuring that future architectural innovations serve societal interests while preserving individual autonomy and trust.

This volume is intended for a broad audience, including architects, engineers, researchers, educators, students, and professionals involved in the design, construction, and operation of the built environment. It aims to provide both foundational knowledge and forward-looking insights that can support innovation, research, and professional practice. Whether readers are exploring smart structures for the first time or seeking to deepen their understanding of emerging technologies, this book offers a framework for understanding how intelligence, adaptability, and sustainability are reshaping architecture.

Ultimately, this book is an invitation to imagine a future in which architecture transcends static form and becomes an active, responsive, and intelligent component of human and ecological systems. It is a future where buildings learn, adapt, communicate, and evolve; where technology and design work in harmony; and where architecture contributes not only to human comfort and functionality but also to environmental stewardship and societal well-being.

It is my hope that the ideas, examples, and perspectives presented in these pages will inspire readers to explore new possibilities, challenge conventional assumptions, and contribute to the creation of a more resilient, sustainable, and intelligent built environment for future generations.

Shahryar Habibi

Introduction: the fusion of architecture and smart structures

1

1.1 Scope

The scope of this book includes a comprehensive journey into the realms of architectural technology, structural engineering, and the revolutionary world of smart buildings. It explores the intricate fusion of architectural esthetics and engineering ingenuity, where the boundaries of form and function blur, resulting in innovative, adaptable, and visually striking architectural solutions. This book explores the spectrum of architectural elements, from kinetic structures that gracefully shift and adapt to tensegrity systems that embody beauty in tension, revealing their applications, principles, and esthetics. The book serves as a bridge between architectural and engineering disciplines and addresses the need for interdisciplinary collaboration in creating sustainable and intelligent architectural environments. In addition, it looks to the future and offers insight into emerging trends and technologies that are poised to shape the architectural landscape in profound ways.

This introductory chapter addresses an exploration of the profound synergy between architecture and smart structures. The fusion of these two domains marks a pivotal moment in architectural history, one where form and function seamlessly intertwine to create intelligent, adaptable, and esthetically captivating environments. The scope of this book is vast, encompassing a comprehensive journey through architectural technology, structural engineering, and the world of smart structures. From the elegance of kinetic structures in motion to the equilibrium of tensegrity systems, the book navigates through a diverse range of architectural elements. It is this comprehensive scope that brings about its significance. The integration of smart structures is not merely a trend but a transformative force that addresses the pressing needs of sustainability, efficiency, and interdisciplinary collaboration in architectural design.

1.2 Significance

It resides in the transformative potential that smart structures bring to the architectural world. It represents a response to the evolving demands of our built environment, propelled by imperatives of sustainability, energy efficiency, and the pursuit of seamless integration between form and function. This book serves as a beacon guiding architects, engineers, students, and design enthusiasts toward a new era of architectural innovation. By delving deeply into the world of smart materials, sensors, and adaptive technologies, it empowers its readers with the knowledge and inspiration needed to push the boundaries of architectural creation. It emphasizes that as the architectural sphere increasingly acknowledges the importance of

Smart Civil and Structural Engineering. DOI: https://doi.org/10.1016/B978-0-443-30223-7.00011-5

environmentally responsible design, the integration of smart structures becomes a pivotal means of crafting sustainable and eco-conscious architectural solutions. Furthermore, it champions the power of collaboration between architects and engineers, nurturing a climate of creativity and cooperative problem-solving, which is at the heart of designing environments that not only respond to evolving conditions but also enhance the well-being of their inhabitants.

The significance of this book reverberates in its potential to inspire, educate, and redefine the way we envision, construct, and experience architectural spaces. In an age where architecture must harmonize with the planet and the evolving needs of society, it acts as a guide and an instigator of transformation, promising a more sustainable and intelligently designed architectural future.

1.3 Purpose and objectives

The purpose of this book is to shed light on the innovative and transformative potential that arises from the fusion of architecture and smart structures. As smart materials, sensors, and adaptive technologies find their place in architectural design, architects, engineers, students, and design enthusiasts stand to benefit from a wealth of knowledge, inspiration, and practical insights. Through this exploration, the book aims to empower readers to push the boundaries of architectural innovation, fostering the creation of sustainable, energy-efficient, and environmentally responsible architectural solutions. It recognizes the significance of collaboration between architects and engineers in shaping the future of architectural design and emphasizes the importance of responding to changing conditions and human needs.

The objectives of this book are multifold:

- to provide a comprehensive understanding of smart structures and their integration into architectural design;
- to explore the esthetics of smart structures and their influence on architectural esthetics;
- to highlight the practical applications and case studies that exemplify the beauty and adaptability of smart structures;
- to underscore the importance of interdisciplinary collaboration between architects and engineers, with real-world examples of successful projects;
- to offer insights into emerging trends and technologies that will shape the future of architectural design and its role in sustainability.

This book is an invitation to embark on a journey through the fusion of architecture and smart structures (a journey that promises to redefine the way we conceive, construct, and experience architectural spaces in a rapidly evolving world).

1.4 Overview of the book

This book is a comprehensive exploration of the evolving landscape of architectural design and engineering, where innovation, sustainability, and esthetics converge. This book delves into the fusion of architectural creativity and structural engineering ingenuity, highlighting the transformative role of smart structures in shaping the future of architecture. It offers a rich tapestry of insights, practical applications, and visionary trends that define this dynamic field.

1.4.1 Key features and themes

Integration of smart structures: The book explores the integration of smart materials, sensors, and adaptive technologies into architectural design. It unveils the potential for structures to respond intelligently to changing conditions, energy needs, and user experiences.

- Esthetic potential: Beyond functionality, the book emphasizes the esthetic potential of smart structures. It showcases how these innovations influence architectural esthetics and presents case studies of visually striking projects that inspire architects, engineers, and designers.
- Diverse architectural elements: From kinetic structures that gracefully shift and adapt to tensegrity systems that exemplify the artistry of tension and compression, the book dissects a spectrum of architectural elements. Each element is examined in terms of its applications, principles, and esthetics.
- Collaboration and interdisciplinarity: Recognizing that the most innovative projects often emerge from collaborative efforts, the book highlights the importance of architects and engineers working together. Real-world examples of successful interdisciplinary collaborations serve as lessons for effective cooperation.
- Sustainability and eco-friendly design: With a growing focus on environmentally responsible design, the book investigates the role of smart structures in creating sustainable and eco-friendly architectural solutions. It underlines how these structures contribute to energy efficiency and reduced environmental impact.
- Future trends: As technology continues to evolve, the book offers insights into emerging trends and technologies that will shape the future of architectural design and structural engineering. It provides a glimpse into the architectural landscape of tomorrow.

This book is designed to appeal to a diverse audience, including architects, structural engineers, students, design enthusiasts, and professionals in the built environment. Its multidisciplinary approach, combined with real-world case studies and a forward-looking perspective, makes it an invaluable resource for those seeking to push the boundaries of architectural innovation while maintaining a strong focus on sustainability and collaboration. This book promises to inspire, educate, and redefine the way we envision and construct architectural spaces, ushering in a future where esthetics and intelligence are in perfect harmony.

1.5 Introduction

Smart structures are the product of the dynamic synthesis of cutting-edge technologies with conventional design concepts in the realm of modern architecture. Through this merging, a new era of buildings is ushered in, one in which they function not just as physical structures but also as sentient creatures that can adjust to their environment and occupants. Architecture, the art and science of building structures, has seen significant transformations over the millennia. Simultaneously, the emergence of intelligent technologies has transformed multiple facets of human existence, encompassing the constructed surroundings. The merging of architecture and smart structures signifies the meeting point of these two fields and a change in perspective in how we envision, build, and engage with our environment.

The development of architecture dates back to ancient civilizations, where basic buildings served utilitarian purposes. Architectural styles evolved over time to reflect cultural, social, and technological progress. Similarly, the development of smart technologies, from simple mechanization to complex digital systems, has affected various industries, including architecture. Smart structures in architecture are characterized by the ability to incorporate intelligent systems that increase functionality, efficiency, and sustainability. Key principles guiding the integration of smart technologies include sustainability, efficiency, and adaptability.

Smart structures are designed to be flexible, able to adjust to shifting demands from users, the environment, and technological advancements. The integration of smart structures into modern architecture has led to innovative applications, including smart façades and smart buildings. Despite numerous advantages, the integration of smart structures with architecture presents several challenges, including complexity in integrating disparate systems, interoperability issues, and cybersecurity issues, which present technical challenges for implementing intelligent architecture. Balancing the functional requirements of smart systems with architectural esthetics also presents design challenges that require seamless integration and visual harmony.

Drawing inspiration from nature, biomorphic design principles inform the development of organic, adaptive structures that mimic natural systems. The future of architecture and smart structures holds immense potential, with anticipated developments such as the integration of artificial intelligence (AI) and machine learning, advancements in sustainable technologies, and the expansion of smart city concepts.

The fusion of architecture and smart structures represents a transformative shift in the built environment, offering unprecedented opportunities for sustainability, efficiency, and innovation. While challenges persist, ongoing advancements and collaborative efforts promise a future where buildings not only serve as shelter but also as intelligent, responsive entities, enriching the lives of occupants and communities.

1.5.1 Defining smart structures

In the realm of engineering and materials science, smart structures represent a transformative approach to design and construction, imbuing traditional structures with the ability to sense, respond, and adapt to external stimuli autonomously. Smart structures, also known as adaptive or intelligent structures, incorporate advanced materials and technologies to augment their functionality beyond conventional static systems. These structures can sense variations in their surroundings, including pressure, temperature, and mechanical stress, and dynamically modify their characteristics or actions in response.

At the heart of smart structures lie sensing elements that enable them to gather data about their surroundings. Real-time feedback on temperature, strain, vibration, and

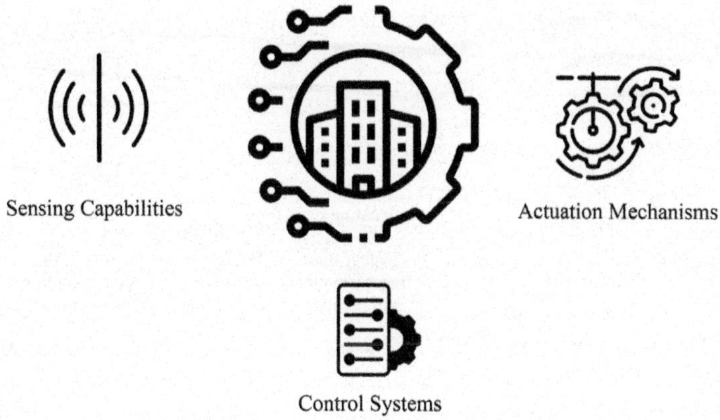

Figure 1.1 Schematic illustration of characteristics of smart structures.

electromagnetic fields is provided by sensors to the structure's control system. In addition to sensing, smart structures are equipped with actuation mechanisms that facilitate responsive actions based on incoming data. Actuators convert electrical, hydraulic, or pneumatic signals into mechanical motion, allowing the structure to adapt its shape, stiffness, or damping properties as needed. Furthermore, control systems serve as the brains of smart structures, processing sensor inputs and issuing commands to actuators to regulate their behavior (Fig. 1.1). These systems may employ algorithms, feedback loops, or AI to optimize performance, enhance stability, and ensure reliability under varying operating conditions.

Active smart structures are dynamic systems that adjust their properties in response to changing external conditions. This adaptability is achieved through mechanisms that allow them to actively control their structural performance, such as adaptive lattice structures (Shen et al., 2024), shape-memory alloys (SMAs) (Liu et al., 2024), and piezoelectric actuators (Uchino, 2017). For example, adaptive trusses are designed to adapt their shape or stiffness and respond to loads to maintain stability and improve durability. When equipped with shape-memory alloys, these structures can be subject to deformation at certain temperatures. SMAs have a unique ability to "remember" their original shape; when heated above their transformation temperature, they return to this form, thereby manipulating the structural configuration. This change is reversible: when the SMA cools below this temperature, it can deform again, allowing repeated adaptive behavior based on thermal conditions. Piezoelectric actuators work on a different principle; they use piezoelectric ceramic materials. These materials exhibit the piezoelectric effect, generating electrical energy in response to applied mechanical forces such as pressure or impact. Conversely, in the inverse piezoelectric effect, these materials produce a mechanical displacement when exposed to electrical energy, making them versatile in applications requiring precise adjustments (Fig. 1.2). In an active grating design, piezoelectric actuators can respond to acoustic waves, voltages, or impact forces and generate specific mechanical outputs, such as displacements, force adjustments, or vibration damping, depending

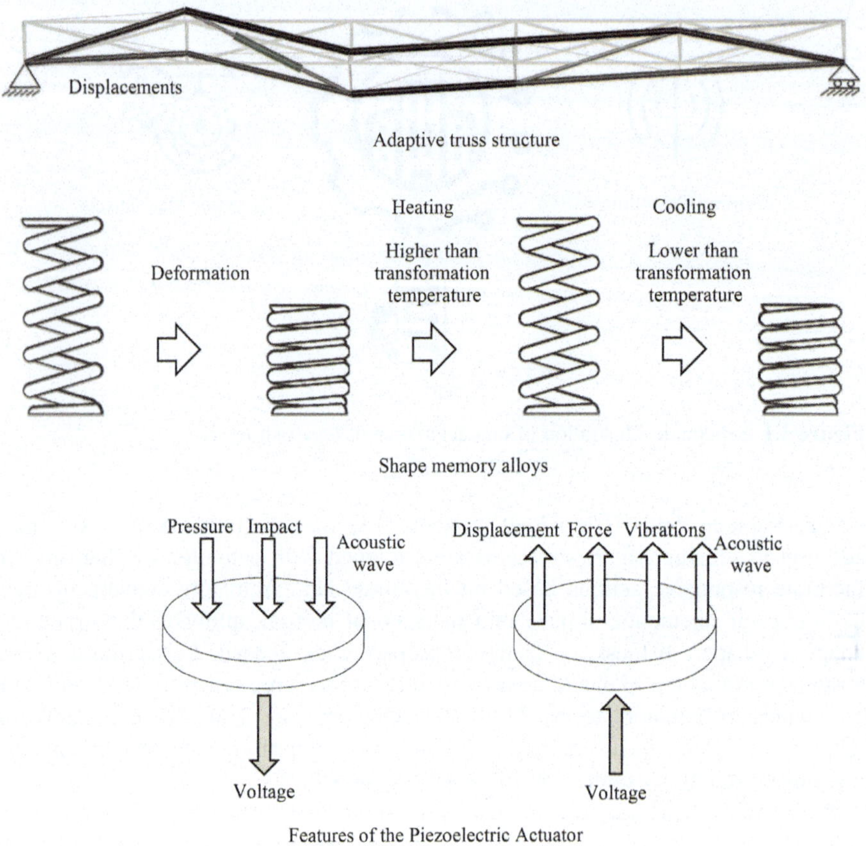

Figure 1.2 Schematic illustration of adaptive truss structures, shape-memory alloys (SMAs), and piezoelectric actuators.

on the input energy. This dual-functional ability allows the structure to effectively manage stressors, optimize its performance, and increase resilience.

Passive smart structures use the intrinsic properties of materials or geometric configurations to adaptively respond to environmental changes and require no external power or active control. Notable examples of these systems are shape-memory polymers (Chen and Andrawes, 2017), self-healing composites (Sanaei Ataabadi et al., 2024), and morphing profiles (Yokota and Barthelat, 2024). For example, shape-memory polymers can return to their original shape after deformation, activated by changes in temperature. In Fig. 1.3, the polymer shown initially holds a specific "original shape" at a low temperature. It assumes a "deformed shape" when bent at low temperatures. After heating, the polymer gradually returns to its original shape once it reaches a high temperature, showing its temperature-responsive properties. Finally, cooling stabilizes the material in its restored form. This cycle illustrates how shape-memory polymers adapt through passive temperature-induced

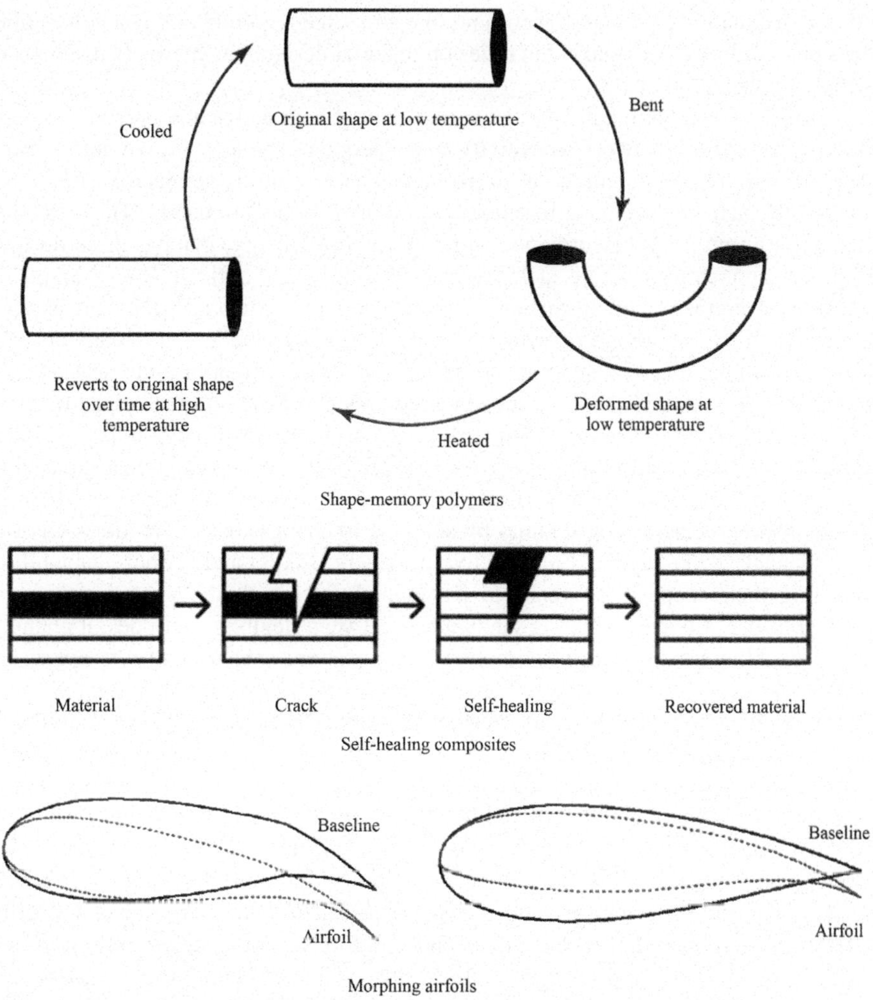

Figure 1.3 Schematic illustration of passive smart structures, showcasing shape-memory polymers, self-healing composites, and morphing airfoils.

changes without requiring additional energy. Self-healing composites, another example, exhibit the ability to autonomously repair internal damage. The figure shows the process where the material initially cracks due to mechanical stress or fatigue. The inherent self-healing ability is activated, leading to a recovery phase where the crack is filled or bridged over time, restoring the integrity of the material. This feature significantly increases the durability and service life of structures by mitigating long-term wear and tear. Morphing airfoils demonstrate passive adaptability in aerodynamic applications. The figure shows the airfoil in two configurations: the "base" shape and a modified shape achieved by rotating or bending part of the structure. This shift changes the airfoil profile, which can optimize lift, drag, or other aerodynamic parameters according to operational requirements. Such morphing does not require active control but relies on material design that passively responds to airflow or structural stress.

Semiactive smart structures integrate elements of both active and passive systems, creating a balance between sensitivity and energy efficiency. Unlike fully active systems that require continuous performance and control, semiactive structures selectively adapt in response to external stimuli or user commands. This selective adaptability optimizes performance while conserving energy, making these designs ideal for applications where stability and flexibility are critical. One example is semiactive dampers, which change their damping properties according to applied forces, such as vibrations or seismic shocks. In these applications, dampers can appear as simple diagonal supports between floors or roofs that provide resistance to movement or as more complex, stepped structures that allow gradual adjustments in stiffness and energy absorption. Similarly, magnetorheological (MR) fluids exhibit a shift from a free-flowing liquid state to a structured, semisolid state when exposed to a magnetic field. This property allows MR fluids to serve as a variable medium in shock absorbers, increasing stability by adapting to changes in impact forces without a permanent power supply. Electroactive polymers (EAPs) add further adaptability, changing their shape in response to an applied electric field. When a voltage is applied to the polymer, which is sandwiched between flexible electrodes, it changes shape, enabling precise control of motion in applications such as actuators or sensors (Fig. 1.4).

Semiactive smart structures are particularly important in the context of civil and structural engineering, as they represent a versatile solution for achieving adaptable performance without the high energy requirements of fully active systems. These structures can fine-tune their response based on real-time data inputs, enabling adaptive adaptation to changing environmental conditions or operational needs. By strategically adjusting parameters such as stiffness and damping, semiactive systems can improve stability and resistance to seismic and wind loads and offer a dynamic response that meets safety and comfort standards (Gutierrez Soto and Adeli, 2019). This approach underlines the value of semi-active systems in infrastructure that faces diverse and unpredictable stressors, such as bridges or high-rise buildings located in seismic zones. Key to the effectiveness of semiactive smart structures is their reliance on advanced materials and actuators. For example, magnetorheological fluids exhibit controllable rheological properties in the presence of a magnetic field, allowing structures to rapidly change their damping characteristics without the need for significant energy input. Such materials are widely used in bridge and building foundations to dampen vibrations and improve occupant comfort (Sternberg et al., 2014). Similarly, EAPs offer a promising solution for adaptive façade systems that respond to environmental changes such as temperature fluctuations and solar exposure, thereby contributing to the energy efficiency of buildings (Bar-Cohen, 2004). These innovations illustrate the potential of semiactive materials and actuators to improve structural adaptability while being consistent with sustainability goals.

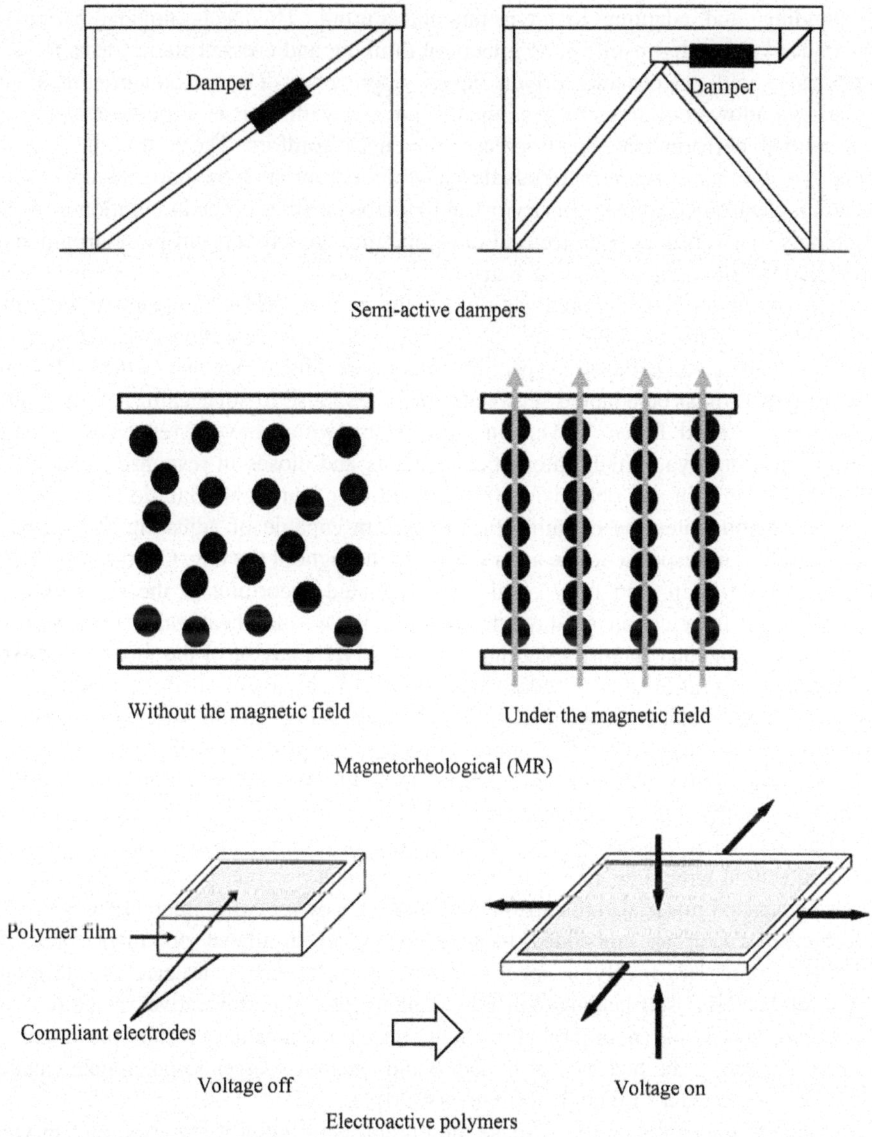

Figure 1.4 Schematic illustration of semiactive smart structures, including semiactive dampers, magnetorheological fluids, and electroactive polymers.

1.5.2 The impact of smart structures on architectural design

Recent progress in material and structural engineering and digital technology has dramatically transformed the field of architecture. Among the strictest developments is the integration of smart structures, materials, and systems capable of sensing, responding, and adapting to environmental stimuli. This technological evolution introduced a paradigm shift in architectural thinking and crossed static forms toward responsive and performance-driven design approaches. The introduction of smart structures allows buildings to dynamically interact with their environment, optimize operational performance, and increase passenger comfort. These structures show adaptive abilities made of materials that have properties such as shape memory, self-healing, electromechanical sensitivity, and variable stiffness. The integration of these materials transforms an architectural envelope into an active agent in environmental modulation rather than a passive barrier.

One of the convincing applications is the use of SMA in adaptive building envelopes and structural elements. SMA can undergo deformation and restore their original shape when exposed to specific temperature sills, which makes them ideal for elements that must be adapted seasonally or in response to solar gain. A remarkable example is the Al Bahar Towers in Abu Dhabi, which have a responsive façade system, Mashrabiya, which automatically opens and closes in response to sunlight, reducing solar gain and optimizing the penetration of daylight (Babilio et al., 2019).

Fig. 1.5 illustrates the dynamic façade system capable of adjusting the aperture configuration in response to the solar exposure throughout the day. Conditions—fully enclosed, half-open, and fully open—are regulated according to the solar path to optimize the intersection of daylight and minimize solar heat gain. The gradient scheme reflects solar intensity and leads to adaptive behavior of the façade elements over time, while the pattern model shows geometric logic and modular repetition of panels. Such systems are an example of how smart architecture can integrate environmental sensitivity with computing design and promote energy efficiency and esthetic articulation. Responsive façade systems like this allow for climate-adaptive skins, improving occupant comfort and building performance.

In another emerging domain, self-healing composites are increasingly being applied to architectural structures, particularly in concrete and polymer-based façade systems. These advanced materials are capable of autonomously repairing microcracks through mechanisms such as embedded healing agents, encapsulated polymers, or even biological processes involving specific strains of bacteria. For instance, the development of bacterial self-healing concrete demonstrates the innovative integration of material science and microbiology, where bacteria embedded within the concrete matrix activate in the presence of moisture and calcium sources to precipitate calcite, thereby sealing cracks (Wiktor and Jonkers, 2011).

Such systems significantly increase the durability, structural integrity, and service life of building components and, at the same time, reduce long-term maintenance and resource consumption costs. In architectural design, it translates into façades and elements carrying a burden that can deal with damage caused by environmental stressors such as thermal cycles, freezing and melting cycles, or seismic movements. The consequence is particularly critical in extreme climatic zones or in areas susceptible to earthquakes, where the integrity of traditional materials is likely to be endangered. By incorporating self-healing technologies, architects and engineers can propose durable structures that respond to damage in a passive but efficient way and correspond to the principles of sustainable and regenerative design (Roos, 2021).

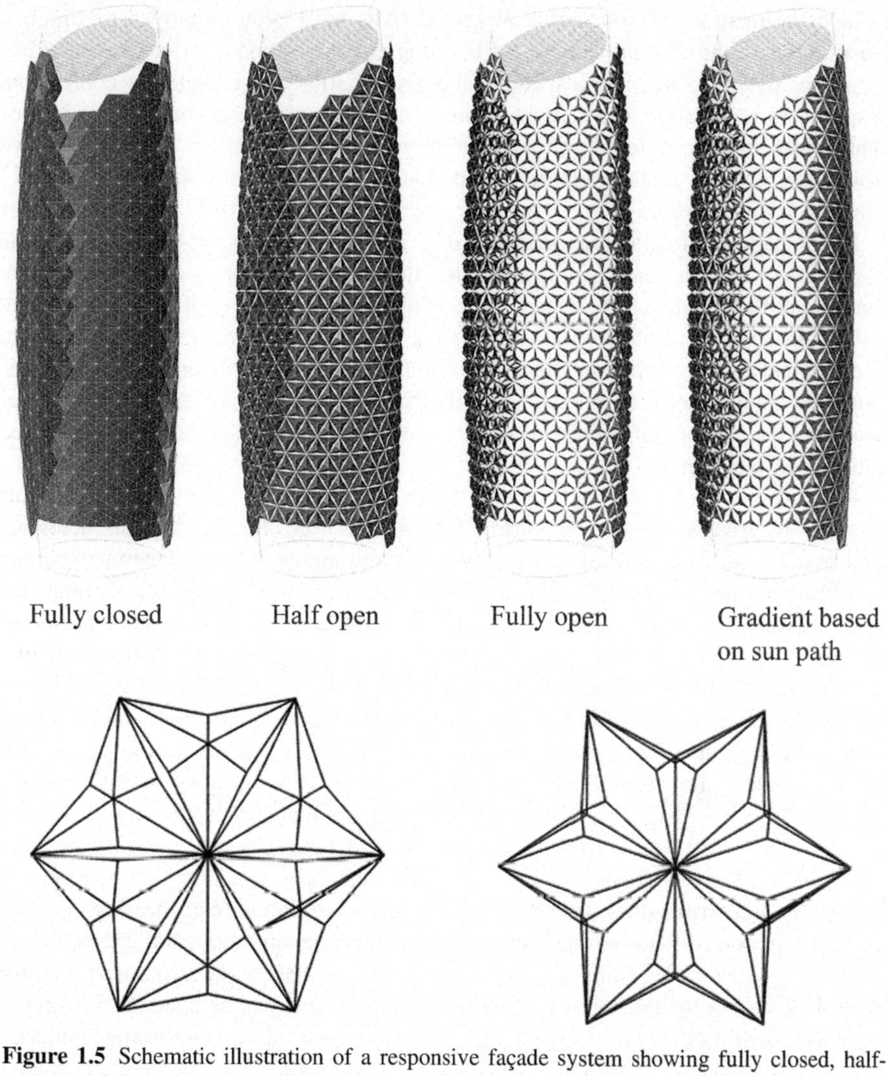

Figure 1.5 Schematic illustration of a responsive façade system showing fully closed, half-open, and fully open states, including a sun-path-based gradient diagram and associated pattern model.

EAPs represent a compelling class of smart materials that offer high levels of adaptability in architectural applications. These materials can change their shape, size, or stiffness when subjected to an electrical stimulus, making them ideal for systems requiring controlled, reversible motion without the complexity of traditional mechanical components (Natamai and Subramanian, 2021). Their lightweight, flexibility, and silent actuation make EAPs especially suitable for kinetic façades, responsive partitions, and adaptive shading systems.

In experimental and conceptual projects, EAPs have been employed to simulate organic, breathing-like movements in building envelopes and interior elements. For example, dynamic membranes embedded with EAPs can expand or contract in response to occupancy levels, light intensity, or thermal variations, creating spaces that gently morph in form and permeability. These applications suggest a move toward responsive, sentient environments that blur the line between static construction and living systems.

By embedding interactive behavior into material systems, EAP-based designs foster new spatial narratives where the built environment becomes an active participant in the human experience. This enables a deeper level of user engagement and environmental responsiveness, advancing architecture from a passive shelter into a dialogical and evolving medium. As computational control and energy systems improve, the architectural integration of EAPs may become more feasible on a larger scale, merging material intelligence with sustainable design principles and digital fabrication techniques.

Researchers are exploring the innovative use of EAPs as dynamic materials capable of transforming their form without the need for mechanical actuators (Vazquez et al., 2019). These ultra-lightweight and highly flexible materials respond to electrical stimulation, typically around 4000 to 5000 volts, enabling controlled deformations such as bending, stretching, or morphing. The system is composed of multiple layers arranged in a precise sequence: a silicone insulation layer, a conductive powder layer, five prestretched acrylic polymer films at the core, followed by another conductive powder layer, a second silicone insulation layer, and an acrylic frame that maintains the structural tension of the entire assembly (Fig. 1.6). This layered composition allows the material to function like an artificial muscle, capable of shifting its geometry in real-time through programmable voltage inputs.

One of the most profound impacts of smart structures lies in the development of adaptive building envelopes. Traditional façades operate as static barriers between the interior and exterior, often requiring supplementary systems (e.g., heating, ventilation, and air conditioning or shading devices) to regulate environmental conditions. In contrast, smart façades embedded with sensors and actuators can dynamically adjust in real time, modulating light, airflow, thermal performance, or opacity in response to external stimuli such as solar radiation, wind speed, or occupancy patterns.

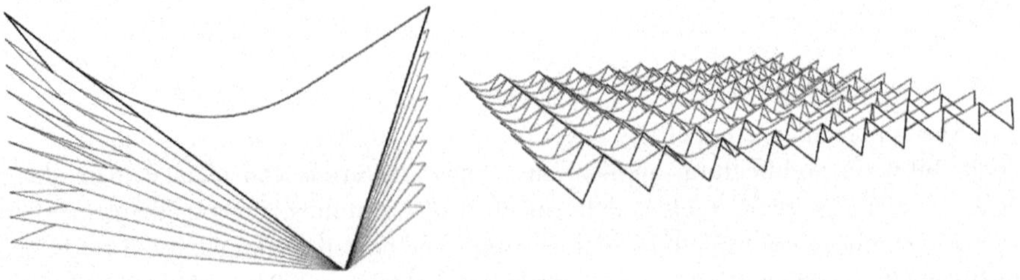

Figure 1.6 A schematic overview of the shapeshift system, a cutting-edge exploration into the application of EAPs. *EAPs*, electroactive polymers.

Figure 1.7 Views of the responsive façade demonstrator developed by ITKE/ITFT, University of Stuttgart. *ITKE*, Institute of Building Structures and Structural Design.

For instance, buildings such as those developed by the Institute of Building Structures and Structural Design (ITKE) at the University of Stuttgart have pioneered the integration of kinetic elements into façade systems that shift in response to solar position, reducing heat gain while maximizing natural daylight. A notable example is ITKE's first large-scale outdoor demonstrator of responsive fiber-reinforced façade elements, which explores the potential of material-driven responsiveness. Rather than relying on motors or external actuators, these façade components are engineered using glass and carbon fiber-reinforced polymers that deform predictably under environmental stimuli, such as changes in temperature, humidity, or solar radiation. Inspired by natural hygroscopic movements, the system responds passively but dynamically, adjusting the façade's openness to control solar exposure and indoor comfort (Fig. 1.7).

This project represents a synergy between technology, material science, esthetics, and environmental stewardship, offering a sustainable alternative to energy-intensive active systems. By embedding intelligence within the material itself, ITKE's approach exemplifies a new frontier in bio-inspired kinetic architecture, where form, function, and responsiveness are seamlessly integrated. Such innovations not only reduce reliance on conventional energy systems but also propose new design paradigms for future-oriented architecture that is adaptive, low-maintenance, and deeply context-responsive (Born et al., 2024).

In the realm of interior architecture and spatial flexibility, smart structures enable the design of transformable spaces (Spitz et al., 2018). Using motion-enabled components, walls, floors, or ceilings can shift position or reconfigure based on programmatic needs. For example, operable partitions equipped with actuators and sensors can open or close to accommodate changing occupancy levels, improving space utilization in offices, cultural centers, or educational facilities. This dynamic reprogramming supports multifunctionality, user agency, and temporal adaptation, aligning with contemporary needs for hybrid and multiuse environments.

Moreover, smart structures introduce a new language of form and interaction in architecture. The integration of movement, responsiveness, and data-driven behavior challenges the conventional notion of buildings as static objects. Instead, it fosters a design culture where the building envelope or structure itself becomes performative, an active interface between human and environment. This shift is closely aligned with concepts of biomimicry (Shashikumar et al., 2024), cyber-physical systems (Li et al., 2023), and interactive architecture (Fox and Kemp, 2016), inspiring a generation of designers to explore kinetic morphologies, sensorial feedback loops, and intelligent control systems that create richer experiential and ecological narratives (Khaddour et al., 2024).

Additionally, the deployment of smart structures enhances sustainability goals by facilitating energy conservation, resource optimization, and building lifecycle management. Responsive systems can reduce reliance on mechanical ventilation, artificial lighting, or static thermal insulation by adapting to real-time environmental changes. As such, smart design strategies align closely with passive design principles but take them further through data integration, machine learning, and predictive modeling. Projects that incorporate building management systems with embedded smart components illustrate how operational energy demand can be minimized through intelligent coordination between architecture and systems engineering.

From a cultural and conceptual perspective, smart structures invite new architectural expressions that resonate with contemporary digital culture, emphasizing interactivity, fluidity, and behavioral performance. In this context, smart structures do not only serve functional purposes; they become media, interfaces, and narrative devices, facilitating a deeper connection between inhabitants, space, and technology.

The integration of smart structures into architectural design represents a shift from static form-making to adaptive systems thinking. It encourages the synthesis of materials science, digital fabrication, embedded intelligence, and user-centric design, offering new possibilities for responsive, efficient, and meaningful built environments. As technologies mature and become more accessible, smart structures are poised to become foundational elements in the next generation of sustainable, intelligent architecture.

From a design perspective, the implementation of smart structures necessitates interdisciplinary collaboration. Architects must work closely with engineers, material scientists, and digital technologists to fully exploit the potential of responsive systems. This collaboration facilitates not only technological feasibility but also enriches the design process, embedding intelligence at every scale—from material behavior to urban interaction. The growing emphasis on sustainability and regenerative design further reinforces the relevance of smart structures. By enabling buildings to actively manage energy, adapt to environmental conditions, and extend their operational life cycles, these systems support a more resilient and resource-conscious built environment. Smart structures challenge conventional dichotomies between form and function, passive and active, or static and kinetic, promoting a more integrated architectural language.

Smart structures are reshaping the conceptual and practical boundaries of architecture. They empower buildings to become dynamic ecosystems, responsive not only to environmental forces but also to human activity and social context. As these technologies mature, they will increasingly define the future of architectural innovation, where performance, adaptability, and user engagement converge to create intelligent, responsive, and sustainable environments.

A compelling example of convergence between smart structural systems and architectural innovation is the Experimenta Science Center in Heilbronn, Germany. This feature gives examples of how a building can be more than a stable shell—it becomes a responsible, protesting, and experienced environment. Architecturally conceived as a spatial spiral, the internal layout is shaped by a helical circulation path that integrates functional zones such as laboratories, exhibitions, and studios. After passing through the foyer, visitors enter the thematic worlds arranged along a central spiral of interconnected rooms and learning spaces that rise floor by floor (Fig. 1.8). This spatial fiction guides the user engagement, aligning with the educational mission of the center and embedding social interactions in the very fabric of architectural design.

Structurally, the building showcases how smart design does not always require movement or automation; instead, it can lie in the precision and logic of the structural system. The interplay of form, load paths, and material articulation forms a quasiintelligent skeleton capable of responding to both static forces and spatial needs. One of the most critical aspects of the load-bearing system is the vertical force transfer mechanism. The majority of the vertical load is channeled through strategically placed columns at key intersections between floor plates, locations optimized for multistory load continuity without lateral deviation. This results in a structurally efficient and architecturally coherent framework where the flow of forces is aligned with the geometry of space.

This integrated intellect is more reflected in the overall steel structure of the building. An example of a principal load-bearing element is a V-truss system, where diagonal members are attached to truss cords, which are attached to truss cords to move bolted joints (Fig. 1.9). In addition to guaranteeing structural rigor, these characteristics enable effective prefabrication and assembly. Meanwhile, the chord-to-column connections employ nonpile bolt plate pairs, which enable controlled

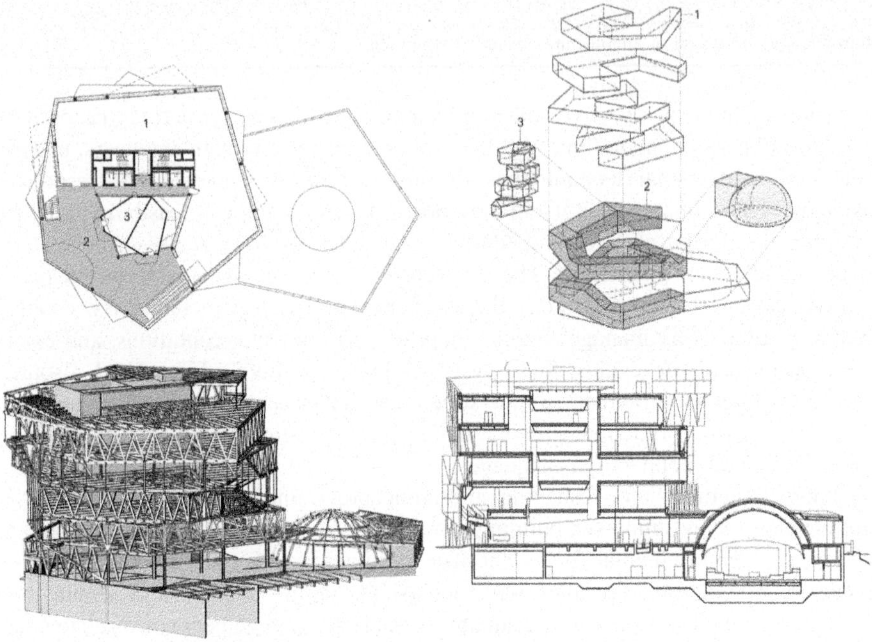

Figure 1.8 Spatial organization and structural logic of the Experimenta Science Center.

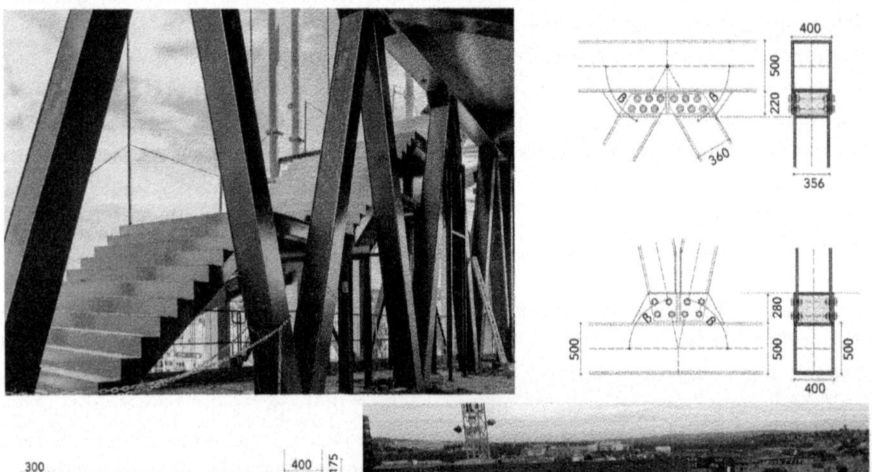

Figure 1.9 V-truss connection details of the Experimenta Science Center.

articulation and modular assembly. The expansion segments of these connections show a fine understanding of load transfer and material behavior, such as the way composite beams are connected to truss chords—a system that reduces torsional stress and improves fabrication scalability.

The structure is not thus “smart” because it morphs or reacts to the sensor but rather because its form, logic, and material integration respond to environmental loads, user movement, and spatial performance. This example shows how architectural and structural intelligence can converge with the environment to create conscious, socially attractive, and technically refined buildings.

1.5.3 Why architects and engineers need to collaborate

As the boundaries between architecture and structural engineering continue to blur in the age of intelligent and adaptive systems, the need for close, sustained collaboration between architects and engineers has never been more critical (Kronenburg, 2007). In traditional practice, architectural vision and structural logic were often treated as sequential or siloed efforts, design first, structure later. But in the context of smart structures, this division becomes not only inefficient but fundamentally counterproductive.

Smart buildings are not merely constructed; they are engineered ecosystems, integrating sensors, adaptive materials, mechanical intelligence, and data-driven operations within a carefully orchestrated spatial and esthetic framework. These systems demand a concurrent design process in which form and function coevolve from the conceptual phase onward. In this context, collaboration is not optional; it is essential to bridge performance-driven design and human-centered spatial experience.

Collaboration begins with a shared design language. Architects are trained to prioritize spatial quality, human experience, and cultural context, while engineers bring deep knowledge of structural mechanics, materials, and environmental performance. To align these skill sets, each discipline must learn to translate its priorities and constraints into a shared design vocabulary. For example, an architect’s desire for a column-free atrium must be evaluated alongside an engineer’s assessment of span limits, material behavior, and load path optimization.

This collaborative negotiation can yield more than compromise; it can drive innovation. In projects like the Experimenta Science Center, structural strategies such as optimized vertical load paths and V-truss systems were developed not in isolation, but in dialog with spatial flow, daylight access, and visitor circulation. These solutions were not imposed by one discipline on another; they emerged through mutual refinement.

The integration of digital design environments, such as parametric modeling and building information modeling platforms, further reinforces the need for cross-disciplinary workflows. Tools like Grasshopper, Karamba3D, and Revit allow for real-time iteration between geometry and performance, enabling both architect and engineer to evaluate structural behavior, material usage, and spatial performance in unison (Menges and Ahlquist, 2011). These tools encourage early-stage collaboration, where structural strategies can inform spatial configuration before design decisions become fixed.

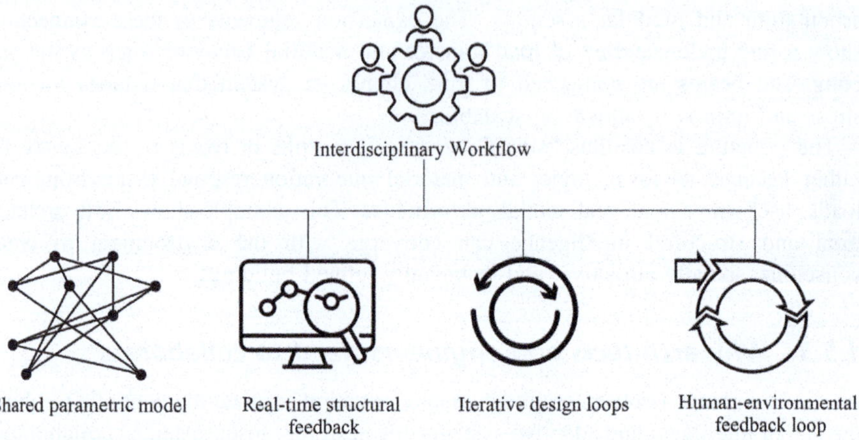

Figure 1.10 Integrated design workflow between architecture and engineering using parametric tools and BIM platforms. *BIM*, building information modeling.

Such platforms act not just as technical utilities but as collaborative media, where both disciplines contribute to a shared parametric model, a dynamic environment that integrates architectural form, structural feedback, and material parameters simultaneously. As illustrated in Fig. 1.10, this workflow diagram highlights four key components of the interdisciplinary design process:

- Shared parametric model: A common digital framework where geometric and performance parameters are embedded into the architectural design from the outset, enabling seamless data exchange between disciplines.
- Real-time structural feedback: Engineering constraints, such as stress distribution, deformation, and load paths, are analyzed and visualized instantly, allowing architects to understand structural implications during form generation.
- Iterative design loops: The design process becomes cyclical and responsive, where architectural modifications are tested against performance metrics and refined accordingly. This iterative logic replaces traditional linear workflows, fostering innovation and integration.
- Human-environmental feedback loop: Performance simulations extend beyond structural behavior to include thermal comfort, daylighting, acoustic response, and user interaction, ensuring that design decisions respond not only to engineering logic but also to occupant experience and environmental context.

By embedding these components into a unified workflow, digital tools can facilitate a new mode of interdisciplinary authorship, where design is not merely an outcome but an evolving negotiation between spatial ambition, structural feasibility, and environmental intelligence (Burry, 2013).

Innovation is a shared responsibility in a smart structure. An ambitious shape must be proposed by the architect and maintained by the engineer; both must coauthor systems that are experience-rich, ecologically responsive, and structurally efficient. This requires early and iterative exchanges (Kolarevic and Malkawi, 2005): identifying load-bearing opportunities within architectural expression, embedding

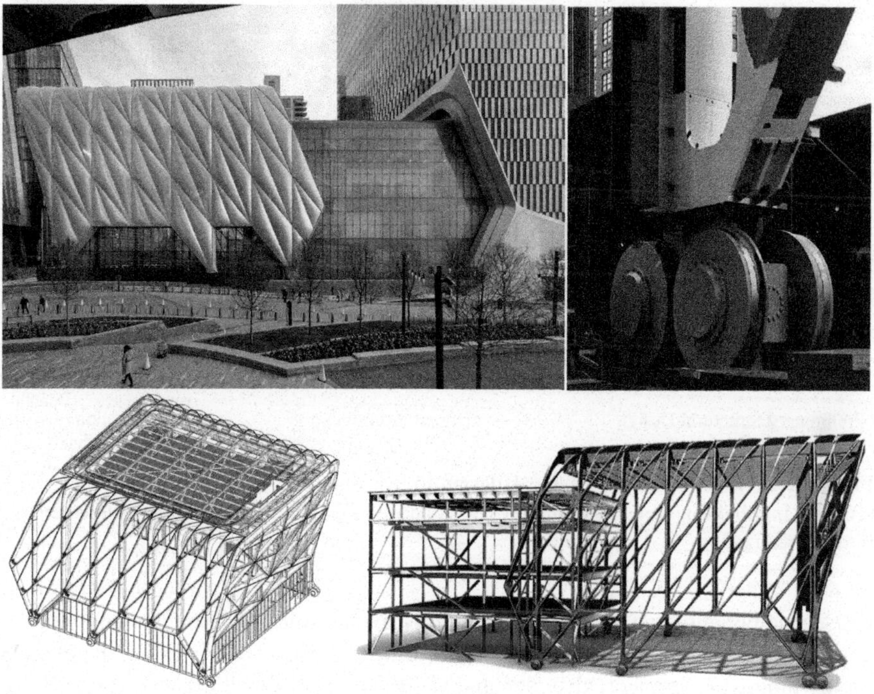

Figure 1.11 Some views of the Shed's kinetic system.

adaptive components within structural logic, and designing building envelopes that perform thermally, acoustically, and visually.

One of the most emblematic examples of such interdisciplinary innovation is The Shed in New York City, designed by Diller Scofidio + Renfro in collaboration with Rockwell Group and structurally engineered by Thornton Tomasetti. Located at Hudson Yards, The Shed exemplifies how a smart structure can merge spatial ambition, technological adaptability, and structural ingenuity into a single architectural organism (Fig. 1.11).

At the heart of The Shed is its kinetic outer shell, a telescoping steel-and-ethylene tetrafluoroethylene (ETFE) enclosure that glides along tracks on massive wheels, allowing the building to physically expand and contract to accommodate a variety of performance and exhibition spaces. This movable shell, weighing over 4000 tons, is not merely a feat of engineering but a result of deeply integrated collaboration between architect and engineer from the very first design conversations. It illustrates precisely the kind of shared responsibility discussed in the previous paragraph; the architect envisioned a flexible cultural venue unconstrained by static walls, and the engineer made this vision structurally, mechanically, and operationally feasible.

The Shed's kinetic system required not only early coordination of spatial and load-bearing strategies but also the embedding of adaptive components—including a complex wheel-and-track system, lightweight cladding, and motorized actuation.

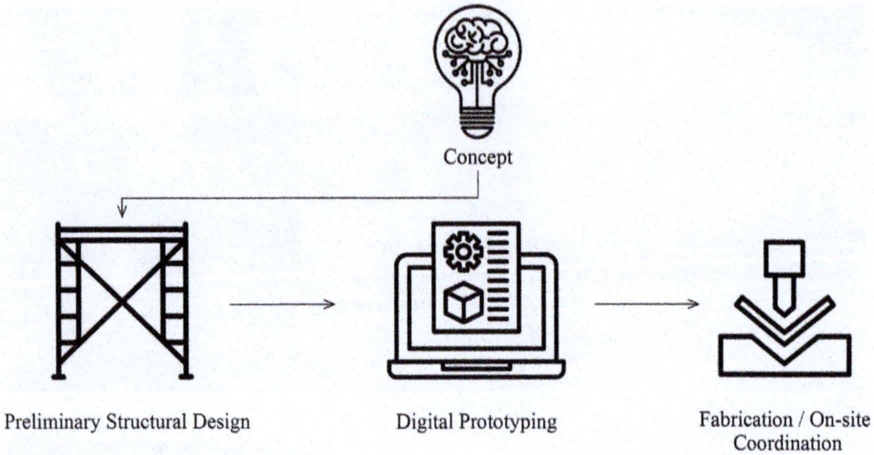

Figure 1.12 Illustration of a typical design-to-construction timeline in smart architecture projects, highlighting key collaboration points between architects and engineers from concept through realization.

These components had to be tested digitally and physically through iterative loops that involved parametric modeling, real-time structural analysis, and full-scale prototyping. The project also demanded attention to envelope performance: the ETFE cladding was chosen for its light weight, transparency, and thermal properties, aligning structural, environmental, and architectural goals.

This example powerfully reinforces the argument that innovation in smart structures is impossible without early, iterative, and equal collaboration. The Shed's responsive envelope is not just a visual or mechanical feature; it's a spatial and social interface that allows the building to adapt to its urban context and programmatic demands in real time. Without the synergistic input of architects and engineers throughout the process, from concept to construction, such a responsive and intelligent design could not have materialized. Such collaboration cannot be confined to isolated project phases; it must be embedded throughout the design-to-construction continuum. As illustrated in Fig. 1.12, the collaborative concept-to-construction timeline breaks this process into four integrated stages, each requiring close coordination between disciplines:

- Concept: At this early stage, architects explore spatial ambitions, typological innovation, and user engagement. Engineers contribute by advising on preliminary structural feasibility and system integration, ensuring that the initial vision is grounded in physical and material logic.
- Preliminary structural design: Here, the structural framework is shaped in response to the architectural concept. Load paths, material strategies, and geometries are tested using computational simulations and physical intuition. The back-and-forth dialog solidifies performance goals while retaining design intent.
- Digital prototyping: Both architects and engineers engage with parametric tools and simulation software to iterate and refine the design. Structural and environmental

simulations are coupled with fabrication constraints, allowing for clash detection, optimization, and detailed coordination before any physical production begins.
- Fabrication/on-site coordination: Digital models are translated into production-ready files for prefabrication or computer numerical control processes. During construction, precise coordination ensures that the built reality adheres to the digital intent. Real-time problem-solving between design and site teams becomes essential in managing tolerance, sequencing, and assembly.

The information above highlights how this sequential yet integrated timeline fosters shared authorship between architects and engineers. Innovation is not confined to the conceptual phase; rather, it evolves collaboratively throughout the project lifecycle, from design intent to on-site realization. This continuity ensures that ambitious architectural visions are structurally and technically viable, with each stage—conceptualization, preliminary structural design, digital prototyping, and fabrication/on-site coordination—acting as a feedback-rich platform for interdisciplinary dialog. Many of today's most forward-looking architectural systems, including kinetic façades, morphing roofs, and structurally active skins, would be unbuildable without this level of cross-disciplinary continuity and mutual accountability. These complex systems demand not only technical precision but also creative negotiation across disciplinary boundaries, underscoring the essential role of integrated workflows in the future of smart, adaptive architecture.

For this collaboration to become embedded in practice, it must begin in education and professional culture. Design studios, engineering labs, and interdisciplinary workshops must promote team-based projects, shared evaluations, and joint critique sessions, allowing future practitioners to develop a culture of mutual respect and fluency in each other's priorities. Cultivating this mindset early helps prevent fragmentation in professional life, where late-stage collaboration often leads to costly rework or esthetic compromise. The challenges of the 21st century, climate resilience, urban densification, and carbon neutrality, are too complex to be addressed within disciplinary silos. Smart structures, with their capacity to adapt, sense, and respond, represent a synthesis of architectural imagination and engineering intelligence. To realize their full potential, architects and engineers must evolve from collaborators into cocreators, jointly responsible for the future of the built environment.

References

Babilio, E., Miranda, R., Fraternali, F., 2019. On the kinematics and actuation of dynamic sunscreens with tensegrity architecture. Frontiers in Materials 6. Available at https://doi.org/10.3389/fmats.2019.00007.

Bar-Cohen, Y.,, 2004. Electroactive Polymer (EAP) Actuators as Artificial Muscles: Reality, Potential, and Challenges, Second Edition. SPIE. Available at https://doi.org/10.1117/3.547465.

Born, L., González San Martín, E.A., Ridder, M., Körner, A.H., Knippers, J., Gresser, G.T., 2024. FlectoSol – a pneumatically activable PV-functionalized façade shading module with bending motion in two directions for solar tracking. Developments in the Built Environment 18, 100372. Available at https://doi.org/10.1016/j.dibe.2024.100372.

Burry, M., 2013. Scripting Cultures. John Wiley & Sons.

Chen, Q., Andrawes, B., 2017. Plasticity modeling of concrete confined with NiTiNb shape memory alloy spirals. Structures 11, 1–10. Available at https://doi.org/10.1016/j.istruc.2017.03.006.

Fox, M., Kemp, M., 2016. Interactive Architecture. Princeton Architectural Press.

Gutierrez Soto, M., Adeli, H., 2019. Semi-active vibration control of smart isolated highway bridge structures using replicator dynamics. Engineering Structures 186, 536–552. Available at https://doi.org/10.1016/j.engstruct.2019.02.031.

Khaddour, L.A., Yeboah, S.K., Dodoo, J.K., 2024. Ecological and carbon footprints of cities. In: Abraham, M.A. (Ed.), Encyclopedia of Sustainable Technologies (Second Edition), Vol. 2. Elsevier, pp. 92–98. Available at https://doi.org/10.1016/B978-0-323-90386-8.00044-9.

Kolarevic, B., Malkawi, A., 2005. Performative Architecture: Beyond Instrumentality. Spon Press.

Kronenburg, R., 2007. Flexible: Architecture That Responds to Change. Laurence King.

Li, G., Ren, L., Fu, Y., Yang, Z., Adetola, V., Wen, J., et al., 2023. A critical review of cyber-physical security for building automation systems. Annual Reviews in Control 55, 237–254. Available at https://doi.org/10.1016/j.arcontrol.2023.02.004.

Liu, X., Wang, W., Hu, D., Qu, J., Cao, C., 2024. Test, design, and parametric analysis of a prefabricated industrial building with shape memory alloy (SMA) cable-restrained bearings for seismic resilience. Engineering Structures 313, 118288. Available at https://doi.org/10.1016/j.engstruct.2024.118288.

Menges, A., Ahlquist, S., 2011. Computational Design Thinking. Chichester.

Natamai, M., Subramanian, 2021. Electroactive Polymers. Walter de Gruyter GmbH & Co KG.

Roos, P.B., 2021. Regenerative-Adaptive Design for Sustainable Development: A Pattern Language Approach. Springer, Cham. Available at https://doi.org/10.1007/978-3-030-53234-5.

Sanaei Ataabadi, H., Liu, Y., Zeng, J.-jie, Zhuge, Y., 2024. Unveiling the impact of absorbent polymers on self-healing efficiency of sludge-derived capsules in cementitious composites. Construction and Building Materials 451, 138803. Available at https://doi.org/10.1016/j.conbuildmat.2024.138803.

Shashikumar, U., Tsai, P.-C., Wang, C.-T., Lay, C.-H., Ponnusamy, V.K., 2024. Beyond biomimicry: innovative bioinspired materials strategies and perspectives for high-performance energy storage devices. Process Safety and Environmental Protection 191, 1193–1217. Available at https://doi.org/10.1016/j.psep.2024.08.123.

Shen, Y., Zhang, X., Xu, W., Ying, X., Tu, Z., Luo, Y., 2024. An adaptive suspended-dome structure: static control methods based on minimum strain energy. Structures 67, 106999. Available at https://doi.org/10.1016/j.istruc.2024.106999.

Spitz, J., Bali, M., Half, D.A., Polle, D., 2018. Smart Building Design. Walter de Gruyter GmbH, pp. 10–67.

Sternberg, A., Zemp, R., de la Llera, J.C., 2014. Multiphysics behavior of a magneto-rheological damper and experimental validation. Engineering Structures 69, 194–205. Available at https://doi.org/10.1016/j.engstruct.2014.03.016.

Uchino, K., 2017. Advanced Piezoelectric Materials: Science and Technology. Elsevier/Wp Woodhead Publishing.

Vazquez, E., Randall, C., Duarte, J.P., 2019. Shape-changing architectural skins: a review on materials, design and fabrication strategies and performance analysis. Journal of Facade Design and Engineering 7 (2), 93–114.

Wiktor, V., Jonkers, H.M., 2011. Quantification of crack-healing in novel bacteria-based self-healing concrete. Cement and Concrete Composites 33 (7), 763–770. Available at https://doi.org/10.1016/j.cemconcomp.2011.03.012.

Yokota, K., Barthelat, F., 2024. Shape morphing of 2D lattice structures from localized contra-rotations. International Journal of Solids and Structures 294, 112771. Available at https://doi.org/10.1016/j.ijsolstr.2024.112771.

Foundations of structural engineering and architectural design

2

2.1 Structural engineering fundamentals

The principles of structural engineering constitute the essential scientific and technical foundation upon which all architectural and infrastructural systems are conceived, analyzed, and realized. These principles establish the critical relationship between architectural design and engineering performance, ensuring that buildings achieve both spatial functionality and structural integrity (Paulino et al., 2026). Structural engineering serves as the bridge between architectural ambition and physical feasibility, allowing complex forms, innovative materials, and advanced building systems to operate safely under a broad range of environmental and operational conditions (Dell'Endice et al., 2023). A thorough understanding of structural behavior enables architects and engineers to collaborate effectively, integrating esthetics, structural stability, and technological innovation into cohesive built solutions. This knowledge is particularly vital in the development of smart structures, where adaptive systems, responsive materials, and digital technologies rely fundamentally on sound structural principles.

Structural engineering is primarily concerned with the ability of a system to safely resist, transfer, and distribute forces without excessive deformation, instability, or collapse. Every structure is subjected to multiple forces throughout its lifecycle, and understanding these forces is essential to ensuring long-term safety, durability, and functionality. Structural loads are broadly classified into static forces and dynamic forces, each of which influences structural behavior differently and requires distinct analytical and design approaches.

2.1.1 Static forces

Static forces are loads that remain relatively constant over time or vary slowly enough to be treated as permanent or predictable. Among these, gravity is the most fundamental force acting continuously on all structures. Gravity generates dead loads, which include the self-weight of structural elements such as beams, columns, slabs, walls, and permanent fixtures. These loads form the baseline structural demand and must be carefully accounted for in the sizing and design of all load-bearing systems (D'Amico and Pomponi, 2020). Structural frameworks must be engineered to transfer these vertical loads safely through columns and foundations to the ground while maintaining dimensional stability and preventing material failure.

Live loads represent another critical category of static force. Unlike dead loads, live loads are variable and arise from occupancy-related factors such as people,

Smart Civil and Structural Engineering. DOI: https://doi.org/10.1016/B978-0-443-30223-7.00016-4

furniture, equipment, and temporary activities within a building. Because live loads fluctuate depending on building function and usage patterns, engineers must design structural systems with sufficient flexibility and reserve capacity to safely accommodate these variations. Residential buildings, offices, industrial facilities, and public structures all require distinct live load considerations according to their operational demands.

Wind and lateral forces further complicate structural design. Wind exerts horizontal pressures on building surfaces, creating shear forces, bending moments, uplift, and overturning effects. Similarly, lateral forces may also arise from seismic influences or uneven soil pressures. To resist these forces, structural systems incorporate specialized strategies such as shear walls, bracing systems, rigid cores, and moment-resisting frames. These elements help distribute horizontal loads efficiently and maintain structural equilibrium under lateral stress.

2.1.2 Dynamic forces

Dynamic forces differ fundamentally from static loads because they vary significantly over time, often involving sudden changes in magnitude, direction, or frequency. These forces introduce complex structural responses, including oscillations, vibrations, resonances, and impacts. Vibrational forces may result from internal sources such as mechanical equipment, occupant movement, or transportation systems, as well as external influences such as seismic activity or wind-induced motion. Excessive vibration can compromise structural serviceability, occupant comfort, and long-term material durability. To mitigate these effects, engineers employ vibration control systems such as tuned mass dampers, base isolators, and damping devices.

Impact forces are sudden, high-intensity loads generated by events such as vehicle collisions, falling objects, blast loads, or accidental strikes. These forces require structural systems capable of absorbing and redistributing energy rapidly to prevent localized or progressive collapse. Materials and structural configurations must therefore be selected with resilience, ductility, and redundancy in mind.

The most significant distinction between static and dynamic forces lies in their temporal behavior and structural consequences. Static forces remain relatively constant and predictable, allowing engineers to perform equilibrium-based calculations with stable load assumptions. Dynamic forces, however, fluctuate over time and often produce complex inertial effects that require advanced analytical techniques, including modal analysis, time-history simulation, and dynamic response modeling.

Fig. 2.1 illustrates the most significant differences between static and dynamic forces, visually demonstrating the constant and predictable nature of static loads compared to the fluctuating and time-dependent behavior of dynamic loads. This figure serves as a foundational conceptual diagram for understanding how different categories of forces influence structural behavior and engineering response.

In static force analysis, moment diagrams play a critical role in understanding how structural members respond to permanent or gradually applied loads. These diagrams illustrate the distribution of bending moments along beams, slabs, and other structural

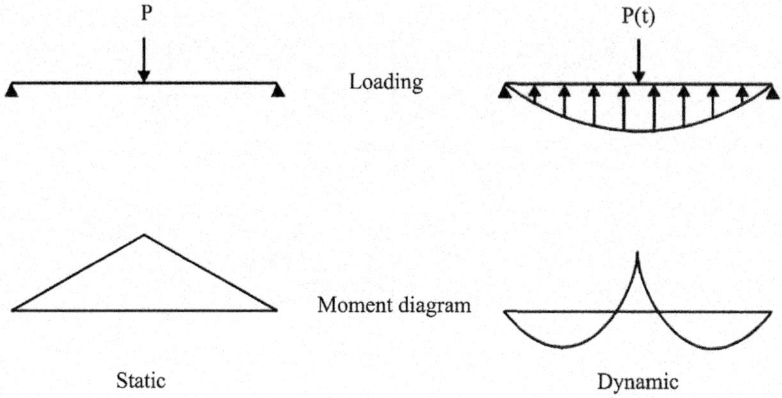

Figure 2.1 Illustration of the most significant difference between a static and dynamic force.

components, enabling engineers to determine reinforcement needs, member dimensions, and support conditions. Because static loads are stable, the resulting moment distributions are generally predictable and allow for relatively straightforward structural optimization (Wu et al., 2022).

Dynamic force analysis, in contrast, examines how moment distributions evolve under changing or sudden forces. Structures subjected to earthquakes, wind gusts, or impact events experience fluctuating internal stresses, resulting in variable bending moments over time. Dynamic moment diagrams therefore provide engineers with essential insights into structural resilience, fatigue behavior, and the capacity to withstand transient or cyclical loading scenarios.

Beyond the distinction between static and dynamic forces, structural engineering fundamentally depends on the selection and organization of appropriate structural systems capable of safely transferring, resisting, and distributing loads under varying physical conditions. Structural systems translate theoretical force analysis into built reality, serving as the physical frameworks through which architectural intent, material performance, and engineering efficiency converge. The selection of a structural typology is never arbitrary; rather, it is determined by a complex interplay of load behavior, span requirements, material properties, environmental conditions, construction methodologies, economic considerations, and spatial or architectural objectives. Different structural systems respond uniquely to gravitational, lateral, dynamic, and environmental forces, making typological understanding essential for both engineers and architects in the design of safe, efficient, and contextually responsive buildings.

2.1.3 Frame systems

Frame systems constitute one of the most widely used structural typologies in contemporary construction (Fig. 2.2). Composed primarily of beams and columns interconnected through rigid or semirigid joints, frame structures create stable load paths that transfer vertical and horizontal forces through bending and axial action.

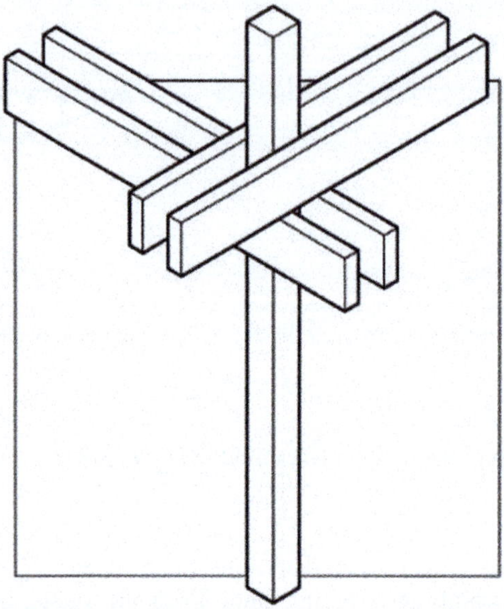

Figure 2.2 Frame systems: Structural load transfer through beams and columns.

These systems are particularly effective in multistory buildings, commercial developments, and institutional projects where flexible interior spaces are required. Rigid frame systems resist lateral forces through moment connections, while braced frames utilize diagonal elements to enhance stability against wind and seismic loads. The adaptability of frame systems allows architects significant freedom in façade articulation and interior planning, making them especially suitable for projects where structural clarity must coexist with functional versatility.

Material selection greatly influences frame system performance. Steel frames offer high tensile strength, long spans, and rapid assembly (Sevugan Rajkannu and Vasantbhai, 2026), whereas reinforced concrete frames provide mass, stiffness, and fire resistance (Wang et al., 2026). Timber framing, increasingly relevant in sustainable design, introduces renewable material strategies while maintaining structural efficiency (Slattery et al., 2026). Thus, frame systems exemplify how engineering principles and architectural flexibility are integrated through structural typology.

2.1.4 Truss systems

Truss systems are engineered assemblies of interconnected triangular elements designed to distribute forces primarily through axial tension and compression rather than bending (Fig. 2.3). This geometric configuration allows trusses to achieve exceptional strength-to-weight ratios, making them ideal for long-span applications such as bridges, stadiums, industrial facilities, and large exhibition halls. By efficiently channeling loads through triangulated members, trusses minimize material use while maximizing structural performance.

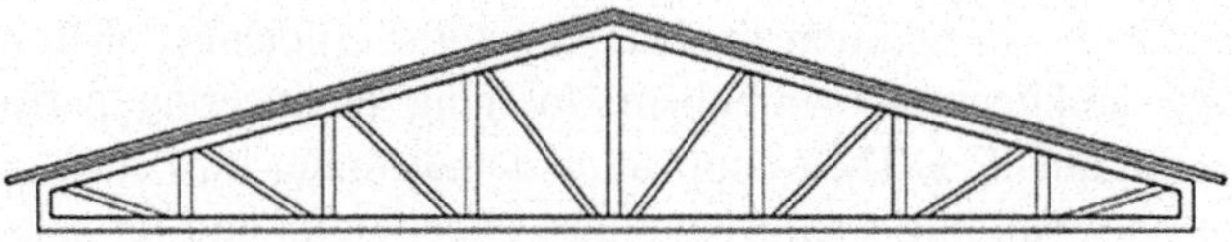

Figure 2.3 Truss systems: Triangulated structural efficiency for long-span applications.

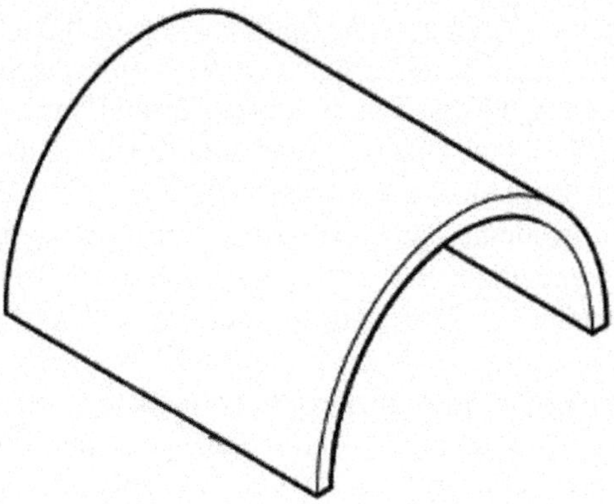

Figure 2.4 Shell structures: Curved surface geometry and membrane load distribution.

The selection of truss systems often arises when large unobstructed spaces are required, where conventional frame systems would be less efficient or economically impractical. Variations such as Pratt, Warren, Howe, and space trusses offer different load distribution characteristics depending on span, loading conditions, and construction requirements. In architectural applications, exposed truss systems can also contribute to visual identity, expressing structural logic as an esthetic feature. This reflects a growing contemporary approach in which engineering systems are not concealed but integrated into architectural language.

2.1.5 Shell structures

Shell structures represent a highly efficient structural typology in which loads are transferred primarily through membrane stresses rather than bending (Fig. 2.4). Typically characterized by curved geometries such as domes, vaults, or folded plates, shell systems achieve remarkable strength through form rather than mass. Their curvature allows forces to be distributed across the surface, reducing localized stresses and enabling lightweight construction with minimal material consumption.

Shell structures are particularly effective for large-span roofs, auditoriums, sports facilities, and transportation hubs, where both structural efficiency and architectural expression are desired. Materials commonly used include reinforced concrete, steel, glass composites, and advanced polymers. Computational modeling has significantly advanced shell design by allowing precise analysis of complex geometries and stress distributions, enabling architects and engineers to optimize both performance and form simultaneously. In addition to their structural efficiency, shell systems often embody strong architectural symbolism, merging engineering performance with expressive spatial qualities. Their application demonstrates how structural typologies can shape not only building stability but also experiential and formal outcomes.

2.1.6 Tensile systems

Tensile structures operate through membranes, cables, or flexible surfaces that resist loads primarily through tension rather than compression or bending (Fig. 2.5). These systems are among the lightest structural typologies and are especially effective for roofs, pavilions, temporary installations, and large-span enclosures where minimal material use and visual lightness are prioritized.

Cable nets, fabric membranes, and pneumatic structures exemplify tensile systems, relying on prestress and curvature to maintain stability. Because tensile systems require careful balancing of forces, they often demand advanced computational analysis and precision engineering. Environmental loads such as wind uplift, snow accumulation, and thermal movement are particularly critical in their design. Architecturally, tensile systems offer opportunities for innovative and fluid forms that challenge traditional structural conventions. Their transparent or semitransparent materials can also enhance daylight penetration and environmental responsiveness (Pronk, 2021). The integration of tensile systems into contemporary architecture reflects the increasing fusion of lightweight engineering, environmental performance, and formal experimentation.

2.1.7 Composite systems

Composite structural systems combine multiple materials or structural behaviors to maximize performance while minimizing weaknesses inherent in individual components. Common examples include steel-concrete composite floors, reinforced

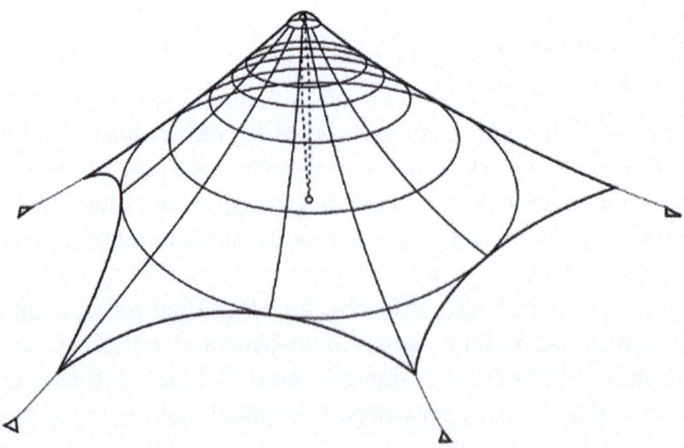

Figure 2.5 Tensile systems: Lightweight structures based on tension and prestress.

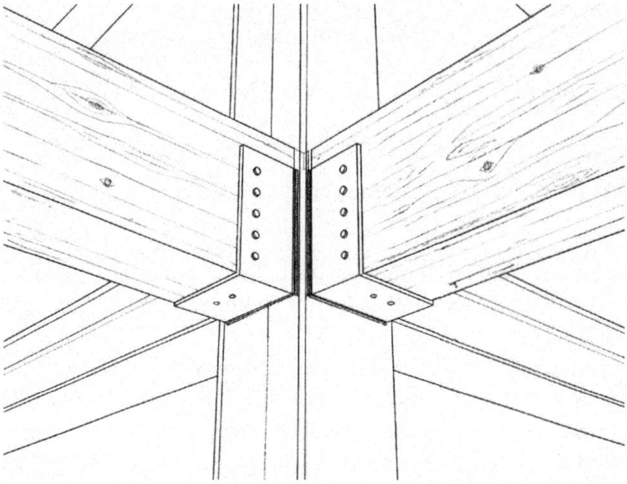

Figure 2.6 Composite systems: Integrated material performance in structural design.

concrete with steel reinforcement, timber-steel hybrids (Fig. 2.6), and fiber-reinforced polymers integrated into conventional systems.

The primary advantage of composite systems lies in their ability to exploit complementary material properties. Steel contributes tensile strength and ductility, concrete offers compressive resistance and thermal mass, while engineered timber introduces sustainability and lightweight performance. Through strategic integration, composite systems achieve greater efficiency, resilience, and adaptability than singular material systems. Composite structures are increasingly important in high-performance buildings, bridges, and adaptive architecture, where structural optimization, environmental performance, and lifecycle considerations must be balanced. Their development reflects broader trends in smart engineering, where structural systems are designed not only for strength but also for multifunctional performance.

2.1.8 Hybrid structural systems

Hybrid structural systems represent the evolution of structural engineering toward integrated, multifunctional frameworks. These systems combine multiple structural typologies—such as frame, truss, shell, tensile, or composite elements—within a single project to address complex architectural, environmental, and operational demands (Fig. 2.7).

For example, contemporary skyscrapers may combine rigid cores, perimeter bracing, outriggers, and composite floor systems to optimize lateral resistance, vertical load transfer, and spatial flexibility. Similarly, large public buildings may integrate shell roofs with frame supports and tensile façade systems to achieve both performance and esthetic goals. Hybrid systems are particularly relevant in smart structures because they support adaptability, digital optimization, and interdisciplinary integration. Through parametric design tools, advanced simulation, and

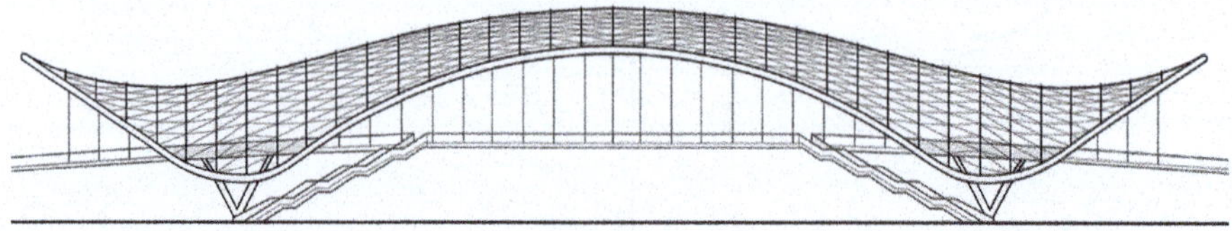

Figure 2.7 Hybrid structural systems: Integrated multisystem structural intelligence.

digital fabrication, hybrid systems can be tailored to highly specific performance objectives, reconciling structural behavior with environmental intelligence and architectural ambition.

The study of structural systems typologies demonstrates that structural engineering extends far beyond force resistance; it is a discipline deeply embedded in spatial organization, material innovation, and architectural possibility (Sandaker et al., 2022). Structural systems are selected not solely for their technical adequacy but for their ability to mediate between engineering demands and design aspirations. Span length, material availability, load complexity, environmental conditions, construction technologies, and programmatic requirements all influence this selection process. As buildings become increasingly complex, sustainable, and technologically integrated, understanding structural typologies becomes essential for creating systems that are not only stable but also adaptive, resource-efficient, and architecturally expressive. Structural systems therefore function as both technical frameworks and design instruments, bridging theoretical engineering principles with real-world applications. This typological foundation prepares the groundwork for more advanced discussions of computational optimization, smart materials, and responsive structures, where structural performance evolves from static resistance toward dynamic intelligence.

Material behavior is another essential aspect of structural engineering fundamentals. Structural systems rely on materials such as steel, reinforced concrete, timber, and advanced composites, each possessing unique properties related to strength, elasticity, ductility, and durability. Engineers must align material characteristics with structural requirements to optimize safety, performance, and sustainability.

Structural engineering fundamentals form the indispensable basis for all subsequent advancements in intelligent and adaptive architecture. By understanding force classifications, load paths, moment behavior, and material performance, designers can create structures that not only meet safety requirements but also support the integration of emerging technologies, sustainable systems, and responsive design strategies. These principles remain central to the evolution of architecture from static construction toward adaptive, performance-driven, and technologically integrated built environments.

2.2 Architectural design principles

Architectural design principles form the intellectual, spatial, and esthetic framework through which the built environment is conceived, organized, and realized. These principles guide the transformation of conceptual ideas into physical spaces that must simultaneously satisfy functional requirements, structural feasibility, environmental responsiveness, and human experience (Elnagar et al., 2024). Far beyond the simple creation of shelter, architectural design shapes how individuals interact with space, how societies express cultural values, and how built systems respond to evolving technological and ecological conditions. As such, architectural design is both an artistic discipline and a strategic methodology, balancing creativity with technical rigor.

At the foundation of architectural design lies the integration of form and function. Form refers to the physical appearance, geometry, proportion, and visual expression of a structure, while function encompasses usability, spatial performance, circulation, and programmatic effectiveness. Successful architecture emerges when these dimensions operate in harmony rather than opposition. Buildings must not only fulfill practical needs but also communicate meaning, evoke emotional responses, and contribute positively to their broader environmental and social contexts. This synthesis of beauty and utility remains central to all architectural practice.

Spatial organization is one of the most essential principles in architecture, as it determines how occupants move through, perceive, and experience the built environment. The arrangement of spaces, transitions, thresholds, circulation paths, vertical connections, and spatial hierarchies directly influences functionality and user behavior. Architects carefully orchestrate these elements to create coherent environments that optimize efficiency, accessibility, privacy, and social interaction. Effective spatial planning transforms buildings from static enclosures into dynamic experiential systems. Human-centered design further expands these principles by prioritizing occupant comfort, well-being, and experiential quality. Architectural environments must address physical ergonomics, thermal comfort, visual access, acoustics, safety, and psychological engagement. Buildings increasingly serve not only as functional infrastructures but also as environments that influence productivity, health, and emotional experience. Consequently, architectural design must carefully align technical performance with human needs.

Esthetic composition remains a defining dimension of architecture. Principles such as proportion, rhythm, balance, contrast, scale, materiality, and visual hierarchy shape how buildings are perceived and understood. These compositional strategies allow architects to create visually coherent structures that resonate culturally and emotionally while contributing to place identity. Importantly, esthetic principles are not isolated from technical considerations; structural systems, environmental performance, and material expression increasingly become integral components of architectural beauty.

Environmental responsiveness is now inseparable from architectural design methodology. Contemporary buildings must respond intelligently to climate, orientation, resource consumption, and ecological context. Passive solar strategies, daylight optimization, natural ventilation, thermal regulation, and sustainable material choices shape both form generation and operational performance. This approach transforms architecture into an environmentally integrated discipline, where buildings act in dialog with natural systems rather than as isolated objects. Material selection and tectonic expression are equally significant. Materials define not only structural capacity but also sensory experience, durability, environmental impact, and construction feasibility. Architects engage with materials as both technical systems and expressive mediums, using them to articulate identity, sustainability, and innovation. Advances in fabrication technologies, digital modeling, and engineered materials have expanded opportunities for architectural experimentation and performance optimization (Esan et al., 2026).

The creative architectural process is inherently iterative and interdisciplinary. Concept development, environmental analysis, structural integration, regulatory compliance, digital simulation, and user needs must all converge throughout the design process. Increasingly, architects collaborate with engineers, computational designers, environmental consultants, and fabrication specialists to ensure that design solutions are both visionary and feasible. This collaborative process reinforces architecture as a synthesis of multiple disciplines.

The evolution of contemporary architectural design has been profoundly shaped by the integration of computational methodologies, parametric systems, and algorithm-driven processes, fundamentally transforming how buildings are conceived, analyzed, and realized (Gu et al., 2021). Traditional architectural design often relied on sequential workflows in which conceptual form generation preceded technical evaluation. In contrast, computational and parametric design methodologies establish a more dynamic and iterative relationship between form, performance, material systems, and environmental responsiveness (Brown et al., 2020). These methodologies enable architects and engineers to simultaneously explore esthetic ambition, structural logic, environmental optimization, and fabrication feasibility within integrated digital ecosystems.

Computational design represents a paradigm shift from static drafting toward data-informed design intelligence. Through advanced software environments, architects can process complex variables, analyze performance criteria, and generate responsive solutions that would be difficult or impossible to achieve through conventional manual methods. Rather than simply representing geometry, computational systems actively participate in shaping it, allowing architecture to evolve as a performance-driven discipline where design decisions are increasingly informed by measurable parameters.

2.2.1 Parametric modeling

Parametric modeling serves as one of the most influential innovations in modern design methodology. In parametric systems, design elements are defined not as isolated geometries but as interconnected parameters governed by rules, relationships, and variable inputs. This approach allows designers to manipulate entire systems by adjusting key parameters, resulting in dynamic models capable of continuous transformation while maintaining internal coherence (Kim et al., 2024).

For example, building height, façade density, solar orientation, structural spacing, and environmental conditions can all be embedded into a parametric model. Adjustments to one parameter automatically influence related components, enabling rapid iteration and systemic optimization. This methodology significantly enhances design flexibility, precision, and adaptability while reducing inefficiencies associated with traditional redesign processes.

Parametric modeling is particularly valuable in projects characterized by geometric complexity, environmental responsiveness, or performance optimization. It supports architects in creating highly customized forms while ensuring that these forms remain technically viable and performance-oriented. As a result, parametric design has become a foundational methodology in smart architecture, where responsiveness and adaptability are essential.

2.2.2 Algorithmic form generation

Building upon parametric systems, algorithmic form generation introduces computational procedures that actively produce geometry based on predefined logical frameworks. Rather than manually shaping form, designers establish algorithmic rules that determine spatial, structural, or environmental outcomes (Garcia-Alvarado et al., 2026). These rules may include mathematical equations, behavioral simulations, environmental data, or structural performance criteria.

Algorithmic processes allow architects to generate highly sophisticated forms, including complex façades, responsive envelopes, optimized structures, and adaptive spatial organizations. This approach shifts design from direct authorship to strategic system definition, where architects design the logic behind form rather than every geometric detail individually.

Algorithmic generation is especially relevant in smart structures because it enables the integration of multiple performance criteria into formal development. Structural loads, daylight optimization, airflow behavior, and fabrication constraints can all influence geometry simultaneously. This creates buildings that are not only visually innovative but also technically optimized and environmentally responsive.

2.2.3 Generative design

Generative design further expands computational methodologies by incorporating artificial intelligence, machine learning, and optimization algorithms into the design process (Zhuang et al., 2025). In generative systems, designers define goals, constraints, and performance criteria, while computational engines autonomously generate multiple design alternatives based on these inputs. For example, objectives may include minimizing structural material, maximizing daylight access, reducing energy consumption, or optimizing occupant circulation. The system evaluates thousands of potential configurations and proposes solutions that balance these often competing priorities. Designers then assess, refine, and select outcomes based on broader qualitative and contextual considerations.

Generative design introduces a new level of design intelligence, where architecture becomes increasingly data-driven, predictive, and performance-based. It allows for unprecedented exploration of design possibilities while significantly accelerating iterative processes. This methodology is particularly important for future smart buildings, where adaptive systems, sustainability, and performance integration require the reconciliation of numerous technical variables.

2.2.4 Digital workflows and integrated design platforms

The effectiveness of computational and parametric methodologies depends heavily on integrated digital workflows. Contemporary architectural practice increasingly relies on interconnected software ecosystems that unify design, structural analysis, environmental simulation, fabrication documentation, and lifecycle management.

Building information modeling (BIM), parametric platforms such as Grasshopper, simulation software, and digital fabrication tools collectively support seamless collaboration among architects, engineers, contractors, and fabricators. These workflows dissolve traditional disciplinary boundaries by enabling real-time data exchange, interdisciplinary coordination, and continuous performance validation (Getuli et al., 2025).

Digital workflows ensure that design intent remains aligned with structural logic, material behavior, and construction feasibility throughout all project phases. Changes made during conceptual development can immediately inform engineering analysis, fabrication strategies, and operational planning. This level of integration is essential for smart structures, where design complexity and performance demands require coordinated decision-making across multiple systems.

2.2.5 Form–performance relationships

Perhaps the most transformative contribution of computational methodologies is the establishment of direct relationships between architectural form and measurable performance. In conventional practice, form and performance were often treated as separate considerations, with esthetic decisions preceding technical optimization. Computational design fundamentally redefines this relationship by embedding performance criteria directly into formal generation. Architectural geometry can now respond dynamically to the following:

- structural efficiency;
- solar exposure;
- thermal behavior;
- ventilation patterns;
- material optimization;
- occupancy flow;
- fabrication constraints.

This creates a performance-driven design paradigm in which form is not merely expressive but also operationally intelligent. Buildings are conceived as systems where geometry, materiality, and technology interact continuously to achieve optimized outcomes. For example, façade configurations may be generated based on solar analysis, structural lattices optimized through load simulations, or spatial layouts informed by circulation analytics. These relationships produce architecture that is simultaneously more efficient, adaptive, and contextually responsive. The integration of computational and parametric methodologies prepares architectural practice for the demands of advanced smart systems by establishing design processes that are:

- data-driven;
- iterative;

- interdisciplinary;
- performance-oriented;
- adaptive;
- Fabrication-aware.

These methodologies are not simply digital tools but represent a redefinition of architectural intelligence itself. They allow designers to address increasing environmental complexity, material innovation, user variability, and technological integration with unprecedented precision. In smart architecture, buildings are no longer conceived as static compositions but as dynamic systems shaped by continuous interaction between data, performance, and human needs. Computational design serves as the intellectual and technical infrastructure that enables this transformation, bridging creativity with engineering rigor and preparing the field for increasingly responsive, sustainable, and intelligent built environments. By embedding algorithmic thinking, parametric relationships, and performance feedback into architectural methodology, contemporary practice moves beyond conventional design limitations toward a future in which form, function, and intelligence are fundamentally interconnected.

A particularly compelling contemporary example of these architectural principles is the Vessel, located in Hudson Yards, New York City, USA, designed by Heatherwick Studio (Fig. 2.8). Located within the Hudson Yards development, Vessel exemplifies how spatial organization, circulation, formal expression, and structural innovation can converge into a singular architectural concept. Rather than functioning as a conventional building, Vessel operates as an experiential public structure composed of an intricate network of interconnected staircases, platforms, and vertical pathways. Its design prioritizes human movement, exploration, and perception, transforming circulation itself into the primary architectural experience.

The project's geometric framework consists of a vertically expanded honeycomb-like structure, where repeated modules generate a visually striking yet highly organized system. This repetition creates rhythm, hierarchy, and formal coherence while encouraging users to engage with space dynamically from multiple vantage points. The structure's design demonstrates the principle that architecture can shape behavior by guiding movement, framing perspectives, and fostering social interaction. Vessel also highlights the relationship between architecture and structural engineering. Its complex interconnected stair system required precise structural coordination to ensure stability while maintaining openness and visual lightness. Structural systems are integrated directly into the architectural language, rather than concealed, reinforcing the concept that structural logic can become a central esthetic and spatial driver.

From an experiential perspective, Vessel challenges traditional distinctions between structure, sculpture, and architecture. It functions simultaneously as infrastructure, landmark, observation platform, and public gathering space. This multifunctionality reflects broader contemporary design values in which architecture increasingly emphasizes participation, interaction, and urban engagement. Moreover, Vessel demonstrates how contemporary architectural design extends beyond enclosed buildings to include civic and experiential environments that enrich public life.

Figure 2.8 Vessel, New York City—Exterior and interior views, sectional analysis, and partial plan illustrating the project's interconnected spatial organization, vertical circulation system, and structural framework.

Through its emphasis on circulation, human-centered exploration, and iconic formal expression, the project illustrates how architectural principles can be applied to create memorable and socially significant spaces. The integration of architectural creativity with engineering precision in Vessel reinforces the importance of interdisciplinary collaboration. The project exemplifies how architectural vision can be realized through structural innovation, digital modeling, and fabrication technologies, ensuring that design ambition remains technically achievable.

Architectural design principles provide the conceptual and practical basis for creating built environments that are functional, meaningful, esthetically compelling, and technically sound. Projects such as Vessel demonstrate that successful architecture emerges from the careful orchestration of space, form, movement, structure, and user experience. As architecture continues to evolve in response to technological innovation, environmental demands, and societal transformation, these foundational principles remain essential in guiding the creation of adaptive, human-centered, and culturally resonant environments.

2.3 Intersections of engineering and design

The intersection of structural engineering and architectural design represents one of the most significant areas of innovation in the contemporary built environment. Historically, architecture and engineering were often approached as complementary yet distinct disciplines, with architects focusing primarily on spatial, esthetic, and cultural dimensions, while engineers concentrated on structural stability, material performance, and technical execution. However, the rapid advancement of smart technologies, computational tools, responsive materials, and integrated digital systems has fundamentally transformed this relationship. Today, the boundaries between these professions are increasingly blurred, giving rise to a collaborative model in which design and engineering operate as interconnected components of a unified process.

This convergence is redefining how buildings are conceived, developed, and experienced. Rather than viewing engineering as a secondary technical process that supports completed architectural concepts, contemporary practice positions structural intelligence, environmental performance, and technological responsiveness as active drivers of design from the earliest conceptual stages. In this integrated paradigm, buildings are no longer static objects; they are adaptive systems capable of sensing, responding, and evolving in relation to environmental conditions, user behavior, and operational demands.

Smart civil and structural engineering does not simply involve adding advanced technologies to conventional systems. Instead, it represents a fundamental rethinking of the relationship between form, function, and performance. Structural systems are increasingly designed not only to provide support but also to participate actively in environmental regulation, energy optimization, and spatial adaptability. Likewise, architectural form is no longer developed independently of engineering constraints but emerges through the integration of structural logic, material capabilities, and technological opportunities.

This shift demands a new interdisciplinary design culture in which architects, engineers, fabricators, computational designers, and clients collaborate from the initial stages of project development. Effective smart integration depends on the early alignment of goals, ensuring that design intent, structural performance, environmental objectives, and user-centered outcomes are addressed holistically. Smart technologies cannot function effectively as postdesign additions; they must be embedded within the conceptual and technical DNA of the project.

The rise of BIM, parametric design platforms, performance simulation tools, and digital fabrication technologies has accelerated this transformation by providing shared collaborative environments. These tools enable real-time coordination between multiple disciplines, allowing design teams to optimize geometry, structural behavior, environmental systems, and construction logistics simultaneously. As a result, design and engineering are increasingly characterized by parallel workflows rather than sequential processes.

A powerful example of this integrated methodology is the W Abu Dhabi—Yas Island, designed by Asymptote Architecture and engineered by Arup (Fig. 2.9). The project's signature feature is its expansive curvilinear steel-and-glass grid-shell canopy, which envelops the hotel and serves simultaneously as an architectural identity, structural system, and environmental interface. The complexity of this canopy required close interdisciplinary coordination between formal design strategies, structural analysis, digital fabrication, and environmental performance objectives. Parametric modeling and iterative simulation were essential to maintaining alignment between geometry, constructability, and operational performance throughout the design and construction process (Hong et al., 2024).

This project shows how design and engineering can no longer be a linear or sequential process. Instead, they should develop as parallel, interconnected workflows, where form-making and performance-tuning are simultaneous and mutual. In its essence, integration of smart systems creates buildings that act as adaptive, living systems. These systems adjust to real-time input, whether from climatic conditions, energy loads, or the presence of a user. Such accountability not only increases performance but also increases health, comfort, and engagement (Veiga et al., 2025).

Another exemplary project is the Council House 2, commonly known as CH2, which illustrates the intersection of environmental engineering and architectural

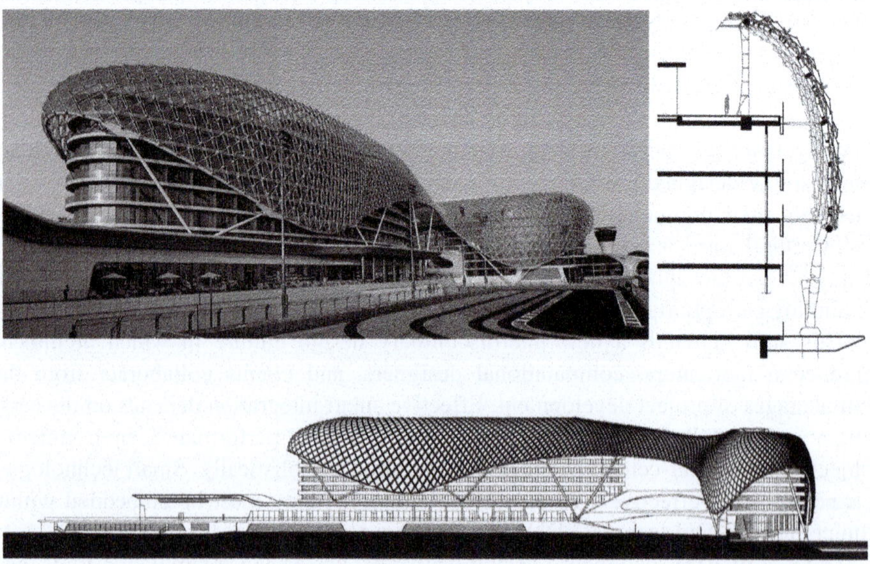

Figure 2.9 W Abu Dhabi—Yas Island Hotel: Parametric envelope and integrated structural design.

Figure 2.10 Council House 2 (CH2), Melbourne: Environmental engineering and adaptive building systems.

design (Fig. 2.10). This building incorporates an integrated array of smart environmental systems, including phase-change materials, thermal chimneys, automated louvers, wind turbines, and passive cooling strategies. These technologies are not merely technical appendages; they are embedded directly into the architectural logic of the building, enabling it to regulate temperature, ventilation, and energy use dynamically. CH2's innovative synthesis of passive and active systems earned it Australia's first 6-Star Green Star rating and established it as a pioneering model for sustainable smart architecture (Kolarevic and Parlac, 2015). The project demonstrates how buildings can function as responsive ecological systems, shifting architecture from passive enclosure toward active environmental participation.

The correct value of integrating smart engineering and design lies in creating an environment which is not only technically advanced but also human-friendly. Future buildings should distribute more than spatial utility; they should increase psychological well-being, support practical flexibility, and reduce ecological footprints (Sternberg, 2009).

Similarly, the Bloomberg European Headquarters by Foster + Partners represents a highly sophisticated integration of engineering intelligence and architectural design (Fig. 2.11). The project incorporates advanced natural ventilation systems, responsive lighting technologies, water conservation systems, and custom-designed bronze façade blades that optimize daylight penetration and airflow. These features are integrated seamlessly into the architectural language of the building, ensuring that technical performance enhances rather than compromises user experience. The headquarters achieved one of the highest sustainability ratings globally, yet its true innovation lies in demonstrating that intelligent performance can be experientially meaningful as well as technically effective (Kanaani, 2019).

These case studies collectively reveal that the future of architecture depends on a profound integration of structural engineering, environmental systems, digital technologies, and human-centered design. The most successful projects are not

Figure 2.11 Bloomberg European Headquarters, London: Smart workplace and sustainable performance integration.

defined solely by technological sophistication but by the coherence with which these systems are unified. Several key principles emerge from this intersection:

- early-stage interdisciplinary collaboration;
- performance-driven form generation;
- real-time adaptability and responsiveness;
- integration of environmental systems into architectural logic;
- user-centered technological implementation;
- digital workflow continuity across design, fabrication, and operation.

Importantly, this convergence also expands the role of architecture itself. Buildings are increasingly conceived not merely as objects of occupation but as living systems that mediate between users, technology, and environment. Through embedded sensing systems, adaptive materials, and computational intelligence, buildings become capable of regulating themselves, enhancing comfort, optimizing energy use, and responding dynamically to external and internal stimuli.

As environmental pressures, urban complexity, and technological capabilities continue to evolve, the integration of engineering and design will become increasingly essential. Future architects and engineers must develop cross-disciplinary literacy, combining creative design thinking with structural understanding, computational competence, and systems integration. This holistic approach will enable the realization of buildings that are technically advanced, environmentally responsible, socially responsive, and experientially rich.

The intersections of engineering and design represent far more than professional collaboration; they signify a paradigm shift in the conception of the built environment. Through this integrated framework, architecture and engineering converge not only in form but also in purpose, producing structures that are adaptive, sustainable, and deeply aligned with the evolving needs of society (Rahim, 2016).

References

Brown, N.C., Jusiega, V., Mueller, C.T., 2020. Implementing data-driven parametric building design with a flexible toolbox. Automation in Construction 118, 103252. Available at https://doi.org/10.1016/j.autcon.2020.103252.

D'Amico, B., Pomponi, F., 2020. On mass quantities of gravity frames in building structures. Journal of Building Engineering 31, 101426. Available at https://doi.org/10.1016/j.jobe.2020.101426.

Dell'Endice, A., Bouten, S., Van Mele, T., Block, P., 2023. Structural design and engineering of Striatus, an unreinforced 3D-concrete-printed masonry arch bridge. Engineering Structures 292, 116534. Available at https://doi.org/10.1016/j.engstruct.2023.116534.

Elnagar, E., Düvier, C., Batra, Z., Christoffersen, J., Mandin, C., Schweiker, M., et al., 2024. Creating a comprehensive framework for design, construction and management of healthy buildings. Energy and Buildings 324, 114883. Available at https://doi.org/10.1016/j.enbuild.2024.114883.

Esan, T.M., Kupolati, W.K., Ackerman, C., 2026. Advances in additive manufacturing for architectural acoustics: design, materials, validation, and implementation challenges in building construction. Additive Manufacturing Letters 17, 100368. Available at https://doi.org/10.1016/j.addlet.2026.100368.

Garcia-Alvarado, R., Soza-Ruiz, P., Valenzuela-Astudillo, E., 2026. Parametric form-finding for 3D-printed housing: a computational workflow from generative exploration to architectural development. Applied Sciences 16 (7), 3527. Available at https://doi.org/10.3390/app16073527.

Getuli, V., Bruttini, A., Rahimian, F., 2025. Parametric design methodology for developing BIM object libraries in construction site modeling. Automation in Construction 170, 105897. Available at https://doi.org/10.1016/j.autcon.2024.105897.

Gu, N., Yu, R., Behbahani, P.A., 2021. Parametric design: theoretical development and algorithmic foundation for design generation in architecture. In: Sriraman, B. (Ed.), Handbook of the Mathematics of the Arts and Sciences. Springer International Publishing, Cham, pp. 1361–1383. Available at https://doi.org/10.1007/978-3-319-57072-3_8.

Hong, S.W., Lee, J., Lee, J.K., 2024. Human behaviour simulation for promoting usefulness and user-centric values in parametric design. Automation in Construction 162, 105386. Available at https://doi.org/10.1016/j.autcon.2024.105386.

Kanaani, M., 2019. The Routledge Companion to Paradigms of Performativity in Design and Architecture. Routledge.

Kim, S., Myint, M.H., Yu, Y., Lee, W., Kim, T.W., Koo, B., 2024. Application of parametric modeling techniques for automated detailing of exterior finishes in architectural bim models: a South Korean case study. Buildings 14 (12), 4013. Available at https://doi.org/10.3390/buildings14124013.

Kolarevic, B., Parlac, V., 2015. Building Dynamics: Exploring Architecture of Change. Routledge, London. Available at https://doi.org/10.4324/9781315763279.

Paulino, D.M.S., Fivet, C., Ligler, H., Napolitano, R., 2026. Integrating spatial and structural considerations in floor plan transformations of historic masonry buildings. Automation in Construction 187, 106878. Available at https://doi.org/10.1016/j.autcon.2026.106878.

Pronk, A., 2021. Flexible Forming for Fluid Architecture. Springer International Publishing, Cham, pp. 1–582. Available at https://doi.org/10.1007/978-3-030-71551-9.

Rahim, A., 2016. Catalytic Formations: Architecture and Digital Design. Routledge.

Sandaker, B.N., Kleven, B., Wang, A.R., 2022. Structural typologies and the architectural space—studies of the relationship between structure and space by application of structural types to multistory buildings. Architecture, Structures and Construction 2 (1), 199–221. Available at https://doi.org/10.1007/s44150-022-00034-z.

Sevugan Rajkannu, J., Vasantbhai, P.J., 2026. Investigation on the composite behaviour of steel-framed glass façade structures. Progress in Engineering Science 3 (2), 100295. Available at https://doi.org/10.1016/j.pes.2026.100295.

Slattery, T.P., Cowled, C.J.L., Ramli Sulong, N.H., Crews, K., Brooke, H., 2026. Structural performance of timber-framed shear walls of various configurations. Construction and Building Materials 523, 146291. Available at https://doi.org/10.1016/j.conbuildmat.2026.146291.

Sternberg, E.M., 2009. Healing Spaces: The Science of Place and Well-Being. Belknap Press of Harvard University Press.

Veiga, I., Oliveira, T., Naranjo-Zolotov, M., Martins, R., Karatzas, S., 2025. Indoor air quality perception: enablers and inhibitors of perceived occupant comfort. Environment International 202, 109690. Available at https://doi.org/10.1016/j.envint.2025.109690.

Wang, Y., Huang, X., Chen, Q., Liu, X., Lei, J., 2026. Dynamic response and sectional forces of reinforced concrete frames subjected to impact loads. Engineering Structures 352, 122160. Available at https://doi.org/10.1016/j.engstruct.2026.122160.

Wu, J.-R., Di Sarno, L., Freddi, F., D'Aniello, M., 2022. Modelling of masonry infills in existing steel moment-resisting frames: nonlinear force-displacement relationship. Engineering Structures 267, 114699. Available at https://doi.org/10.1016/j.engstruct.2022.114699.

Zhuang, X., Zhu, P., Yang, A., Caldas, L., 2025. Machine learning for generative architectural design: advancements, opportunities, and challenges. Automation in Construction 174, 106129. Available at https://doi.org/10.1016/j.autcon.2025.106129.

Smart structures: concepts and innovations

3

3.1 What makes a structure "Smart"

A smart structure represents a profound evolution beyond the conventional understanding of buildings as static physical entities. Rather than functioning solely as shelters designed to resist environmental forces, smart structures are conceived as dynamic, responsive, and adaptive systems that continuously interact with their surroundings, occupants, and operational conditions. They embody the convergence of architectural design, structural engineering, material science, digital technologies, and computational intelligence, transforming the built environment into an active participant in environmental regulation, resource management, and user experience. In this paradigm, intelligence is not defined by the mere presence of isolated technological components such as sensors or automation devices; instead, it emerges through the systemic integration of sensing, interpretation, decision-making, and adaptive response mechanisms across multiple scales of the building system.

The defining characteristic of a smart structure lies in its ability to perceive change, process information, and respond effectively. This transforms architecture from a passive form into an operational ecosystem capable of maintaining performance, optimizing efficiency, and improving resilience over time. Much like biological organisms that detect stimuli, process signals, and initiate responses, smart structures operate through interconnected feedback systems that allow them to sense environmental, structural, and human conditions and adjust accordingly. This cyber-physical integration redefines buildings as living systems that are capable not only of reacting but also of learning, predicting, and evolving in response to changing circumstances.

At the foundation of every smart structure is the capacity for sensing. Sensing technologies provide the building with awareness by continuously monitoring physical, environmental, and behavioral variables. Embedded structural health monitoring systems can detect strain, stress, deformation, vibration, and material fatigue, while environmental sensors monitor temperature, humidity, daylight intensity, air quality, and acoustic conditions. Occupancy sensors further expand this intelligence by tracking movement patterns, density, and user presence. Technologies such as fiber-optic sensors, piezoelectric materials, microelectromechanical systems, and wireless sensor networks enable highly precise data acquisition, forming the informational basis upon which smart performance depends. Without this sensory layer, a building cannot achieve meaningful adaptability, as intelligence begins with awareness.

However, sensing alone does not constitute smartness. Data must be interpreted and transformed into actionable knowledge. Thus, the second critical component of smart structures is data processing and computational feedback. Through microcontrollers, cloud platforms, internet of things (IoT) infrastructures, building information

Smart Civil and Structural Engineering. DOI: https://doi.org/10.1016/B978-0-443-30223-7.00005-X

modeling (BIM), artificial intelligence (AI), and machine learning algorithms, collected data is analyzed to identify patterns, predict future conditions, and optimize operational decisions. This computational layer serves as the cognitive system of the structure, converting raw environmental or structural data into strategic responses. Predictive maintenance systems, for example, can identify emerging material degradation before visible damage occurs, while machine learning platforms can optimize thermal comfort by analyzing occupancy schedules and climatic trends. In this way, buildings move from reactive systems to proactive environments, where future performance is anticipated rather than merely corrected after failure.

The third defining capability of smart structures is actuation and adaptive response. Once data is processed, the structure must possess mechanisms through which physical or operational adjustments can occur. These responses may be mechanical, material, spatial, or environmental. Adaptive façades may alter geometry or transparency in response to solar exposure; heating, ventilation, and air conditioning (HVAC) systems can recalibrate airflow based on occupancy; structural dampers may respond dynamically to seismic forces; and smart materials such as shape memory alloys (SMAs), electrochromic glass, and phase change materials can modify their behavior without external mechanical intervention. Actuation transforms intelligence into measurable performance by enabling the structure to physically adapt to fluctuating conditions. Through this capacity, buildings maintain equilibrium among comfort, safety, energy efficiency, and operational continuity.

Equally important is the principle of systems integration. True smartness cannot emerge from disconnected subsystems operating independently. Instead, structural, mechanical, environmental, digital, and user interaction systems must function as a coordinated whole. In smart structures, HVAC systems communicate with façade controls, structural monitoring interfaces connect with maintenance protocols, and occupancy analytics influence energy distribution strategies. This interconnectivity creates a cyber-physical ecosystem in which intelligence is distributed yet unified. The building behaves not as a collection of technological fragments but as an integrated organism capable of multidimensional responsiveness. Systems integration is therefore essential in transforming technological complexity into coherent operational intelligence.

A central operational model within smart structures is the closed-loop feedback system, which consists of sensing, processing, response, and recalibration. This loop continuously cycles as conditions change, ensuring that the building remains aligned with performance objectives. For example, increased solar radiation may be detected by façade sensors, analyzed through control algorithms, and responded to through automated shading systems, while subsequent environmental changes are reassessed to confirm effectiveness. This feedback loop allows buildings to self-regulate, maintain balance, and improve continuously over time. Such operational logic is what distinguishes genuinely smart systems from static automation.

Beyond technical performance, smart structures fundamentally redefine sustainability. Traditional sustainable design often focuses on static efficiency targets established during design or certification stages. Smart structures, by contrast, embed sustainability within operational behavior, allowing environmental performance to be continuously refined through adaptation. Energy conservation is achieved through demand-responsive systems; water management is enhanced through intelligent monitoring; material longevity is supported by predictive maintenance; and occupant well-being is improved through personalized environmental adjustments. Sustainability thus becomes dynamic, data-informed, and continuously optimized rather than fixed or prescriptive.

The human dimension is equally critical. Smart structures are not solely technological artifacts; they are environments designed to enhance human experience. Through user-centered interfaces, adaptive comfort systems, behavioral analytics, and personalized environmental controls, occupants become active participants within the building ecosystem. This shifts architectural functionality from predetermined standardization toward behavioral responsiveness, where spaces evolve according to user needs, preferences, and activities. In this context, intelligence serves not only operational efficiency but also social and experiential quality.

A compelling real-world example of these principles is the Asahi Glass Company (AGC) Headquarters in Louvain-la-Neuve, Belgium (Fig. 3.1). This project demonstrates how smart architectural intelligence can be embedded directly into the building envelope through adaptive façade technology (Samyn and De Coninck, 2014).

The façade system integrates advanced glazing technologies, environmental sensors, computational controls, and responsive shading mechanisms to regulate daylight, thermal gain, and occupant comfort dynamically. Sensors continuously monitor sunlight intensity, glare potential, internal lighting conditions, and solar orientation, transmitting

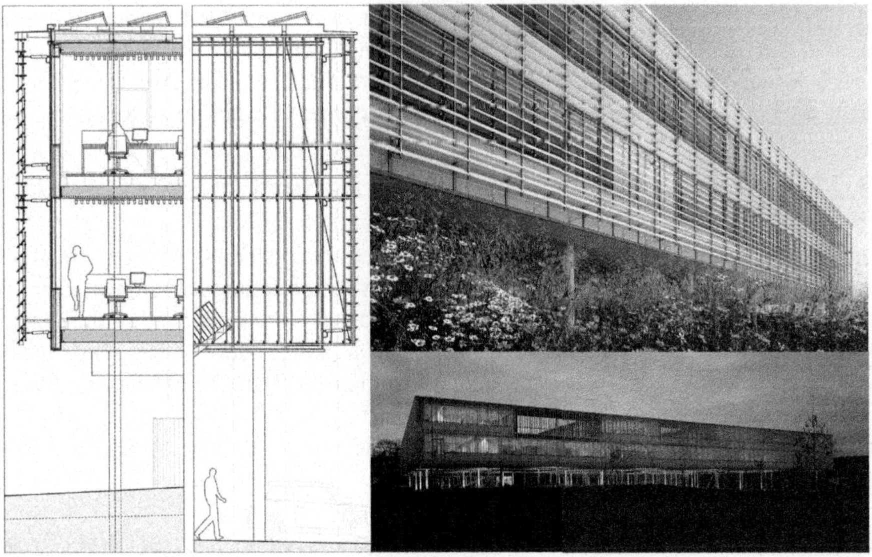

Figure 3.1 Views of the AGC Headquarters building.

real-time data to centralized management systems. The computational infrastructure interprets this environmental information using control algorithms that determine optimal façade responses throughout the day. Movable glazing elements, insulated façade layers, and silk-screened solar control devices adjust automatically to minimize overheating, maximize daylight penetration, and preserve visual comfort. These adaptive interventions not only improve occupant satisfaction but also reduce energy consumption associated with artificial lighting and mechanical cooling.

Importantly, the AGC Headquarters demonstrates that smartness is not confined to a single adaptive feature. The façade operates as part of a broader integrated environmental system, coordinating with lighting management and HVAC systems to ensure holistic performance. This systems-level coordination exemplifies the core principles of sensing, computational intelligence, adaptive actuation, and systemic integration. Rather than serving as a passive enclosure, the building envelope becomes an active environmental mediator, continuously optimizing building performance while reinforcing architectural expression.

The AGC Headquarters case highlights a critical lesson: smart structures derive intelligence not from isolated technologies but from the meaningful orchestration of architecture, engineering, materials, and digital systems into coherent adaptive frameworks. This approach redefines the role of buildings within contemporary society. They are no longer inert objects designed solely for shelter or symbolic expression but intelligent systems capable of sensing, learning, responding, and evolving.

Ultimately, what defines a structure as truly "smart" is its operational capacity to coordinate sensing, interpretation, and adaptive performance within a unified framework. Rather than emphasizing isolated technological novelty, smart structures distinguish themselves through continuous responsiveness, systemic coordination, and performance refinement across changing environmental and functional conditions. Their significance lies in redefining building operation as an evolving process, one in which structural systems actively enhance efficiency, adaptability, and long-term performance in response to changing environmental and functional conditions.

3.2 Materials and technologies behind smart structures

Smart structures emerge from the convergence of materials science, structural engineering, and computational intelligence, where building components evolve from passive physical elements into performance-active systems. Within this framework, materials and technologies operate as interdependent layers that enable structures to sense, process, and respond to environmental and operational conditions. This transformation shifts architectural and structural design toward systems capable of continuous adaptation, extending performance beyond static assumptions into dynamic operational behavior.

3.2.1 Smart materials: functional responsiveness in architecture

Smart materials have transformed contemporary architectural design by enabling building components to respond dynamically to changing environmental conditions.

Unlike conventional construction materials that maintain fixed physical properties throughout their service life, smart materials possess the ability to adapt their behavior in response to external stimuli such as temperature, light, humidity, pressure, or electrical input. This functional responsiveness enhances building performance, improves occupant comfort, and contributes significantly to energy efficiency and environmental sustainability. Among the most influential smart materials in architecture are SMAs, phase change materials (PCMs), and electrochromic glazing systems, each contributing to the development of adaptive and intelligent building envelopes.

SMAs are metallic materials capable of recovering their original shape after deformation when exposed to thermal or electrical activation. This unique behavior results from reversible phase transformations within the alloy structure. In architectural applications, SMAs are increasingly used in adaptive façades, responsive shading systems, deployable structures, and kinetic building components. Their ability to generate controlled movement without complex mechanical systems makes them particularly suitable for lightweight and energy-efficient responsive architecture. SMAs enable façades and structural systems to react autonomously to environmental changes, such as solar radiation or temperature fluctuations, thereby enhancing the adaptability and resilience of building envelopes.

PCMs represent another major category of smart materials widely applied in sustainable architecture. PCMs absorb, store, and release thermal energy during phase transitions, typically between solid and liquid states. This latent heat storage capability allows PCMs to moderate indoor temperature fluctuations and stabilize thermal conditions within buildings. During periods of elevated temperature, PCMs absorb excess heat as they melt, while stored heat is released when temperatures decrease, and the material solidifies. Through this process, PCMs reduce heating and cooling demands, improve thermal comfort, and contribute to lower operational energy consumption (Jayathunga et al., 2024).

The integration of PCMs into building envelopes has become increasingly common in wall systems, roof assemblies, ceiling panels, glazing systems, and façade cladding materials. Their passive thermal regulation capability provides an effective strategy for enhancing building energy performance without increasing mechanical system dependency. Research on PCM-enhanced façade panels has demonstrated substantial improvements in energy efficiency, including reductions in heating coil power demand of up to 11.3%, highlighting the effectiveness of PCMs in minimizing thermal loads while maintaining occupant comfort (Bigaila and Athienitis, 2017). Furthermore, PCM-integrated façades contribute to peak load reduction and improve the overall thermal inertia of buildings, particularly in climates characterized by significant day–night temperature variations (Novais et al., 2015). Despite their advantages, several challenges remain associated with PCM implementation in architecture. The selection of appropriate PCM compositions requires careful consideration of melting temperature ranges, thermal conductivity, encapsulation methods, fire performance, durability, and economic feasibility (Zhang et al., 2024).

Electrochromic glass is another important innovation in responsive façade technology. Electrochromic glazing systems can dynamically modify their transparency or tint level when subjected to low-voltage electrical signals, allowing precise control over daylight penetration and solar heat gain. This adaptive capability enables façades to respond continuously to external environmental conditions, improving visual comfort while reducing building energy consumption. In cold conditions, electrochromic glass remains transparent to maximize passive solar heat gain and daylight access, thereby reducing heating requirements (Fig. 3.2). Conversely, during hot periods, the glazing darkens automatically to limit excessive solar radiation, decrease cooling loads, and prevent indoor overheating.

The performance of electrochromic glazing can be further enhanced through integration with ventilated façade systems. Ventilated façades incorporate an air cavity between the external cladding layer and the insulated building envelope, enabling natural or mechanical airflow that improves thermal regulation. In warm climates, excess heat accumulated within the cavity is dissipated through convection, while tinted electrochromic glazing simultaneously minimizes solar heat transfer into interior spaces. In colder conditions, the cavity functions as an additional insulating layer, while the glazing remains clear to support passive heating strategies. The combination of ventilated façades and electrochromic glazing therefore creates highly adaptive building envelopes capable of optimizing thermal and visual performance under varying climatic conditions.

Electrochromic glazing systems provide several architectural and environmental benefits. By dynamically regulating visible light transmission and solar heat gain, these systems reduce glare, enhance occupant comfort, and decrease reliance on artificial lighting and mechanical cooling systems (Li et al., 2023). Electrochromic

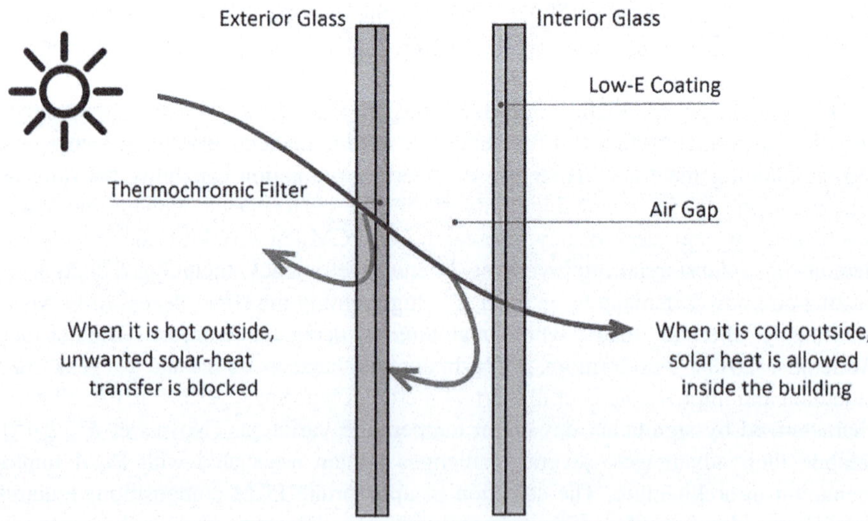

Figure 3.2 Schematic illustration of the functionality of electrochromic glass.

technologies are increasingly applied in windows, skylights, curtain walls, and double-skin façades, offering architects greater flexibility in balancing aesthetics, transparency, and energy performance (Bui et al., 2021). Studies conducted on commercial buildings equipped with electrochromic glazing have reported primary energy savings ranging from 7% to 16%, together with reductions in lifecycle operational costs of up to 18% when compared with conventional glazing systems (Detsi et al., 2024).

Nevertheless, the widespread adoption of electrochromic glazing remains constrained by factors such as higher initial costs, manufacturing complexity, and integration requirements with building management systems. Ongoing research aims to improve switching speed, optical performance, durability, and affordability while enabling more efficient integration with intelligent building automation systems (Villa et al., 2024). Future developments are expected to incorporate artificial intelligence and real-time environmental sensing, allowing electrochromic façades to operate autonomously according to occupancy patterns, climatic data, and energy optimization strategies (Es-sakali et al., 2024). The emergence of smart materials has fundamentally transformed the role of the building envelope from a static protective barrier into an adaptive and responsive environmental mediator. Through the integration of SMAs, PCMs, electrochromic glazing, and other intelligent materials, contemporary façades can actively regulate heat transfer, daylight penetration, ventilation, and structural movement in response to changing environmental conditions. These innovations contribute significantly to sustainable architecture by enhancing energy efficiency, reducing environmental impact, and improving occupant well-being. As research and technological development continue to advance, smart materials are expected to play an increasingly central role in the evolution of high-performance, resilient, and environmentally responsive buildings.

3.2.2 Enabling technologies: sensing and structural intelligence

The operational intelligence of smart structures is enabled through integrated sensing, actuation, and control systems that translate environmental conditions into measurable and actionable data. These systems form a cyber-physical framework that supports continuous structural awareness. A significant example is One World Trade Center, which integrates a comprehensive structural health monitoring system embedded within its high-rise steel and concrete framework (Fig. 3.3). Fiber-optic strain sensors, accelerometers, and temperature monitoring devices continuously record structural behavior related to wind loads, vibration patterns, and thermal effects. This data is transmitted to centralized computational platforms that analyze performance trends and support predictive maintenance strategies.

This monitoring system enables early detection of structural anomalies, ensuring long-term performance stability and operational safety in one of the tallest buildings in the United States. The feedback and control layer employs computational platforms ranging from microcontrollers and IoT-based systems to AI-assisted digital twins to interpret sensor data and determine optimal responses. This integration allows predictive operations, where buildings "learn" from historical data to predict

Figure 3.3 View and architectural section of One World Trade Center, illustrating the tower's tapered geometry, structural core, and spatial organization from base to spire.

future conditions. For example, AI-powered control systems can optimize HVAC performance by correlating occupancy patterns, outdoor weather, and indoor comfort goals (Kim et al., 2025). The resulting architecture operates less as a static structure and more as a responsive ecosystem guided by continuous information flow.

Beyond monitoring, smart structures incorporate actuation mechanisms that enable physical transformation in response to environmental input (Bakker, 2020). These include mechanical actuators, piezoelectric elements, and motorized systems that allow façades and structural components to adjust geometry, permeability, or stiffness. When combined with sensor networks, these systems establish a closed-loop interaction between environmental conditions and structural response. This enables façades to regulate shading, ventilation, and thermal performance dynamically, reducing reliance on energy-intensive mechanical systems while improving environmental stability.

3.2.3 Biological-structural integration: Bosco Verticale

A distinct trajectory in smart material systems is represented by Bosco Verticale, where vegetation is integrated as an active environmental and material component of the architectural system (Fig. 3.4). Unlike digitally driven smart systems, this project introduces biological intelligence as a performance layer within the built environment. Trees, shrubs, and vegetation are structurally embedded into façade balconies, forming a distributed ecological system that contributes to environmental regulation.

Figure 3.4 View and architectural section of Bosco Verticale, Milan, Italy.

This integration modifies structural behavior by introducing variable loads associated with soil, water retention, and plant growth. Reinforced balcony systems and carefully calibrated load transfer mechanisms ensure stability while supporting ecological functionality. The building envelope thus operates as a hybrid system in which architectural structure and biological growth coexist.

Vegetation contributes to environmental performance through shading, evapotranspiration, air filtration, and microclimatic regulation. This reduces solar heat gain, improves local air quality, and stabilizes external thermal conditions, demonstrating that environmental adaptation can be achieved through biological processes rather than mechanical systems.

Smart structures operate through interconnected layers of sensing, computation, and response. In One World Trade Center, this intelligence is digitally driven through sensor-based monitoring systems, while in Bosco Verticale, adaptation occurs through biological growth and ecological interaction. Despite their differences, both systems operate through continuous feedback between structure and environment. This reveals a shift toward distributed intelligence, where responsiveness is not centralized but embedded across multiple layers—material, structural, environmental, and biological. Buildings become adaptive systems capable of maintaining performance through continuous interaction with external conditions.

The convergence of smart materials, structural systems, and environmental mechanisms leads to integrated material-technology frameworks in which performance emerges from system interaction rather than isolated components. In such systems:

- materials actively respond to environmental stimuli;
- structural systems support both load and environmental regulation;
- environmental processes are embedded within architectural form.

This integration transforms buildings into adaptive systems that operate across multiple temporal scales, from real-time response (sensor-driven behavior) to long-term ecological adaptation (biological systems such as vegetation growth).

Smart materials and enabling technologies redefine the boundaries of structural engineering by transforming buildings into responsive systems capable of continuous adaptation. Through the integration of sensor networks in One World Trade Center and biological systems in Bosco Verticale, two complementary models of intelligence emerge: one digital and computational, the other ecological and biological. Together, these case studies demonstrate that structural intelligence is not limited to electronic systems but can also emerge through living processes embedded within architecture. This dual perspective establishes a broader material paradigm in which buildings function as adaptive environments that continuously negotiate between structure, environment, and life.

3.3 Sensors and their role in smart design

The integration of sensing technologies constitutes one of the fundamental operational pillars of smart structures, establishing the critical interface through which architecture perceives, interprets, and responds to dynamic physical, environmental, and human conditions. Sensors serve as the perceptual infrastructure of intelligent buildings by continuously transforming real-world variables—including structural stress, thermal fluctuations, humidity, air quality, light intensity, occupancy, and vibration—into measurable digital data. Through this continuous data acquisition, buildings evolve from static physical enclosures into adaptive cyber-physical systems capable of real-time situational awareness, predictive analysis, and responsive control.

Within smart architectural ecosystems, sensors function not merely as isolated measurement tools but as interconnected nodes embedded within broader computational and operational frameworks (Makanadar, 2024). Their integration with building management systems, IoT networks, BIM, and digital twin platforms enables closed-loop feedback cycles in which sensed conditions are analyzed, interpreted, and translated into operational decisions. This process fundamentally transforms buildings into responsive systems capable of optimizing performance, enhancing resilience, and continuously refining environmental and structural efficiency (Mohammadi et al., 2024).

Sensing technologies in smart design can be broadly classified into four principal domains: structural sensing, environmental sensing, occupant-centered sensing, and acoustic sensing. Each category contributes distinct layers of intelligence while collectively reinforcing comprehensive building awareness. Structural sensing systems are primarily dedicated to monitoring physical integrity, load behavior, and long-term material performance. Technologies such as strain gauges, displacement sensors, accelerometers, and fiber-optic monitoring systems continuously assess deformation, vibration, stress distribution, and fatigue accumulation across structural systems. Among these technologies, Fiber Bragg Grating sensors have become particularly significant due to their precision, durability, and adaptability under demanding operational conditions (Martínez Vicente et al., 2023). Embedded within beams, façades, bridges, and complex structural envelopes, these sensors enable early anomaly detection, predictive maintenance, and improved lifecycle resilience, particularly in large-span or structurally unconventional buildings.

Environmental sensing systems focus on monitoring the relationship between building envelopes and surrounding climatic conditions. These systems are especially critical in smart façades, ventilated envelopes, and adaptive building skins, where empirical performance validation is essential. Surface temperature sensors, thermocouples, resistance temperature detectors (RTDs), heat flux sensors, humidity sensors, and ambient environmental detectors continuously monitor thermal transfer, moisture behavior, airflow, and climatic variation. Such sensor networks provide indispensable real-world validation for simulation models, allowing architects and engineers to calibrate digital predictions and optimize long-term building performance.

Empirical monitoring frameworks significantly enhance the reliability of predictive design by bridging the gap between computational simulation and operational reality. Laboratory experimentation and in situ testing methods—including infrared thermography, heat flux analysis, and environmental logging—reveal discrepancies between modeled assumptions and actual performance, ensuring that smart façades achieve intended energy and environmental objectives (Fig. 3.5).

Surface temperature monitoring systems, including RTDs and thermocouples, are directly integrated into façade assemblies to track thermal behavior across internal and external material layers (Fig. 3.6). These systems provide critical data regarding solar heat gain, thermal lag, airflow effectiveness, and seasonal temperature variability, enabling precise assessment of façade responsiveness under changing

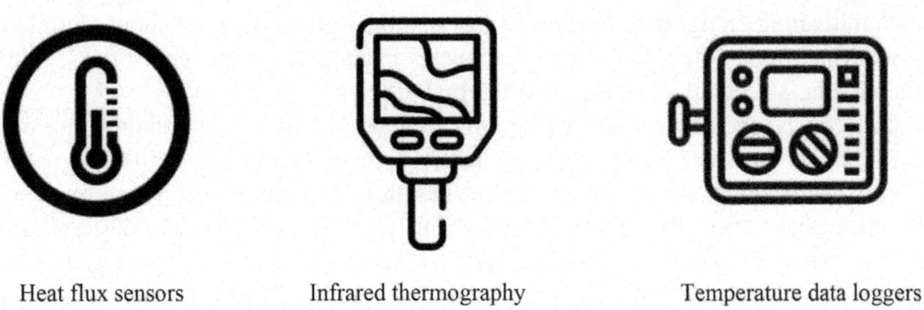

Figure 3.5 Empirical validation and simulation calibration framework.

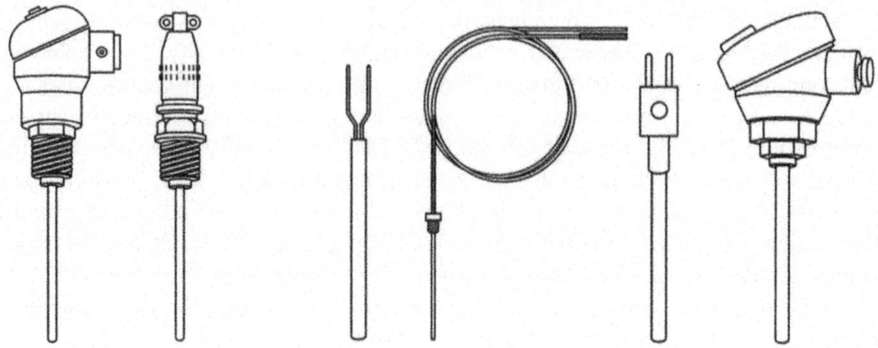

Figure 3.6 Schematic illustration of surface temperature sensors: RTDs and thermocouples. *RTDs*, resistance temperature detectors.

environmental conditions. In ventilated façades, this information is particularly valuable for evaluating cavity performance and optimizing thermal buffering strategies.

Heat flux sensors embedded within envelope assemblies measure conductive heat transfer through façade systems, providing direct data on insulation performance, U-values, and thermal bridge formation (Fig. 3.7). By quantifying thermal resistance empirically, these sensors allow for targeted material refinement and envelope optimization, supporting more energy-efficient building designs.

Ambient temperature sensors positioned strategically around building exteriors establish external climatic baselines that contextualize façade behavior. These sensors are essential for understanding thermal gradients, seasonal fluctuations, and façade-environment interactions, thereby strengthening comprehensive thermal ecosystem analysis (Fig. 3.8).

Humidity sensors further expand environmental intelligence by monitoring moisture accumulation within façade cavities and material layers. Moisture management is essential for preserving insulation effectiveness, preventing condensation, and mitigating long-term material degradation. Through continuous humidity monitoring, smart façades can proactively manage environmental risks while maintaining structural durability and thermal efficiency (Fig. 3.9).

Collectively, these environmental sensing technologies transform building envelopes from passive barriers into monitored, adaptive systems capable of continuous performance optimization. Sensor-integrated façades thereby become active environmental mediators that contribute significantly to sustainable design objectives. Acoustic sensing systems introduce another critical dimension of smart performance by monitoring sound transmission, vibration patterns, and occupant comfort. Acoustic accelerometers and vibration sensors (Fig. 3.10) mounted on façade assemblies detect structural resonance, sound leakage, and external noise penetration. These systems are essential for evaluating

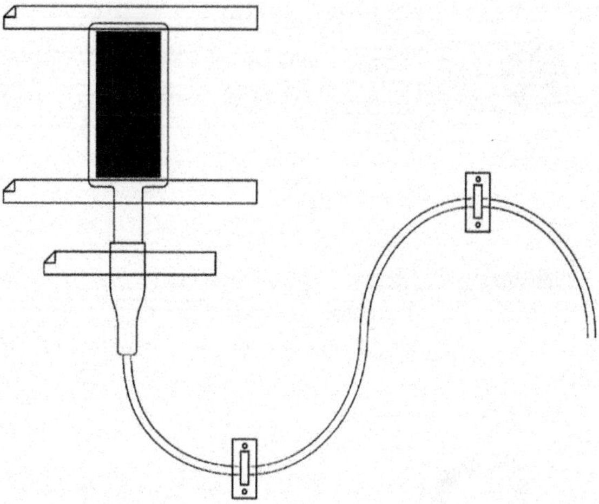

Figure 3.7 Schematic illustration of the installation of a heat flux sensor using tape to fixate the sensor and the connection block.

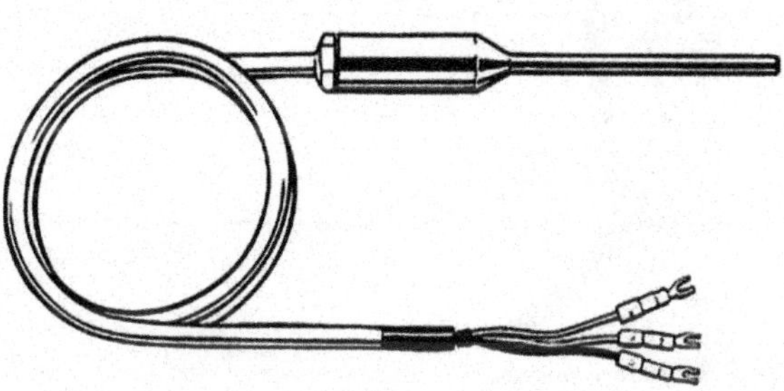

Figure 3.8 Schematic illustration of a temperature sensor dedicated type.

acoustic insulation performance and ensuring compliance with standards such as ISO 140-3 (Montes González et al., 2020). By identifying vulnerabilities in acoustic performance, these sensors support the optimization of façade systems for urban environments where noise mitigation is a significant concern.

Occupant-centered sensing systems extend smart design into the realm of human-responsive architecture. Motion detectors, infrared occupancy sensors, biometric technologies, and behavioral analytics platforms monitor occupancy patterns, movement, and user interaction. These technologies enable buildings to dynamically adjust lighting, HVAC systems, ventilation, acoustics, and security conditions based on actual user behavior rather than static schedules. IoT-integrated occupancy sensing significantly enhances energy efficiency, occupant comfort, and operational adaptability while reinforcing human-centered design priorities (Gao et al., 2025).

The Taipei Performing Arts Center (TPAC) exemplifies the advanced integration of sensing technologies within contemporary smart architecture (Fig. 3.11). Through embedded structural sensors, environmental monitoring systems, and occupancy-

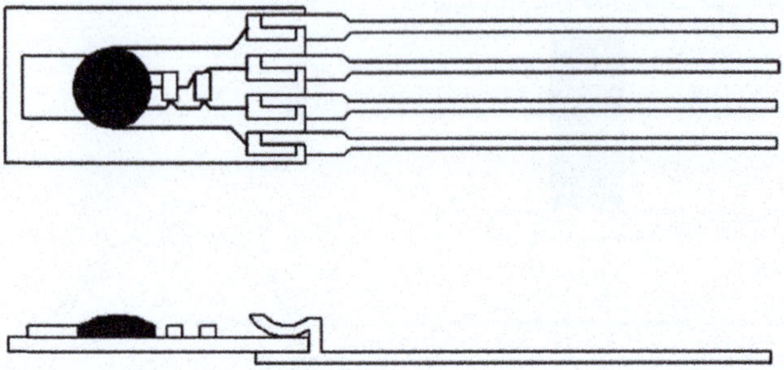

Figure 3.9 Schematic illustration of a digital humidity sensor optimal for all general-purpose humidity applications.

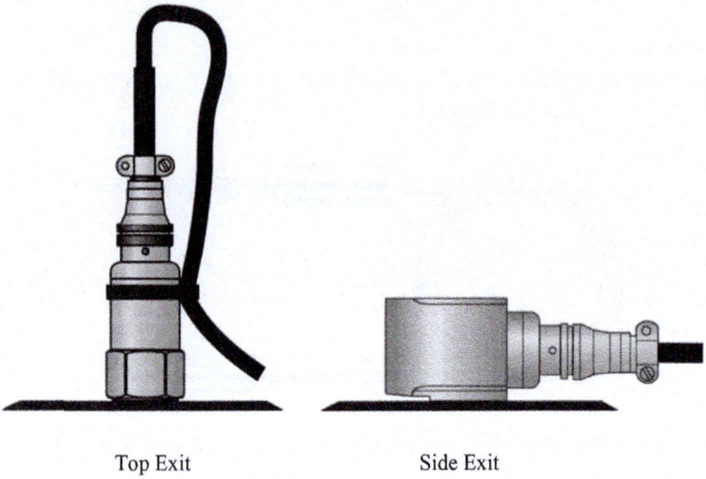

Figure 3.10 Schematic illustration of the mounting of a vibration sensor for horizontal measurement and axial measurement.

responsive controls, TPAC continuously compares real-world performance against predictive digital models, enabling adaptive operational adjustments and predictive maintenance. This case illustrates how sensor ecosystems can unify structural resilience, environmental responsiveness, and occupant-centered intelligence within a single integrated architectural framework.

The future trajectory of sensing technologies in smart design increasingly emphasizes embedded intelligence, autonomy, and predictive capability. Emerging innovations such as self-powered sensors, sensor fusion networks, and AI-integrated analytics are expanding the role of sensors beyond passive monitoring toward autonomous operational forecasting and adaptive control (Rafi et al., 2024; Kuo et al., 2025). Self-powered systems capable of harvesting energy from solar exposure or structural vibration promise enhanced deployment longevity, while machine learning algorithms strengthen predictive maintenance, occupant adaptation, and environmental optimization.

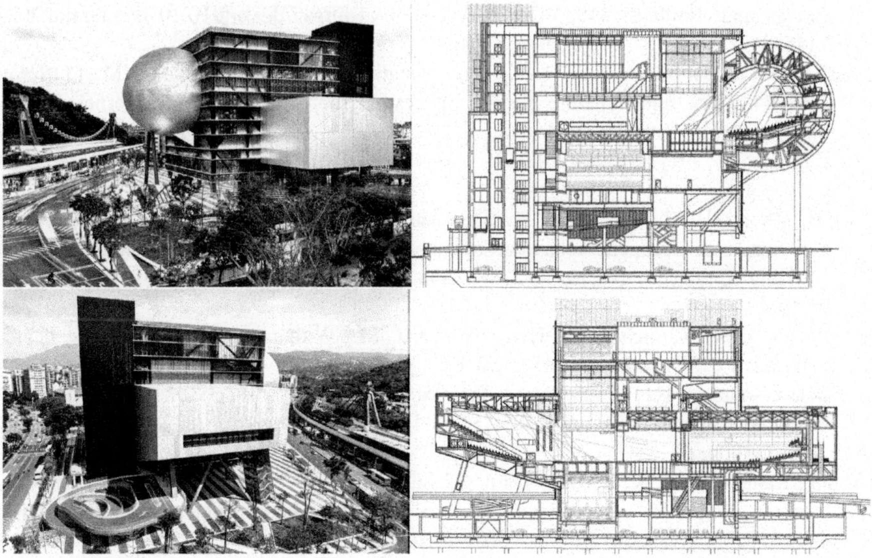

Figure 3.11 View and architectural sections of the Taipei Performing Arts Center, illustrating its structural configuration, sensor integration zones, and adaptive environmental systems.

Sensors function as the foundational intelligence infrastructure of smart architecture. They mediate the relationship between material systems and computational decision-making, allowing buildings to perceive, evaluate, and continuously adapt to evolving conditions. Through structural, environmental, acoustic, and occupant-responsive sensing, architecture is redefined as a dynamic, self-regulating ecosystem where safety, sustainability, comfort, and resilience are continuously optimized. As smart structures continue to evolve, sensing technologies will remain central to transforming buildings into perceptive, adaptive organisms embedded within complex environmental and technological networks.

References

Bakker, R., 2020. Smart Buildings: Technology and the Design of the Built Environment. Riba Publications. Available at https://www.perlego.com/book/1518526/smart-buildings-technology-and-the-design-of-the-built-environment-pdf.

Bigaila, E., Athienitis, A.K., 2017. Modeling and simulation of a photovoltaic/thermal air collector assisting a façade integrated small scale heat pump with radiant PCM panel. Energy and Buildings 149, 298–309. Available at https://doi.org/10.1016/j.enbuild.2017.05.045.

Bui, D.-K., Nguyen, T.N., Ghazlan, A., Ngo, T.D., 2021. Biomimetic adaptive electrochromic windows for enhancing building energy efficiency. Applied Energy 300, 117341. Available at https://doi.org/10.1016/j.apenergy.2021.117341.

Detsi, M., Atsonios, I., Mandilaras, I., Founti, M., 2024. Effect of smart glazing with electrochromic and thermochromic layers on building's energy efficiency and cost.

Energy and Buildings 319, 114553. Available at https://doi.org/10.1016/j.enbuild.2024.114553.

Es-sakali, N., Idrissi Kaitouni, S., Ait Laasri, I., Oualid Mghazli, M., Cherkaoui, M., Pfafferott, J., 2024. Static and dynamic glazing integration for enhanced building efficiency and indoor comfort with thermochromic and electrochromic windows. Thermal Science and Engineering Progress 52, 102681. Available at https://doi.org/10.1016/j.tsep.2024.102681.

Gao, H., Zhong, H., Zou, L., 2025. Human-centric IoT control: a framework for quantifying the impact of occupant behaviour on energy efficiency in shared offices. Journal of Building Engineering 108, 112784. Available at https://doi.org/10.1016/j.jobe.2025.112784.

Jayathunga, D.S., Karunathilake, H.P., Narayana, M., Witharana, S., 2024. Phase change material (PCM) candidates for latent heat thermal energy storage (LHTES) in concentrated solar power (CSP) based thermal applications - a review. Renewable and Sustainable Energy Reviews 189, 113904. Available at https://doi.org/10.1016/j.rser.2023.113904.

Kim, S., Kang, B., Lee, S., Lee, T., Yoon, G., Baek, Y., et al., 2025. AI-optimized energy and comfort control for carbon-neutral tropical buildings. Journal of Building Engineering 111, 113415. Available at https://doi.org/10.1016/j.jobe.2025.113415.

Kuo, T., Manikkan, S., Liu, X., Devarapalli, H., Bilionis, I., Karava, P., 2025. AI-enabled building energy model generation and domain randomization for control policy learning. Building and Environment 113815. Available at https://doi.org/10.1016/j.buildenv.2025.113815.

Li, W., Shah, K.W., Li, Y., 2023. Effects of electrochromic glass window on energy efficiency, thermal and visual performances under tropical climate. Energy for Sustainable Development 76, 101300. Available at https://doi.org/10.1016/j.esd.2023.101300.

Makanadar, A., 2024. Neuro-adaptive architecture: buildings and city design that respond to human emotions, cognitive states. Research in Globalization 8, 100222. Available at https://doi.org/10.1016/j.resglo.2024.100222.

Martínez Vicente, J.L., González-Gallego, M., Terroba Ramírez, F., Frövel, M., López Cela, J.J., 2023. Study of the transverse strain effect on the fiber bragg grating sensor (FBGS) response with polyimide coating under experimental biaxial tests. Composite Structures 326, 117621. Available at https://doi.org/10.1016/j.compstruct.2023.117621.

Mohammadi, M., Assaf, G., Assaad, R.H., 2024. Real-time spatial-temporal mapping and visualization of thermal comfort and HVAC control by integrating immersive augmented reality technologies and IoT-enabled wireless sensor networks: towards immersive human-building interactions. Journal of Building Engineering 94. Available at https://doi.org/10.1016/j.jobe.2024.109887, https://www.sciencedirect.com/science/journal/23527102.

Montes González, D., Barrigón Morillas, J.M., Rey Gozalo, G., Atanasio Moraga, P., 2020. Microphone position and noise exposure assessment of building façades. Applied Acoustics 160, 107157. Available at https://doi.org/10.1016/j.apacoust.2019.107157.

Novais, R.M., Ascensão, G., Seabra, M.P., Labrincha, J.A., 2015. Lightweight dense/porous PCM-ceramic tiles for indoor temperature control. Energy and Buildings 108, 205–214. Available at https://doi.org/10.1016/j.enbuild.2015.09.019.

Rafi, M.R., Hu, F., Li, S., Song, A., Zhang, X., O'Neill, Z., 2024. Deep weighted fusion learning (DWFL)-based multi-sensor fusion model for accurate building occupancy detection. Energy and AI 17, 100379. Available at https://doi.org/10.1016/j.egyai.2024.100379.

Samyn, P., De Coninck, J., 2014. AGC Glass Building.
Villa, D.L., Hahn, N.T., Grey, J.K., Pavich, F., 2024. Futures for electrochromic windows on high performance houses in arid, cold climates. Energy and Buildings 315, 114293. Available at https://doi.org/10.1016/j.enbuild.2024.114293.
Zhang, W., Shi, Y., Yang, H., Zou, Y., Cao, X., Cui, H., 2024. Design optimization of passive PCM-enhanced hospital buildings for efficient energy-saving. Journal of Energy Storage 91, 112037. Available at https://doi.org/10.1016/j.est.2024.112037.

Architectural aesthetics and smart structures

4

4.1 The unity of form and function

The relationship between form and function has long served as one of the foundational principles of architectural theory, traditionally rooted in the belief that the physical form of a building should emerge directly from its intended use, structural requirements, and spatial purpose. Historically associated with modernist thought, this principle emphasized that architectural expression should not be arbitrary or purely decorative but rather a coherent manifestation of practical necessity (Lucchi, 2023). In the age of smart structures, however, this concept has evolved far beyond its conventional interpretation (Yiannoudes, 2016). Form is no longer simply the result of programmatic organization or structural efficiency; instead, it becomes an adaptive, responsive, and technologically enhanced medium through which performance, environmental intelligence, and human experience are simultaneously articulated.

Within smart architecture, the unity of form and function is significantly expanded through the integration of advanced computational methodologies, responsive materials, sensing technologies, and real-time control systems. Buildings increasingly operate as dynamic systems rather than static compositions, allowing architectural form to embody environmental responsiveness, structural intelligence, and operational adaptability (Kolarevic and Parlac, 2015). This transformation positions form not merely as a physical container but as an active participant in performance optimization (Ramadan et al., 2026). Architectural geometry, façade articulation, structural configuration, and spatial organization are continuously informed by environmental data, user behavior, and technological feedback loops, creating structures where appearance and operation are inseparable (Cocho-Bermejo and Vogiatzaki, 2022).

Traditionally, architecture and engineering often functioned as parallel but distinct disciplines. Architects frequently emphasized symbolic, cultural, and esthetic ambitions, while engineers prioritized stability, safety, and efficiency. This separation often resulted in compromises between visual expression and technical performance. However, the rise of computational design, parametric modeling, building information modeling, and smart technologies has fundamentally altered this relationship. These tools enable a deeply integrated design process in which structural logic, environmental analysis, and formal exploration occur simultaneously (Napps et al., 2026). In this new paradigm, architectural form emerges through iterative collaboration, where esthetic decisions are shaped by performance criteria, and technical systems contribute directly to visual and spatial identity.

This convergence has generated a new architectural language in which structural systems, environmental strategies, and technological intelligence actively shape

Smart Civil and Structural Engineering. DOI: https://doi.org/10.1016/B978-0-443-30223-7.00007-3

design expression. Smart structures exemplify an inseparable fusion of artistic ambition and engineering precision, where buildings are conceived as responsive ecosystems rather than inert objects. Esthetic qualities increasingly arise from performance-driven objectives rather than superficial stylistic choices, demonstrating how beauty can emerge from efficiency, adaptability, and technical sophistication (Abramovich, 2016). In this context, form becomes performative—an embodiment of structural, environmental, and experiential intelligence.

A compelling example of this integration is Foster + Partners' Apple Park Visitor Center in Cupertino (Fig. 4.1). The project illustrates how formal identity can emerge directly from environmental and structural optimization. Its circular geometry, expansive overhangs, high-performance glazing systems, and carefully calibrated orientation are not solely esthetic decisions but environmental strategies designed to reduce solar heat gain, enhance daylight penetration, and optimize natural ventilation. The building's sleek, minimalistic appearance is therefore inseparable from its operational logic. Architectural expression, environmental control, and structural innovation function collectively, producing a design in which performance and visual identity reinforce one another.

Figure 4.1 Views of the Apple Park Visitor Center.

The Apple Park Visitor Center demonstrates that contemporary architectural beauty can be generated through optimization rather than ornamentation. Its design confirms that sustainable and intelligent systems need not constrain creativity; rather, they can become drivers of innovation. This reflects a broader transformation in architectural thinking, where the building envelope, structure, and environmental systems collectively define esthetic identity.

Smart structures extend this principle even further by embedding adaptive behavior directly into formal systems. Responsive façades, kinetic envelopes, morphing roofs, and programmable shading systems represent a significant departure from static architecture. In these systems, form is not fixed but evolves continuously in response to changing climatic conditions, occupancy patterns, or operational requirements. The building envelope itself becomes an active interface between internal and external environments, capable of altering transparency, geometry, or thermal properties in real time. Consequently, formal qualities emerge not solely from geometry but from behavioral intelligence.

This shift introduces the concept of architecture as a living system, in which buildings visually communicate performance through motion, transformation, and environmental responsiveness. Kinetic façades that adjust according to solar exposure, adaptive shading systems that recalibrate based on daylight levels, and responsive structural systems that accommodate varying loads all demonstrate how functionality can actively shape appearance over time. The building's visual identity becomes dynamic, changing throughout the day or across seasons as it responds to external stimuli.

Such responsiveness fundamentally redefines architectural esthetics. Rather than pursuing permanence and static perfection, smart structures embrace adaptability, temporality, and operational expression. Form becomes an evolving manifestation of environmental intelligence, revealing performance through visible adaptation. This approach challenges traditional architectural assumptions by positioning buildings as active agents within ecological and technological systems.

Moreover, the unity of form and function in smart structures extends beyond technical performance to encompass human experience. Spatial quality, comfort, visual engagement, and psychological well-being are increasingly integrated into performance-based design strategies. Responsive systems can enhance daylight quality, thermal comfort, acoustic conditions, and occupant interaction, ensuring that technological sophistication directly supports user experience. In this sense, function is not limited to structural or mechanical performance but includes the broader experiential dimensions of architecture.

The convergence of design intelligence, environmental responsiveness, and structural innovation ultimately establishes a new paradigm in which form and function are mutually generative rather than hierarchically related. Smart structures embody this transformation by demonstrating that esthetic expression, environmental sustainability, structural performance, and technological adaptability can coexist within unified systems. Buildings are no longer conceived as static solutions to predefined needs but as adaptive environments capable of continuous evolution.

As architecture advances into increasingly complex environmental and technological contexts, the unity of form and function becomes not only a design principle but also a strategic necessity. Smart structures reveal that true innovation lies not in prioritizing form over function or function over form, but in achieving their seamless integration. Through this holistic synthesis, architecture evolves into a discipline capable of producing environments that are visually compelling, operationally intelligent, environmentally responsive, and deeply aligned with human needs. The future of architectural design thus resides in the continued exploration of this unity, where buildings function as living systems whose beauty emerges directly from their intelligence, adaptability, and performance.

A notable example of this integrated design philosophy is the Serpentine Pavilion by Snøhetta, which powerfully demonstrates how computational methodologies can unify structural logic, material efficiency, environmental responsiveness, and experiential quality within a single architectural expression (Fig. 4.2). Rather than treating form as an independent esthetic gesture, the pavilion's design emerges from a deeply performance-driven process in which geometry, structural optimization, and public interaction are developed simultaneously. Through advanced computational modeling and parametric analysis, the project achieves a sophisticated balance between lightweight construction, structural precision, and immersive spatial experience.

The pavilion's sweeping lattice roof structure is composed of carefully calibrated timber members and tensioned cable systems, configured according to optimized load

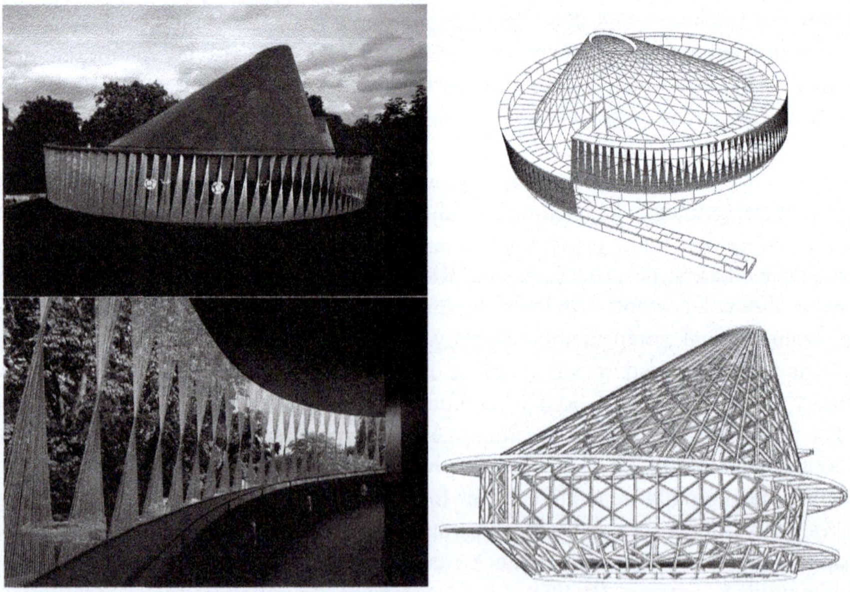

Figure 4.2 Views of the Serpentine Pavilion.

distribution pathways. This approach minimizes material use while maximizing structural performance, illustrating how digital tools can refine complex geometries to achieve both efficiency and elegance (Weber, 2024). The resulting form is not arbitrary but generated through the precise alignment of structural forces, material behavior, and spatial functionality. By translating engineering requirements directly into architectural language, the pavilion exemplifies how structural systems themselves can become primary drivers of visual identity.

Beyond structural performance, the pavilion also reflects the importance of user-centered design within smart architectural practice (Li et al., 2020). Its fluid geometry shapes patterns of movement, gathering, visual connection, and environmental interaction, ensuring that structural logic supports social engagement and sensory experience. The interplay of transparency, shading, and spatial enclosure creates a dynamic public environment that responds to changing light conditions and occupancy patterns. In this way, computational design extends beyond technical problem-solving to facilitate experiential richness, demonstrating that intelligent form-making can simultaneously enhance performance and human interaction (Economidou et al., 2024).

The Serpentine Pavilion further underscores the transformative role of digital workflows in contemporary architecture (Caetano and Leitão, 2020). Computational modeling allows architects and engineers to simulate structural behavior, environmental conditions, and circulation dynamics in parallel, enabling iterative refinement that strengthens both technical and experiential outcomes (Gerber and Lin, 2014). This process dissolves traditional boundaries between architecture and engineering, replacing linear design hierarchies with integrated systems thinking (Knippers et al., 2021). Form is therefore conceived not as a static object but as a responsive outcome of multiple performance parameters working in concert.

The pavilion illustrates how computational intelligence can serve as a bridge between structural rationality and creative ambition. Its design reflects a broader paradigm within smart architecture, where architectural beauty is derived from measurable performance rather than superficial styling. Material selection, force calibration, environmental responsiveness, and spatial quality collectively shape the final form, reinforcing the principle that design excellence emerges from interdisciplinary synthesis.

Ultimately, the Serpentine Pavilion demonstrates that in smart architecture, form and function are not separate considerations but mutually generative forces. Structural systems do not merely support architecture; they actively shape its identity, spatial qualities, and environmental behavior. Likewise, architectural expression enhances structural performance by organizing material systems into coherent, adaptive, and meaningful forms. This reciprocal relationship exemplifies the evolving nature of intelligent design, where computational tools enable architecture to operate as an integrated ecosystem of esthetics, performance, and human experience.

Through projects such as the Serpentine Pavilion, the unity of form and function is redefined for the digital age. Architecture becomes a responsive synthesis of creativity and calculation, where structural optimization, environmental strategy, and public engagement converge to produce spaces that are simultaneously efficient, adaptive, and visually compelling. This model reflects the future trajectory of smart structures, in which computational intelligence transforms architectural design into an evolving dialog between technology, performance, and spatial expression.

A significant illustration of this integrated approach is The Shed in New York, designed by Diller Scofidio + Renfro (2019), which demonstrates how kinetic systems can extend architectural performance beyond fixed spatial boundaries (Fig. 4.3). The building is defined by a movable outer shell that travels along a rail-based infrastructure, enabling the structure to expand or contract in response to programmatic demands and event-specific requirements. This transformation is not merely formal but operational, allowing the building to shift between enclosed and open configurations without compromising structural stability or environmental control. The telescopic mechanism is supported by a robust steel exoskeleton and integrated wheel assemblies, which ensure precise movement while maintaining alignment and load continuity throughout transformation phases.

Figure 4.3 Views of the Shed's Bloomberg Building.

In this configuration, spatial adaptability becomes an intrinsic property of the architectural system rather than an external addition. The envelope functions as an active mediator between interior activities and external urban space, recalibrating boundaries according to use intensity, climatic conditions, and temporal programming. This behavior positions the building as a variable infrastructure rather than a fixed container, where architectural identity is continuously redefined through motion and reconfiguration.

The project further demonstrates how structural engineering and architectural intent converge through mechanical intelligence. The kinetic envelope operates as both enclosure and apparatus, embedding movement as a design principle rather than a mechanical feature. As the shell extends or retracts, the spatial envelope reorganizes lighting conditions, ventilation potential, and perceptual openness, producing distinct environmental atmospheres within a single architectural system. This integration allows the building to negotiate between enclosure and exposure, stability and transformation, without compromising functional continuity.

More broadly, this approach reflects a fundamental reorientation in contemporary architectural thinking, where form is understood as a variable condition rather than a fixed outcome. Instead of treating structure, program, and appearance as separate domains, smart architectural systems merge these dimensions into a unified operational logic. In this framework, formal expression is directly informed by performance criteria, including movement capacity, spatial flexibility, and environmental responsiveness. As a result, architectural identity emerges through behavior, timing, and transformation rather than static composition. This evolving paradigm is supported by advances in computational design, simulation-based modeling, and responsive system integration, which allow architects to test and refine dynamic behaviors prior to construction. Material systems and mechanical components are no longer passive executors of design but active participants in spatial formation. The integration of these elements enables buildings to function as adaptive infrastructures capable of responding to shifting contextual conditions.

Smart structures redefine architecture as a time-based and performance-driven discipline. Buildings are no longer conceived solely as spatial objects but as systems that evolve through interaction with users, the environment, and operational demands. Within this framework, kinetic and adaptive architectures such as The Shed illustrate how responsiveness can be embedded directly into form, producing environments that continuously adjust, negotiate, and reorganize themselves while maintaining coherence between structural logic, functional requirements, and experiential quality.

4.2 Designing for visual impact

Designing for visual impact within smart structural systems requires a redefinition of architectural expression, where appearance is no longer treated as an independent compositional outcome but as the visible consequence of underlying system behavior. In this framework, esthetics are produced through operational logic rather than applied ornamentation (Cecilia Russo and Ferrara, 2017), and the building's visual identity is continuously shaped by its interaction with environmental forces, occupancy patterns, and technological feedback systems.

Unlike conventional approaches in which visual composition is resolved through static geometries and predefined material arrangements, smart structures generate form through iterative computational processes that integrate structural efficiency, environmental adaptation, and spatial performance. As a result, visual impact emerges from the articulation of dynamic relationships between load distribution, energy exchange, and adaptive responses (Kanaani, 2024), rather than from isolated formal decisions. The building becomes legible as a system in motion, where variation, transformation, and responsiveness contribute to its visual character.

A key dimension of this shift lies in the use of computational modeling and simulation-driven design environments, which allow architectural forms to be continuously tested against multiple performance criteria. These tools enable designers to explore complex geometric configurations that respond simultaneously to structural demands, daylight optimization, airflow dynamics, and spatial usability. The resulting architectural outcomes often exhibit a high degree of formal complexity, not as an esthetic objective in itself, but as the spatial byproduct of resolving competing performance conditions (Meagher, 2015).

Within this context, visual impact is closely tied to the perception of intelligence embedded in form. Facades that modulate transparency, structures that reveal their load logic, and envelopes that adjust in response to environmental stimuli communicate a sense of architectural awareness. This perceptual quality transforms buildings into expressive systems, where change over time becomes part of the visual language. Rather than presenting a fixed image, smart structures generate evolving visual conditions that reflect real-time interaction between internal systems and external environments. Material systems further contribute to this dynamic expression. Adaptive surfaces, variable translucency materials, and responsive coatings allow the building envelope to shift its appearance according to light, temperature, or user activity. These transformations reinforce the idea that visual impact in smart architecture is not static but temporal, defined by variation across different conditions and moments of occupancy.

Designing for visual impact in smart structures represents a convergence of performance and perception. Architectural expression is no longer separated from functional logic but is instead embedded within it, producing environments where esthetics arise from measurable processes and adaptive behaviors. This integration enables buildings to communicate their internal intelligence externally, creating architectures that are visually distinctive not through stylistic emphasis, but through the visible articulation of responsiveness, efficiency, and systemic coherence.

A significant example is the Morocco Pavilion at Expo 2020 Dubai, which demonstrates how a climate-responsive façade can create a strong architectural identity (Fig. 4.4). The building's rammed-earth exterior, combined with passive cooling strategies, produces a visually striking monolithic form that simultaneously reflects regional traditions and contemporary performance optimization. The pavilion's spatial and material expression emerges from thermal mass behavior, proving that performance itself can be the foundation of esthetic impact.

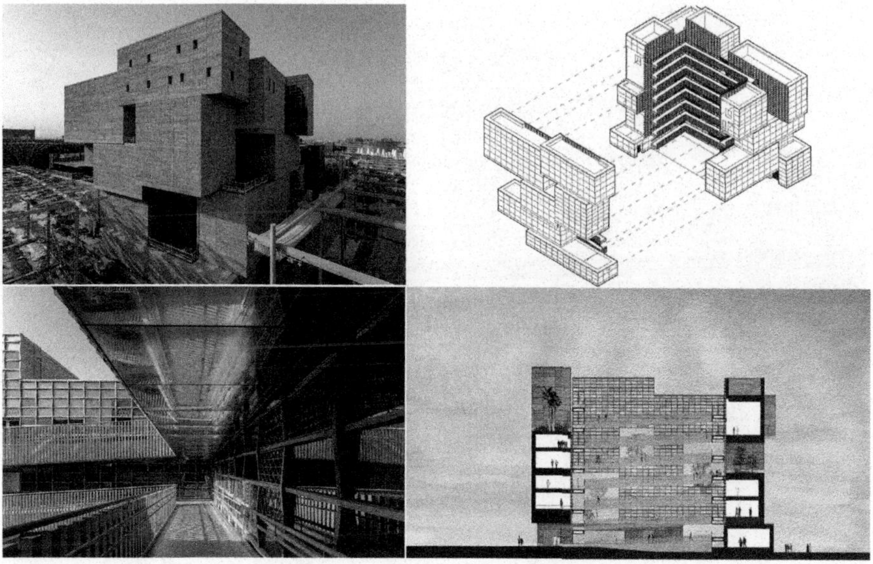

Figure 4.4 Views of the Morocco Pavilion at Expo 2020 Dubai.

Another exemplary project is the Bund Finance Center in Shanghai, which features a kinetic façade composed of layered bronze tubes that move to regulate light, visibility, and environmental exposure (Fig. 4.5). Inspired by traditional Chinese theater curtains, the façade's operable screens slide along tracks to create continuously shifting patterns of light and shadow. This dynamic envelope not only produces a visually striking and culturally resonant form but also performs as a climate-responsive interface that reduces glare and solar heat gain. The project demonstrates how computational design, environmental simulation, and advanced fabrication techniques can merge to generate a façade system where visual identity is inseparable from functional adaptability and energy performance. The building's ability to transform its appearance in real time illustrates how smart technologies elevate esthetic expression through movement, environmental intelligence, and cultural symbolism.

Smart structures capable of movement or transformation offer new avenues for visual impact by producing dynamic esthetics that change throughout the day, season, or occupancy cycle. Here, visual impact is not static but temporal, as the building expresses its intelligence through motion, shading adjustments, or environmental modulation.

Smart structures achieve visual impact not through formal spectacle but through the legibility of their environmental intelligence. In this context, the Dynafit Headquarters, also referred to as the "Glass Peaks for Mountaineering Gear," offers a particularly compelling illustration (Fig. 4.6). Designed by Barozzi Veiga, the building is defined by its steeply inclined, faceted glass façade. The architects and

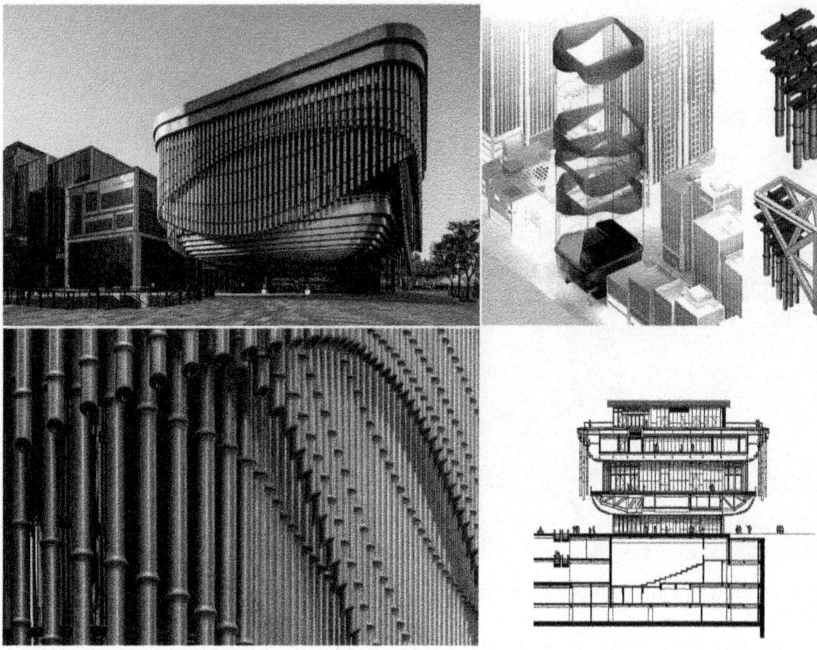

Figure 4.5 Views of the Bund Finance Center.

engineers developed an energy- and cost-efficient shading strategy of remarkable simplicity, integrating optimized glass geometries with calibrated panel inclination. The result is an envelope that reduces solar heat gain while maintaining high daylight penetration and panoramic views.

As solar conditions shift throughout the day, the prismatic façade refracts light in constantly changing patterns, producing a crystalline expression that appears both sharp and atmospheric. This visual dynamism is not an esthetic afterthought: it emerges directly from the envelope's environmental logic, demonstrating how a performance-driven façade can generate a distinct architectural identity rooted in climatic responsiveness.

The Olympic Aquatics Center in Paris similarly exemplifies how geometric precision and environmental performance coalesce into visual coherence (Fig. 4.7). Its large-span timber–steel roof, optimized through parametric modeling, distributes daylight uniformly across interior pools while minimizing heat gain and structural weight. The resulting undulating profile evokes hydrodynamic flow, translating environmental moderation and structural efficiency into a unified formal language.

Material systems play a central role in the expressive capacity of smart structures, particularly when performance characteristics are made visible. At the Dynafit Headquarters, the high-transmittance glazing, combined with the façade's steep inclinations, acts as both a shading mechanism and a visual medium. The simplicity of the solution is noteworthy: rather than relying on mechanically actuated screens,

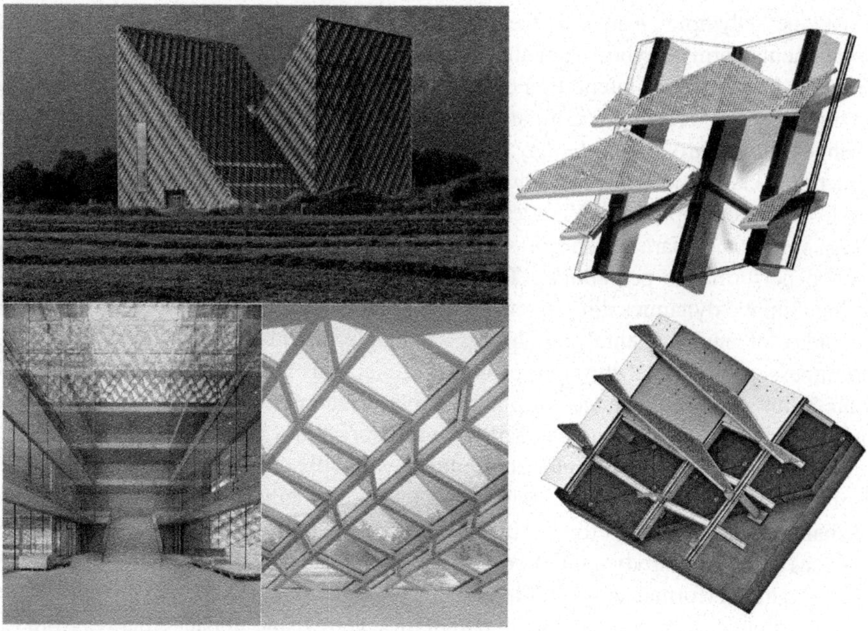

Figure 4.6 Views of the Dynafit Headquarters.

Figure 4.7 Views of the Olympic Aquatics Center.

the design employs geometry and material properties as passive regulators of solar exposure (Wijewardane et al., 2025). Variations in light intensity, weather conditions, and seasonal angle changes are visually expressed across the façade, allowing material performance to become an esthetic agent.

At the Olympic Aquatics Center, laminated timber and lightweight metal components form a hybrid roof that reveals structural rationality while optimizing embodied energy. The integration of translucent roof bands diffuses daylight into the swimming halls, creating a calm and luminous interior environment whose character changes subtly throughout the day. These material strategies demonstrate how smart structures can produce a continuously evolving visual experience grounded in environmental adaptation (Wang et al., 2024).

An important characteristic of smart structures is their ability to make environmental performance perceptible. The Dynafit Headquarters achieves this by registering climatic dynamics directly onto its inclined glass surfaces. The façade becomes an index of environmental variability; sun angle, cloud density, and atmospheric brightness are legible through changing reflections and refractions. Form and materiality thus operate as communicative systems, conveying environmental intelligence without mechanical articulation (Bibri et al., 2024).

The Olympic Aquatics Center communicates intelligence through its aerodynamic roof geometry, engineered to resist wind uplift and enhance thermal stability. Its formal articulation is therefore inseparable from its environmental logic: the building reads as a fluid, hydrodynamic body whose visual identity arises from its structural and climatic performance. Both the Dynafit Headquarters and the Olympic Aquatics Center articulate a maturing esthetic paradigm where visual impact emerges from the inherent logic of smart performance. This paradigm is defined by:

- Legible performance: Adaptations and environmental moderations made visible in form and material.
- Computational precision: Optimized geometries that reduce material use while enhancing efficiency.
- Material expressiveness: Envelopes whose visual qualities evolve with atmospheric conditions.
- Temporal dynamism: Identities shaped by changing light, weather, and seasonal patterns.

These projects demonstrate that contemporary architectural esthetics increasingly derive from the interplay of structure, climate, and material intelligence. In this light, the visual impact of smart structures is not an application or ornament.

4.3 Case studies in aesthetic excellence

Esthetic excellence in the context of smart structures emerges when architectural form is conceived not as an autonomous visual artifact but as the perceptible outcome of integrated environmental intelligence, structural reasoning, and technological mediation. Rather than relying on applied ornamentation or purely symbolic gestures, such architecture derives its expressive power from the logic of performance, how a building senses, responds to, and interacts with its surroundings and users. In this framework, beauty is inseparable from behavior, and visual identity becomes a manifestation of underlying systems rather than an imposed stylistic layer.

Contemporary smart structures demonstrate that esthetic impact can be generated through advanced materials, computational design methodologies, and performance-driven structural optimization (Kathiravel et al., 2026). Parametric modeling, environmental simulation, and digital fabrication allow architects and engineers to align formal expression with climatic responsiveness, structural efficiency, and spatial experience. As a result, architectural form is increasingly shaped by data-informed processes in which geometry, material behavior, and environmental flows are calibrated simultaneously. These processes enable buildings to articulate complexity, variation, and responsiveness in ways that were previously unattainable through conventional design methods.

Within this paradigm, the esthetic quality of a building is closely linked to its capacity to communicate intelligence, both visually and experientially. Façades may register changes in light and climate, structures may express load paths through refined geometries, and spaces may adapt subtly to patterns of use and movement. The resulting architecture is not static, but dynamic in perception, offering continuously evolving visual and spatial conditions. Such qualities are particularly evident in projects where structure itself becomes the primary generator of spatial and esthetic meaning.

A seminal example of this approach is the KAIT Pavilion and Plaza in Yokohama, designed by Junya Ishigami. The KAIT Pavilion is characterized by an expansive field of ultra-slender steel columns, more than 300 in number, distributed across the plan according to a rigorously calibrated structural logic (Fig. 4.8). Eschewing conventional grids, axial hierarchies, or compositional symmetry, the columns are arranged in a seemingly irregular yet precisely engineered pattern. This configuration produces a spatial condition marked by ambiguity, openness, and visual indeterminacy, wherein architectural boundaries are dissolved rather than explicitly articulated.

The resulting "forest" of columns blurs distinctions between interior and exterior, structure and space, enclosure and openness. Spatial limits are not defined by walls or envelopes but are instead perceived through shifting densities, changing sightlines, and subtle variations in proximity. As users move through the pavilion, the space continuously reconstitutes itself, generating layered depth and a heightened awareness of bodily movement and perception. This deliberate indeterminacy constitutes the core of the project's esthetic excellence, transforming structural necessity into an experiential and perceptual framework.

Complementing the pavilion, the KAIT Plaza operates as a minimalist, semiopen, column-free space that foregrounds the architectural potential of overlooked interstitial environments (Fig. 4.9). Its apparent simplicity is underpinned by advanced structural optimization, allowing for an unobstructed spatial field that reinforces openness and collective use. Together, the Pavilion and Plaza articulate two complementary strategies: one defined by multiplicity and spatial fragmentation, the other by clarity and openness, both unified by a commitment to structural precision and material restraint.

Crucially, the intelligence of the KAIT projects does not reside in visible technological apparatuses, responsive systems, or sensor-driven adaptability. Instead, smartness is embedded in the optimization of structure itself, where computational analysis and precision fabrication enable each element to be reduced to its essential structural function. The extreme slenderness of the columns results in

Figure 4.8 Views of the KAIT Pavilion.

an almost immaterial presence, allowing structure to recede visually while remaining performatively indispensable. In this sense, material efficiency, spatial perception, and esthetic expression are inseparably linked.

The KAIT Pavilion and Plaza exemplify how smart structural thinking can generate esthetic value through restraint rather than spectacle. Architectural intelligence is expressed quietly and implicitly, through the careful balance between structural necessity and experiential richness. The absence of dominant elements or rigid enclosures fosters a fluid and human-centered relationship between users and space, positioning perception, movement, and atmosphere as primary design drivers.

Taken together, this case study demonstrates that esthetic excellence in smart structures emerges from the convergence of form, performance, and intelligence. Rather than treating technology as an additive feature, the KAIT projects reveal a paradigm in which intelligence is intrinsic to spatial conception and structural logic. Beauty, in this context, arises from coherence between what a building is, how it performs, and how it is experienced, marking a fundamental shift from static

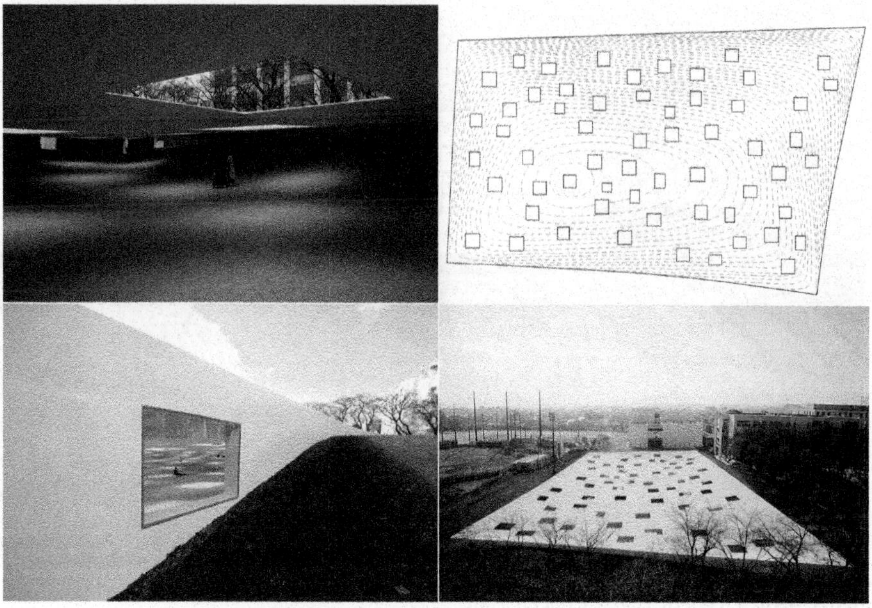

Figure 4.9 Views of the KAIT Plaza.

architectural composition toward dynamically perceived and structurally informed spatial environments.

Harpa Concert Hall in Reykjavík, designed by Henning Larsen Architects in collaboration with Batteríið Architects and artist Olafur Eliasson, exemplifies how computationally generated geometry can simultaneously define the esthetic identity and environmental performance of an architectural façade (Fig. 4.10). The building's crystalline envelope, composed of hexagonal and quasibrick glass modules, was developed through an iterative design process integrating parametric modeling, solar radiation analysis, and optical simulation. Through this workflow, each geometric unit was systematically calibrated in relation to façade orientation, daylight penetration requirements, and thermal performance criteria, enabling the envelope to function as a responsive environmental mediator while sustaining a highly articulated visual expression.

What distinguishes Harpa as a smart structural system is the deliberate synthesis of optical intelligence and structural logic. The prismatic geometry of the façade refracts, reflects, and diffuses Iceland's rapidly changing daylight conditions, generating a continuously evolving interplay of light, color, and shadow. This dynamic visual behavior is not a purely esthetic effect but the outcome of a performance-driven design strategy enabled by advanced environmental modeling and digital fabrication techniques. Variations in glass thickness, angular configuration, surface treatment, and coatings contribute directly to passive solar regulation, glare mitigation, and thermal modulation, demonstrating how esthetic qualities can emerge as a direct consequence of environmental responsiveness.

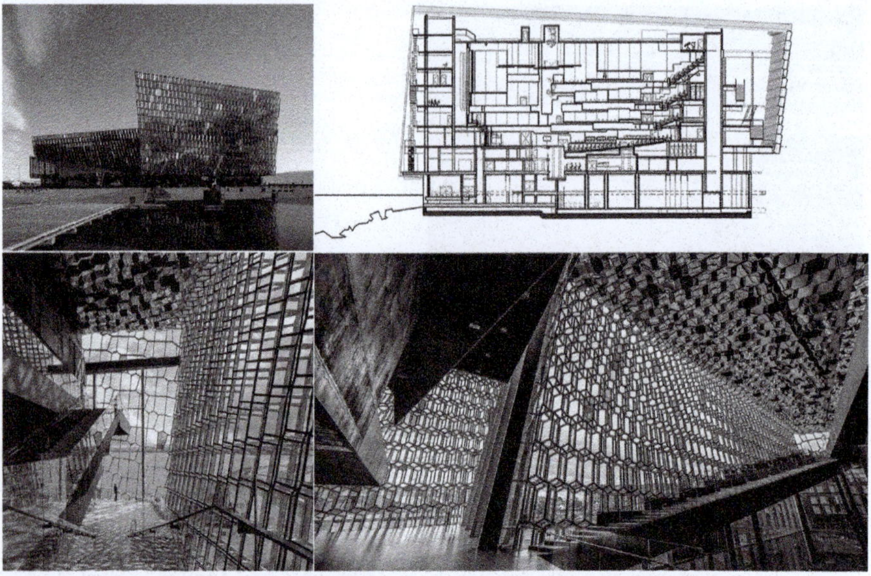

Figure 4.10 Views of Harpa Concert Hall.

Furthermore, the steel substructure supporting the façade is precisely coordinated with the modular geometric system, allowing surface articulation, structural anchorage, and environmental performance to operate as an integrated whole. Rather than treating structure, envelope, and appearance as independent layers, Harpa unifies these dimensions into a single, coherent architectural system. In this regard, the project stands as a compelling example of smart design in which artistic intent, computational precision, and environmental intelligence converge. Harpa's esthetic impact is inseparable from its geometric and environmental performance, illustrating the capacity of smart structures to produce architecture that is simultaneously expressive, functional, and technologically advanced.

References

Abramovich, H., 2016. Intelligent Materials and Structures. De Gruyter. Available at http://doi.org/10.1515/9783110338027.

Bibri, S.E., Huang, J., Jagatheesaperumal, S.K., Krogstie, J., 2024. The synergistic interplay of artificial intelligence and digital twin in environmentally planning sustainable smart cities: a comprehensive systematic review. Environmental Science and Ecotechnology 20. Available at https://doi.org/10.1016/j.ese.2024.100433.

Caetano, I., Leitão, A., 2020. Architecture meets computation: an overview of the evolution of computational design approaches in architecture. Architectural Science Review 63 (2), 165–174. Available at https://doi.org/10.1080/00038628.2019.1680524.

Cecilia Russo, A., Ferrara, M., 2017. Smart solutions, "smart aesthetics"? Design Journal 20 (1), S342–S353. Available at https://doi.org/10.1080/14606925.2017.1352872.

Cocho-Bermejo, A., Vogiatzaki, M., 2022. Phenotype variability mimicking as a process for the test and optimization of dynamic facade systems. Biomimetics 7 (3). Available at https://doi.org/10.3390/biomimetics7030085.

Economidou, E., Itzlinger, A., Frauenberger, C., 2024. Lived experience in human-building interaction (HBI): an initial framework. Frontiers in Computer Science 5. Available at https://doi.org/10.3389/fcomp.2023.1233904.

Gerber, D.J., Lin, S.H.E., 2014. Designing in complexity: simulation, integration, and multidisciplinary design optimization for architecture. Simulation 90 (8), 936–959. Available at https://doi.org/10.1177/0037549713482027.

Kanaani, M., 2024. The Routledge companion to smart design thinking in architecture & urbanism for a sustainable. Living Planet. Routledge. Available at http://doi.org/10.4324/9781003384113.

Kathiravel, R., Hou, D., Feng, H., 2026. Optimizing structural and environmental performance in mass timber hybrid high-rise buildings through parametric design. Journal of Building Engineering 118, 115005. Available at https://doi.org/10.1016/j.jobe.2025.115005.

Knippers, J., Kropp, C., Menges, A., Sawodny, O., Weiskopf, D., 2021. Integrative computational design and construction: rethinking architecture digitally. Civil Engineering Design 3 (4), 123–135. Available at https://doi.org/10.1002/cend.202100027.

Kolarevic, B., Parlac, V., 2015. Building Dynamics: Exploring Architecture of Change Building Dynamics: Exploring Architecture of Change. Taylor and Francis Inc, Canada, pp. 1–298.

Li, Z., Zhang, J., Li, M., Huang, J., Wang, X., 2020. A review of smart design based on interactive experience in building systems. Sustainability 12 (17), 6760. Available at https://doi.org/10.3390/su12176760.

Lucchi, E., 2023. Energy and climatic performances of modern architecture: a complete overview of building physics implications. Engineering Proceedings 53 (1). https://doi.org/10.3390/IOCBD2023-15203.

Meagher, M., 2015. Designing for change: the poetic potential of responsive architecture. Frontiers of Architectural Research 4 (2), 159–165. Available at https://doi.org/10.1016/j.foar.2015.03.002.

Napps, D., Staudt, J., Saluz, U., Chen, X., Steiner, D., Zong, C., Deghim, F., Lang, W., Geyer, P., Schnellenbach-Held, M., Petzold, F., Borrmann, A., König, M., 2026. Evaluation of building design variants in early phases on the basis of adaptive detailing strategies. Buildings 16 (4). Available at https://doi.org/10.3390/buildings16040685.

Ramadan, L.A., Mokadem, A.E., Badawy, N., 2026. Advancing architecture: form-finding of architectural forms using comprehensive multi-objective optimization. International Journal of Architectural Computing 24 (1), 181–214. Available at https://doi.org/10.1177/14780771251316123.

Wang, M., Jia, Z., Tao, L., Xiang, C., 2024. Review of dynamic façade typologies, physical performance and control methods: towards smarter and cleaner zero-energy buildings. Journal of Building Engineering (98), 111310. Available at https://doi.org/10.1016/j.jobe.2024.111310.

Weber, R., 2024. On the Aesthetics of Architecture: A Psychological Approach to the Structure and the Order of Perceived Architectural Space. Routledge. Available at http://doi.org/10.4324/9781003502883.

Wijewardane, S., Vasilakopoulou, K., Santamouris, M., 2025. Dynamic solar shadings: redefining the building envelope. Solar Compass 16. Available at https://doi.org/10.1016/j.solcom.2025.100139.

Yiannoudes, S., 2016. Architecture and Adaptation: From Cybernetics to Tangible Computing. Taylor and Francis Inc., Greece, pp. 1–231.

Tensegrity systems in architecture: beauty in tension

5

5.1 Understanding tensegrity: tension and compression in equilibrium

Tensegrity systems represent one of the most elegant syntheses of structural efficiency, architectural expression, and systemic intelligence. Derived from the concept of tensional integrity, tensegrity structures achieve stability not through continuous mass, rigid frames, or hierarchical load paths, but through a finely tuned equilibrium between discontinuous compression elements and a continuous network of tension (Vilnay et al., 2023). This structural paradigm fundamentally challenges conventional load-bearing architecture by demonstrating that minimal material deployment, when strategically organized through force balance, can produce systems of remarkable strength, adaptability, and visual lightness.

Rather than relying on gravity-driven compression stacks or rigid enclosures, tensegrity systems operate as self-stabilizing assemblies, where structural integrity emerges from internal force equilibrium (Mirats Tur and Juan, 2009). As a result, tensegrity has evolved from an experimental structural curiosity into a powerful design logic capable of informing architectural form-making, performance optimization, and smart structural behavior.

Tensegrity is defined as a structural system in which isolated compression elements (struts) are suspended within a continuous tensile network (cables or tendons), such that the entire system stabilizes itself through a state of self-equilibrium. Unlike conventional framed or load-bearing structures, tensegrity systems do not rely on continuous compressive paths extending from roof to foundation. Instead, stability arises from the prestressed interaction between tension and compression, allowing the structure to maintain its geometry under load with exceptional material efficiency (Wang et al., 2024a).

The term "tensegrity" was popularized by R. Buckminster Fuller (Calladine, 1978), who famously described these systems as "islands of compression in an ocean of tension." The compression members in this arrangement are spatially isolated and kept in place by constantly tensioned cables, so they never come into contact with one another. This information appears overly automated. A closed force system that is stable even under deformation is created when the struts push outward, and the cables pull inward. At the core of tensegrity lies the principle of static and dynamic equilibrium. Every internal force is counterbalanced by an opposing force, producing a stable yet flexible configuration without the need for rigid joints or massive structural components. Importantly, this equilibrium is actively maintained through prestress, which imparts stiffness and structural integrity even before external loads, such as gravity or wind, are applied (Bieniek, 2017).

Smart Civil and Structural Engineering. DOI: https://doi.org/10.1016/B978-0-443-30223-7.00009-7

The simplest and most fundamental tensegrity configuration is the simplex structure, which is geometrically based on a triangular arrangement. In this system, a top triangular plane is connected to a bottom triangular plane through a combination of cables and struts. As illustrated in Fig. 5.1, the simplex tensegrity consists of three compression struts and nine tension cables, forming the minimum stable three-dimensional tensegrity module.

This elementary configuration clearly demonstrates the core principle of tensegrity: compression elements remain discontinuous, while the tensile network provides global stability and geometric control. Despite its simplicity, the simplex structure exhibits high stiffness, redundancy, and spatial clarity, making it a foundational typology for more complex tensegrity assemblies.

More advanced configurations include quadrangular cylindrical tensegrity systems, which introduce orthogonal symmetry and enhanced modular clarity. Fig. 5.2 illustrates the initial configuration and the corresponding self-equilibrium state of a

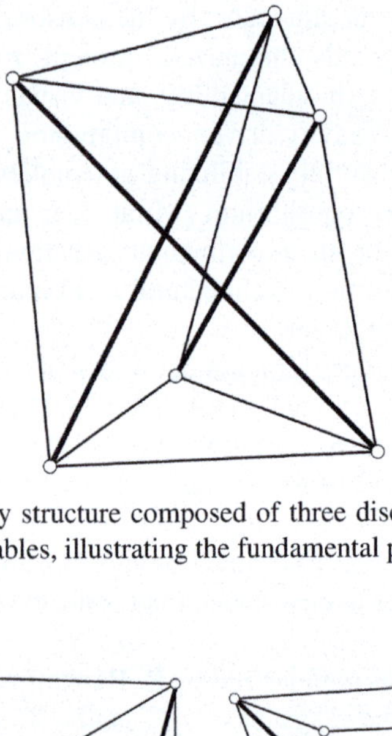

Figure 5.1 Simplex tensegrity structure composed of three discontinuous compression struts and nine continuous tension cables, illustrating the fundamental principle of tensional integrity.

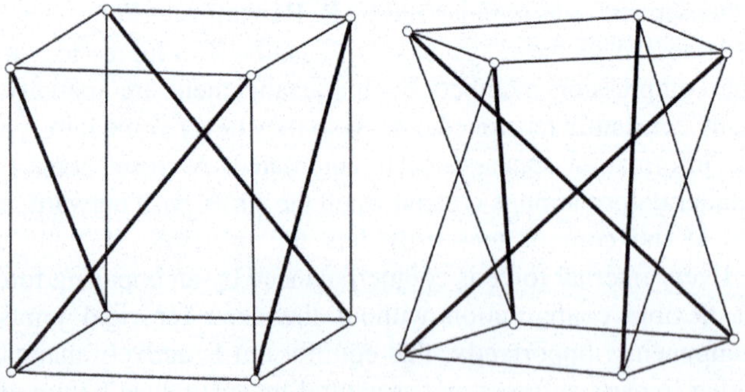

Figure 5.2 Initial configuration and self-equilibrium state of a quadrangular cylindrical tensegrity structure, illustrating stabilization through prestress and internal force balance.

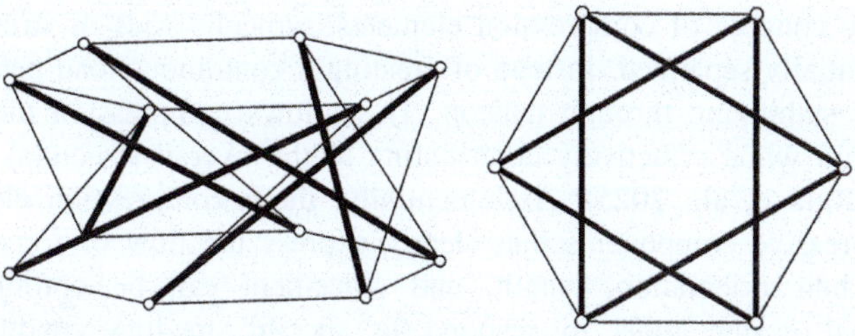

Figure 5.3 Hexagonal tensegrity module shown in three-dimensional perspective (left) and corresponding top view (right), highlighting the spatial separation of compression members and continuity of the tensile network.

quadrangular cylindrical tensegrity structure, clearly demonstrating how global stability is achieved through prestress rather than external support. The transformation from the initial geometry to the self-equilibrated state highlights the critical role of internal force redistribution in defining the final structural form (Chen et al., 2024).

Building upon this logic, Fig. 5.3 presents a hexagonal tensegrity module in both a three-dimensional perspective view and a corresponding top view. This representation clarifies the spatial separation of compression members and the continuity of the tensile network, offering insight into how complex three-dimensional forms emerge from simple force relationships.

In architectural applications, the equilibrium-driven behavior of tensegrity systems enables them to:

- absorb and redistribute loads efficiently;
- achieve exceptionally high strength-to-weight ratios;
- respond elastically to deformation rather than failing catastrophically.

Because loads are transmitted throughout the entire tensile network, localized stress concentrations are significantly reduced. This distributed load behavior aligns closely with principles of structural resilience and redundancy, which are central to contemporary smart and adaptive design strategies. When subjected to external forces, tensegrity systems adjust internally, redistributing stresses without compromising global stability. In tensegrity structures, tension plays the dominant organizational role, acting as the connective tissue that binds the system together (Ma et al., 2026). Cables, tendons, or tensile membranes form a continuous network that defines geometry, maintains spatial relationships, and stabilizes compression elements. The uninterrupted nature of this tensile network allows forces to travel along multiple paths, enhancing robustness and fault tolerance. From an architectural perspective, tension introduces several key properties such as visual transparency and perceived lightness, explicit visualization of force flow, and the ability to span large distances with minimal material. Tension-based systems are particularly suitable for smart and responsive architecture, as tensile elements can be easily integrated with sensors, actuators, or shape-memory materials. By adjusting prestress levels in response to environmental conditions, such as wind, temperature, or occupancy, tensegrity structures can dynamically change stiffness, shape, or load-carrying capacity.

Tensegrity consists of compression elements, typically rods or struts, discontinuous and spatially separated. Instead of creating a continuous load path, they resist buckling by stabilizing through tension. This allows compression members to be extremely thin while effectively contributing to the overall structural performance (Fernández-Ruiz et al., 2025). Architecturally, these compression elements often become expressive components that clearly express the flow of forces within the structure. Their orientation, length, and placement can be optimized through computational design tools to respond to specific loading conditions, spatial constraints, or esthetic intentions. As a result, the structural logic becomes legible, reinforcing the unity of form, force, and function (Wang et al., 2022).

Understanding the equilibrium between tension and compression in tensegrity systems has profound implications for architectural design:

- Material efficiency: maximum performance with minimal mass;
- Adaptability: adjustable prestress enables changes in stiffness or geometry;
- Deployability: structures can be folded, transported, and rapidly deployed;
- Esthetic clarity: structural behavior is made visible and intelligible.

These characteristics make tensegrity particularly relevant for smart and responsive architectural environments, where adaptability, efficiency, and expressive performance are paramount. While originally conceived as a structural concept, tensegrity can be understood as a precursor to contemporary smart structural systems (Al Sabouni-Zawadzka and Gilewski, 2018). Its reliance on distributed intelligence, feedback-driven equilibrium, and systemic interaction closely mirrors the logic of sensor-based and adaptive architectural frameworks. When augmented with digital control systems, responsive materials, and real-time monitoring, tensegrity structures can evolve into intelligent frameworks capable of learning, adapting, and responding over time.

Tensegrity systems redefine architectural stability as a condition of balance rather than mass. By mastering the interplay between tension and compression, architects and engineers gain access to a structural language that is lightweight, expressive, and inherently adaptable. This understanding lays the foundation for subsequent sections of the chapter, which will explore tensegrity's architectural applications, smart material integration, and future potential within adaptive and responsive built environments.

5.2 Tensegrity in architectural design

While tensegrity originated as a structural concept grounded in engineering mechanics, its relevance to architecture lies in its capacity to operate simultaneously as a structure, form generator, and spatial organizer. In architectural design, tensegrity transcends its role as a load-bearing system and becomes a design strategy that informs geometry, material expression, and spatial experience. By distributing forces through a network of tension and isolated compression, tensegrity enables architectures that are lightweight, transparent, and formally expressive while remaining structurally efficient.

Unlike conventional structural systems, where architecture is often constrained by beams, columns, and rigid load paths, tensegrity allows form to emerge from force equilibrium rather than static hierarchy. This shift aligns closely with contemporary architectural paradigms that prioritize performance-driven design, material efficiency, and adaptability. In tensegrity-based architecture, form is not imposed but discovered through the balancing of forces. The geometry of a tensegrity system emerges from the spatial configuration of struts and tendons, often resulting in nonorthogonal, curvilinear, or lattice-like forms that would be difficult to achieve using conventional construction logic.

A compelling extension of tensegrity-based architectural design can be found in biotensegrity models inspired by biological systems, where form emerges through force equilibrium and controlled deformation. An illustrative example is the spine-inspired biotensegrity model (Chai Lian et al., 2020), which demonstrates how tensegrity principles can be employed as a computational form-finding strategy rather than a fixed structural typology (Fig. 5.4). Their work explores three variations of a k=1 biotensegrity spine model, in which compression elements emulate vertebrae while tensile components replicate muscular and connective tissue behavior.

In order to allow the structure to seamlessly switch between configurations while preserving global equilibrium, a method that can simulate controlled shape change across four distinct stages was proposed in this spine-inspired biotensegrity model. In order to keep the cables under tension and prevent them from loosening during geometric reconfiguration, the form-finding procedure looks for the ideal forced elongation of the tension elements. This method shows how tensegrity systems can withstand ongoing deformation without losing structural integrity, a characteristic that is especially important for responsive and adaptive architectural systems.

From an architectural perspective, this model demonstrates how tensegrity can function as a design intelligence, one that integrates geometry, force distribution, and transformation logic into a unified framework. Rather than prescribing form a priori, the final geometry emerges from algorithmic negotiation between tension, compression, and

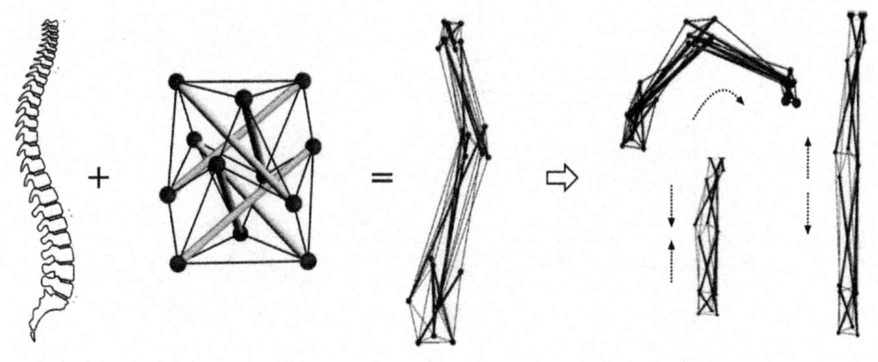

Figure 5.4 Form-finding process of a spine-inspired biotensegrity model showing multistate shape transformation through controlled elongation of tension elements, ensuring continuous equilibrium and preventing cable slack.

boundary conditions, closely aligning with performance-driven and biomimetic design paradigms in contemporary architecture. Such methodologies suggest new possibilities for kinetic structures, adaptive envelopes, and responsive spatial frameworks, where architectural form evolves in response to internal control parameters or external stimuli.

Computational design tools have significantly expanded the architectural applicability of tensegrity. Parametric modeling environments allow architects to manipulate node positions, strut lengths, cable tensions, and degrees of prestress. These parameters directly influence both structural performance and spatial expression. As a result, tensegrity becomes a computationally informed form-finding process, where architectural esthetics and structural optimization are inseparable.

An explicit example of computational tensegrity form-finding is provided by symmetry-adapted force density methods, where equilibrium configurations are generated through mathematical encoding of symmetry constraints (Chen et al., 2018). Fig. 5.5 illustrates a family of tensegrity structures with dihedral symmetries ranging from D_3 to D_{12}, produced by introducing additional cables and controlling the force density ratio between standard and auxiliary tensile elements ($q_a=-0.75q_s$). In symmetry-adapted force density form-finding, the parameter q represents the force density of a structural element, defined as the ratio of axial force to element length. q_s denotes the force density of the primary (standard) tensile elements, while q_a corresponds to the force density assigned to additional cables introduced to refine structural symmetry and stability. The relation $q_a=-0.75q_s$ defines the force density of auxiliary tensile elements as 75% of that of the primary cables, allowing additional members to stabilize and symmetrize the tensegrity structure without overpowering

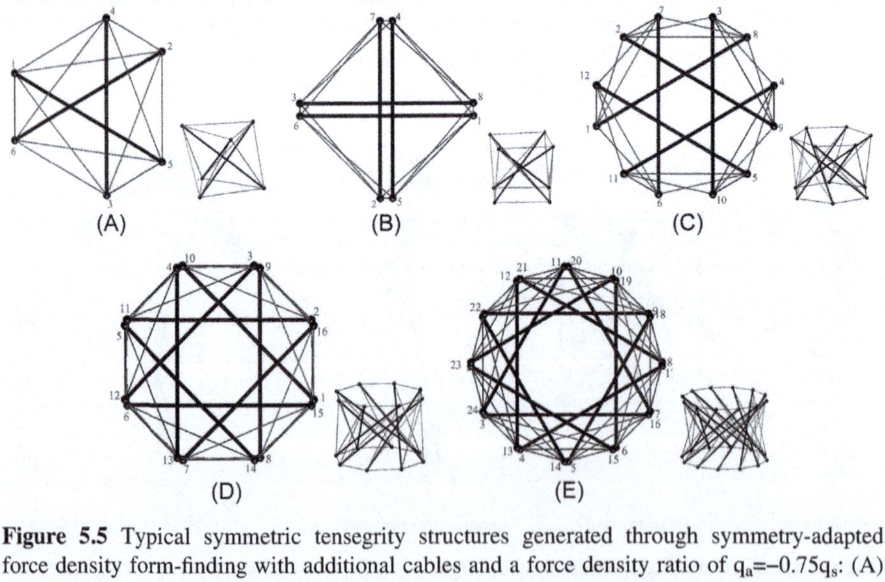

Figure 5.5 Typical symmetric tensegrity structures generated through symmetry-adapted force density form-finding with additional cables and a force density ratio of $q_a=-0.75q_s$: (A) D_3 symmetry; (B) D_4 symmetry; (C) D_6 symmetry; (D) D_8 symmetry; (E) D_{12} symmetry. Forms emerge computationally from equilibrium conditions rather than predefined geometry.

the global equilibrium, thereby acting as a computational control parameter in the form-finding process. From a form-finding perspective, this ratio becomes a design parameter that directly influences the resulting geometry. By calibrating the relative force density of auxiliary cables, the designer can regulate structural stiffness, spatial curvature, and the balance between redundancy and efficiency. In architectural terms, $q_a=-0.75q_s$ enables the generation of symmetric tensegrity forms in which added tensile elements enhance stability and geometric clarity while preserving the lightweight character of the system. Rather than prescribing geometry, this approach allows architectural form to emerge from equilibrium conditions governed by symmetry and prestress, demonstrating how tensegrity can function as a parametric, computation-driven design system.

Beyond structural efficiency, tensegrity systems exert a profound influence on spatial quality and architectural perception by redefining the relationship between structure, enclosure, and occupation. The deliberate separation of compressive and tensile elements eliminates the need for continuous load-bearing frameworks, enabling the creation of large, unobstructed volumes characterized by visual permeability and structural lightness. This spatial condition is exemplified by Tensegrity Späti (Barkow Leibinger, Concéntrico 09), a temporary pavilion installed at the entrance of the Iglesia de Santiago el Real in Logroño (Fig. 5.6). In this project, discontinuous wooden compression struts are suspended within a continuous network of tensioned cables, producing a configuration that appears to hover above the ground while maintaining

Figure 5.6 Tensegrity Späti, Barkow Leibinger, Concéntrico 09—View of the temporary tensegrity pavilion showing suspended wooden compression elements, tensile cable network, and textile membrane enclosure.

global equilibrium. The addition of a laser-cut textile membrane transforms the structural system into an inhabitable architectural space, providing enclosure, shading, and programmatic definition without obscuring the underlying structural logic.

By exposing the mechanisms of equilibrium and self-stress, Tensegrity Späti invites occupants to perceive architecture not as a static assemblage of mass but as a dynamic system in continuous balance. The legibility of force transmission, visible in the contrast between compressive timber elements and tensile cables, enhances the experiential understanding of structural behavior and reinforces contemporary architectural values of transparency and structural honesty. Moreover, the integration of circulation, seating, and social program shifts tensegrity from its historical association with sculptural or experimental artifacts toward a fully architectural condition. Through its lightweight construction, minimal ground impact, and use of renewable and recyclable materials, the project demonstrates how tensegrity can simultaneously support spatial quality, environmental responsibility, and urban activation. In this context, tensegrity emerges not merely as an efficient structural strategy but as a generative architectural system capable of shaping perception, interaction, and meaning within the built environment.

Tensegrity systems inherently promote material efficiency, encouraging the use of high-performance materials such as high-strength steel cables, carbon fiber or composite struts, and aluminum or lightweight steel members (Santos, 2025). In architectural design, the contrast between tension and compression becomes an esthetic language in itself. Slender struts appear to float within a web of cables, visually expressing the invisible forces that stabilize the structure. This expressive clarity resonates strongly with architectural theories that advocate for the visibility of structure as a source of esthetic value. Furthermore, the modularity of tensegrity components allows for prefabrication and precise assembly, reinforcing the alignment between architectural design, fabrication logic, and construction efficiency (Dong et al., 2025).

One of the most compelling architectural advantages of tensegrity systems is their inherent adaptability. Because stability depends on prestress rather than rigid connections, tensegrity structures can be:

- folded or collapsed for transportation;
- deployed on-site with minimal intervention;
- reconfigured by adjusting tension levels.

These characteristics make tensegrity particularly relevant for temporary architecture, emergency shelters, mobile installations, and space-efficient construction. When augmented with sensors and actuators, tensegrity systems can evolve into smart architectural frameworks, capable of adjusting stiffness, geometry, or load distribution in response to environmental or occupancy changes.

In this context, tensegrity aligns directly with smart structure principles discussed in earlier chapters, acting as a physical embodiment of responsive, adaptive design. Tensegrity has been explored across a range of architectural scales and typologies, including lightweight roofs and canopies, pedestrian bridges and walkways, pavilions and exhibition structures, and kinetic and deployable façades. In each application, tensegrity enables designers to minimize material usage while maximizing spatial and visual impact. Although large-scale permanent buildings employing pure tensegrity remain relatively rare due to construction complexity, hybrid systems, combining tensegrity with conventional framing, have proven highly effective.

Despite its advantages, the architectural adoption of tensegrity systems presents several challenges:

- high sensitivity to geometric precision;
- complexity of analysis and construction;
- dependence on accurate prestress control;
- limited familiarity among practitioners.

Addressing these challenges requires early collaboration between architects, structural engineers, and fabrication specialists. Advanced simulation tools and digital fabrication techniques play a critical role in translating tensegrity concepts from theoretical models into buildable architectural systems.

Beyond its technical characteristics, tensegrity represents a philosophical shift in architectural thinking. It encourages designers to conceive buildings as systems of relationships rather than assemblies of parts. This relational approach resonates with contemporary concerns about sustainability, resilience, and systemic intelligence in the built environment. In tensegrity-based architecture, stability arises from balance, efficiency from interdependence, and beauty from structural clarity. These qualities position tensegrity as both a technical solution and an architectural ideal within the evolving discourse of smart and adaptive design.

Tensegrity in architectural design offers a powerful synthesis of structural logic, spatial innovation, and esthetic expression. By leveraging the equilibrium between tension and compression, architects and engineers can create environments that are lightweight, adaptable, and visually compelling. As digital tools and smart technologies continue to advance, tensegrity systems are poised to play an increasingly influential role in shaping future architectural forms that are not only efficient and resilient but also expressive of the invisible forces that sustain them.

5.3 Practical applications and case studies

The translation of tensegrity from theoretical construct to architectural and infrastructural application has been driven largely by advances in computational design, material engineering, and structural monitoring. Contemporary practice demonstrates that tensegrity is no longer confined to experimental prototypes or artistic installations but can operate as a viable structural strategy across multiple scales. In permanent infrastructure, temporary architecture, and adaptive systems, tensegrity principles have enabled lightweight construction, long-span performance, and expressive structural clarity while supporting evolving demands for resilience and sustainability.

This section examines real-world applications of tensegrity systems, highlighting how the fundamental principles of tension–compression equilibrium are adapted to address functional, environmental, and constructional requirements. The selected case studies demonstrate the versatility of tensegrity across scales and typologies, as well as its compatibility with smart structural strategies.

One of the most prominent large-scale applications of tensegrity-inspired design is the Kurilpa Bridge in Brisbane, designed by Cox Rayner Architects in collaboration with Arup (Fig. 5.7). The bridge employs a hybrid structural system in which inclined steel masts, tensioned cables, and deck elements distribute loads through a network of forces rather than through continuous compressive members (Bel Hadj Ali et al., 2010). This strategy enables long spans with reduced material mass while providing multiple load paths that enhance redundancy and structural robustness. The exposed cable network not only communicates structural behavior visually but also integrates structural health monitoring systems that record responses to wind, temperature variation, and pedestrian loading. In this context, tensegrity operates simultaneously as a structural system, an architectural language, and a platform for smart infrastructure.

At the scale of temporary and experimental architecture, the Underwood Pavilion in Muncie, Indiana (2014) constitutes a significant contemporary application of tensegrity principles mediated through computational design and modular construction. Conceived as a lightweight, demountable installation, the pavilion is composed of 56 parametrically differentiated tensegrity modules, each operating as a self-equilibrated structural unit. As illustrated in Fig. 5.8, each module consists of isolated compression struts held in position by a continuous tensile system formed by polymer fabric and tension elements. The compression members do not form continuous load paths; instead, global stability emerges

Figure 5.7 Kurilpa Bridge, Brisbane. Perspective views and sectional drawings illustrating the tensegrity-inspired cable–mast system, highlighting the interplay between tensioned cable networks and inclined compression masts in achieving long-span structural equilibrium and spatial transparency.

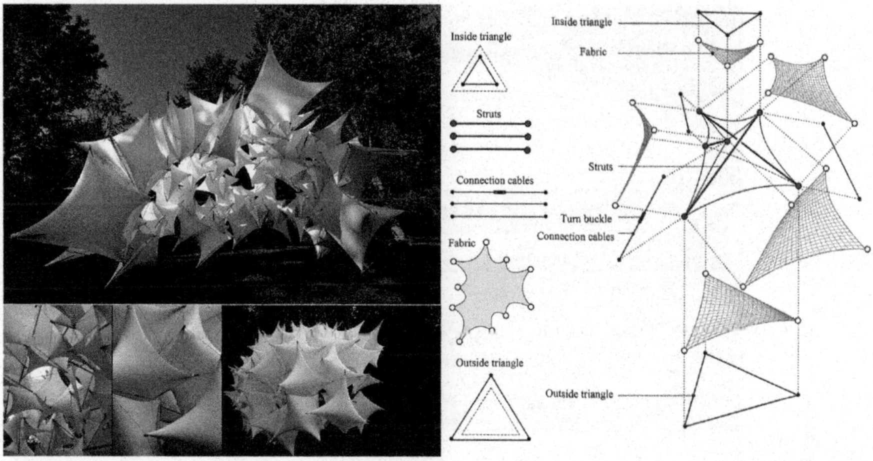

Figure 5.8 Underwood Pavilion, Muncie, USA (2014). Left: views of the assembled pavilion illustrating the spatial continuity generated by the aggregation of parametrically differentiated tensegrity modules. Right: step-by-step assembly of a single tensegrity module, showing isolated compression struts stabilized by a continuous tensile polymer fabric network.

from prestress and the geometric interdependence of tension and compression, satisfying the fundamental criteria of tensegrity behavior.

Parametric design tools played a central role in the development of the pavilion. By controlling parameters such as strut length, nodal geometry, and membrane tension, the designers were able to generate a family of geometrically distinct modules while preserving structural equilibrium across the system. This approach enabled local variation and curvature without compromising global stability, demonstrating how computational form-finding can reconcile architectural expression with strict structural constraints. The aggregation of these differentiated modules produces a spatially continuous enclosure whose form emerges from the repetition and transformation of a single tensegrity logic rather than from a predefined architectural geometry. From an architectural perspective, the pavilion illustrates how tensegrity can operate simultaneously as structure, enclosure, and spatial generator. The tensile polymer fabric performs multiple roles: it stabilizes the compression struts, defines the pavilion's envelope, and modulates light and permeability. The resulting space is neither fully enclosed nor purely skeletal but instead occupies an intermediate condition that is particularly suited to pavilions, exhibition spaces, and temporary installations. The modular nature of the system allows for rapid assembly and disassembly, minimal ground intervention, and efficient material use, aligning the project with principles of sustainable and reversible construction.

Importantly, the Underwood Pavilion demonstrates that tensegrity systems can move beyond isolated experimental prototypes toward architecturally legible and inhabitable structures when supported by precise digital workflows. The clear correspondence between the assembly of individual modules and the emergent global form reinforces the didactic value of tensegrity, making structural behavior

Figure 5.9 Georgia Dome, Atlanta, USA. Exterior and interior views of the cable-supported roof structure, illustrating the large-span enclosure achieved through the equilibrium of tensioned cables and compressive elements in a tensegrity-inspired structural system.

visible and intelligible to occupants. In this sense, the project exemplifies how tensegrity-based architecture can integrate computational design, material efficiency, and spatial experience, reinforcing its relevance within contemporary performance-driven and lightweight architectural practices.

Large-span enclosures further illustrate the adaptability of tensegrity concepts when hybridized with conventional structural systems. A well-documented example is the Georgia Dome in Atlanta, whose roof structure was designed by Weidlinger Associates using a cable-supported system informed by tensegrity principles. As shown in Fig. 5.9, the exterior view reveals the lightweight and elevated roof profile, while the interior view highlights the expansive, column-free volume made possible by the cable network. Although the system does not constitute a pure tensegrity structure—since compression elements are not fully isolated—the roof relies fundamentally on the equilibrium between tensioned cables and compressive components to achieve structural stability. Loads are distributed through a prestressed cable network rather than through continuous compressive load paths, allowing the roof to span large distances with significantly reduced self-weight.

From an architectural standpoint, this structural strategy enables a high degree of spatial openness and flexibility, accommodating large spectator capacities without visual obstruction. The exposed cable system also enhances the legibility of structural behavior, allowing occupants to perceive the roof as an active system of forces rather than a massive enclosure. Computational analysis and prestress optimization were essential to the design and construction of the roof, ensuring controlled deformation under wind, seismic, and live loading conditions. In this sense, the Georgia Dome demonstrates how tensegrity logic can be selectively integrated into permanent architectural structures to address the demands of large-span enclosures, bridging the gap between experimental structural theory and practical architectural application.

More recently, research-driven projects have extended tensegrity into the domain of adaptive and intelligent structural systems, positioning it at the intersection of architecture, robotics, and smart materials. Leading research institutions such as MIT, NASA Ames Research Center, and ETH Zürich have developed active tensegrity prototypes in which sensors, actuators, and control algorithms are integrated directly into the tensile network. These systems are capable of dynamically adjusting prestress levels, enabling controlled changes in stiffness, geometry, and load distribution in real time. Such capabilities allow tensegrity structures to respond elastically to environmental stimuli, mitigate vibrations induced by wind or seismic activity, and maintain stability under variable loading conditions.

A notable example is NASA's Super Ball Bot and subsequent tensegrity habitat prototypes, which explore deployable tensegrity frameworks for extraterrestrial environments (Fig. 5.10). While developed for space exploration, these projects demonstrate architectural-scale principles such as rapid deployment, structural redundancy, and adaptability to unpredictable ground conditions. Similarly, active tensegrity structures explore feedback-controlled cable networks in which localized actuation enables continuous adjustment of geometry and internal force distribution (Wang et al., 2024b). Rather than functioning as static load-bearing assemblies, these systems behave as distributed structural organisms, where sensing, actuation, and

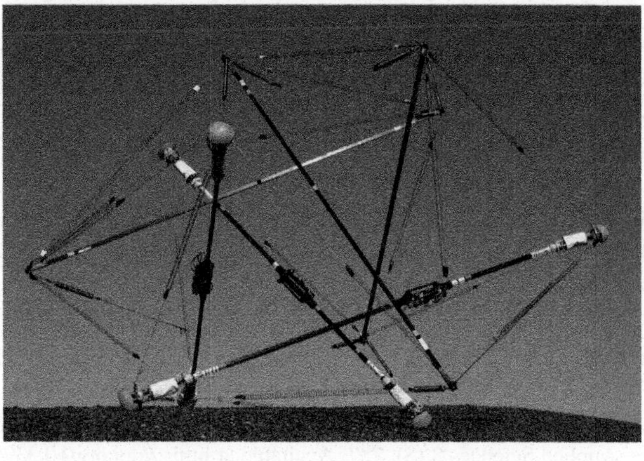

Figure 5.10 A prototype of NASA's Super Ball Bot is inspired by the flexibility of a baby toy.

control are embedded throughout the network. This mode of operation allows the structure to respond autonomously to changes in loading, environmental conditions, or programmatic requirements. For architectural applications, such behavior offers valuable insights into the design of adaptive load-bearing frameworks, responsive envelopes, and reconfigurable spatial systems, in which structural performance and spatial configuration can evolve dynamically over time.

In parallel, architectural research groups have begun to explore tensegrity as a framework for responsive envelopes and reconfigurable spatial systems. Prototypes developed at ETH Zürich and TU Delft integrate tensegrity logic with lightweight membranes and kinetic components, enabling building skins that adjust porosity, curvature, or shading behavior in response to solar exposure and wind loads (Pugnale et al., 2026). Although these systems remain largely experimental, they illustrate how tensegrity can support adaptive architectural elements that blur the boundary between structure and environmental control.

These projects suggest future architectural scenarios in which tensegrity frameworks enable deployable shelters, climate-responsive pavilions, and reconfigurable interior or exterior spaces (Gomez-Jauregui et al., 2023). By combining distributed load-bearing capacity with embedded sensing and actuation, tensegrity systems offer a compelling model for architecture conceived as an intelligent, responsive system capable of learning, adapting, and evolving over time rather than remaining fixed and inert.

Tensegrity principles have also begun to influence façade design and lightweight enclosure systems, particularly in contexts where flexibility and environmental responsiveness are critical. By combining tensegrity frameworks with membranes, kinetic panels, or shading devices, designers can create envelopes that modulate porosity, orientation, or stiffness in response to solar exposure and wind forces. These applications position tensegrity as an intermediary between structure and environmental control, reinforcing its relevance within performance-driven architectural design.

Collectively, these built projects and research prototypes reveal that tensegrity systems are most effective when developed through interdisciplinary workflows that integrate form-finding, structural analysis, and digital modeling from the earliest design stages. Hybridization with conventional construction methods often enhances feasibility, while the integration of monitoring and control technologies improves reliability and adaptability. As computational design tools, digital fabrication techniques, and smart materials continue to advance, tensegrity is increasingly positioned not only as an efficient structural solution (Zhang and Ohsaki, 2025) but also as a conceptual framework for architecture conceived as a dynamic, responsive, and resource-conscious system (Niazy et al., 2024).

References

Al Sabouni-Zawadzka, A., Gilewski, W., 2018. Inherent properties of smart tensegrity structures. Applied Sciences 8 (5), 787. Available at https://doi.org/10.3390/app8050787.

Bel Hadj Ali, N., Rhode-Barbarigos, L., Pascual Albi, A.A., Smith, I.F.C., 2010. Design optimization and dynamic analysis of a tensegrity-based footbridge. Engineering Structures 32 (11), 3650–3659. Available at https://doi.org/10.1016/j.engstruct.2010.08.009.

Bieniek, Z., 2017. The self-equilibrium problem of the class-theta tetrahedral tensegrity module. Composites Part B: Engineering 115, 21–29. Available at https://doi.org/10.1016/j.compositesb.2016.10.054.

Calladine, C.R., 1978. Buckminster Fuller's "Tensegrity" structures and Clerk Maxwell's rules for the construction of stiff frames. International Journal of Solids and Structures 14 (2), 161–172. Available at https://doi.org/10.1016/0020-7683(78)90052-5.

Chai Lian, O., Kok Keong, C., Nishimura, T., Jae-Yeol, K., 2020. Form-finding of spine inspired biotensegrity model. Applied Sciences 10 (18), 6344. Available at https://doi.org/10.3390/app10186344.

Chen, Y., Sun, Q., Feng, J., 2018. Improved form-finding of tensegrity structures using blocks of symmetry-adapted force density matrix. Journal of Structural Engineering 144 (10). Available at https://doi.org/10.1061/(asce)st.1943-541x.0002172.

Chen, Y., Dong, Y., Yuan, X., Ma, S., Dong, S., 2024. Modular assembly of tensegrity structures with diverse mesh division forms. Engineering Structures 315, 118491. Available at https://doi.org/10.1016/j.engstruct.2024.118491.

Dong, Y., Yuan, X., Wang, X., Samy, A., Ma, S., Dong, S., 2025. Simultaneous topology, configuration, and prestress optimization for lightweight design of modular tensegrity chain structures. Thin-Walled Structures 212. Available at https://doi.org/10.1016/j.tws.2025.113184.

Fernández-Ruiz, M.A., Huerta-Gómez-Merodio, M., López-Jiménez, P., Carbonell-Márquez, J.F., 2025. Design and construction of a prototype tensegrity structure with integrated pretensioning and instrumentation features. Buildings 15 (24). Available at https://doi.org/10.3390/buildings15244506.

Gomez-Jauregui, V., Carrillo-Rodriguez, A., Manchado, C., Lastra-Gonzalez, P., 2023. Tensegrity applications to architecture, engineering and robotics: a review. Applied Sciences 13 (15), 8669. Available at https://doi.org/10.3390/app13158669.

Ma, S., Swai, D.R., Tang, J., Dong, Y., Yuan, X., 2026. Equilibrium theory and matrix analysis of tensegrity-beam structures. International Journal of Solids and Structures 327. Available at https://doi.org/10.1016/j.ijsolstr.2025.113805.

Mirats Tur, J.M., Juan, S.H., 2009. Tensegrity frameworks: dynamic analysis review and open problems. Mechanism and Machine Theory 44 (1), 1–18. Available at https://doi.org/10.1016/j.mechmachtheory.2008.06.008.

Niazy, D., Ashraf, M., Bodaghi, M., Zolfagharian, A., 2024. Resilient city perspective: 4D printing in art, architecture and construction. Materials Today Sustainability 26, 100708. Available at https://doi.org/10.1016/j.mtsust.2024.100708.

Pugnale, A., Mirra, G., Halls, J., Park, M.M., Mack, M., Colabella, S., 2026. Deploying FloaTree: computational design-to-construction workflow of a low-tech tensegrity system. Architectural Intelligence 5 (1). Available at https://doi.org/10.1007/s44223-025-00112-4.

Santos, F.A., 2025. Tensegrity metastructure with tunable stiffness, strength, and energy dissipation. Extreme Mechanics Letters 81. Available at https://doi.org/10.1016/j.eml.2025.102420.

Vilnay, O., Chernin, L., Vilnay, M., 2023. Tensegrity Structures Design Methods. CRC Press, Boca Raton, pp. 1–238. Available at https://doi.org/10.1201/9781003370093.

Wang, S., Kotnik, T., Schwartz, J., Cao, T., 2022. Equilibrium as the common ground: introducing embodied perception into structural design with graphic statics. Frontiers of Architectural Research 11 (3), 574–589. Available at https://doi.org/10.1016/j.foar.2022.01.001.

Wang, T., Post, M.A., Tyrrell, A.M., 2024a. TWrist: an agile compliant 3-DoF tensegrity joint. Biomimetic Intelligence and Robotics 4 (3). Available at https://doi.org/10.1016/j.birob.2024.100170.

Wang, Y., Han, Z., Xu, X., Luo, Y., 2024. Topology optimization of active tensegrity structures. Computers & Structures 305, 107513. Available at https://doi.org/10.1016/j.compstruc.2024.107513.

Zhang, J.Y., Ohsaki, M., 2025. A mechanical approach to generation of planar polygonal mesh for grid shells. Journal of Constructional Steel Research 234, 109736. Available at https://doi.org/10.1016/j.jcsr.2025.109736.

Kinetic structures: engineering motion in architecture

6

6.1 Engineering movement in architecture

For centuries, architecture has been fundamentally conceived as static—structures designed to resist forces, maintain stability, and preserve form over time. However, the increasing integration of advanced materials, digital control systems, sensing technologies, and mechanical actuation has transformed this paradigm (Lee et al., 2021). Contemporary architecture is no longer confined to immobility; it can now move, adapt, and respond (Askar et al., 2021). Kinetic structures represent this evolution, embodying a shift from fixed spatial configurations to dynamic, reconfigurable environments (Matheou et al., 2018). Engineering movement in architecture involves the deliberate incorporation of mechanical, structural, and computational systems that allow buildings or their components to change position, shape, orientation, or performance in response to environmental, functional, or user-driven inputs. Unlike traditional structures that rely solely on passive resistance, kinetic systems actively engage with external forces and internal demands, transforming motion into a fundamental architectural strategy.

A kinetic structure is an architectural system designed to incorporate controlled physical movement as an integral part of its performance. Movement may occur at various scales and speeds, including:

- rotating or sliding façade panels;
- folding roofs or retractable enclosures;
- adjustable shading devices;
- expanding or deployable spatial modules;
- shape-changing structural systems.

The motion may be periodic, responsive, programmable, or user-controlled. In all cases, movement is not ornamental but purposeful, serving environmental regulation, spatial flexibility, energy optimization, or experiential enhancement. Kinetic architecture thus integrates principles from structural engineering, mechanical systems, robotics, and smart control technologies, requiring interdisciplinary collaboration from the earliest stages of design (Soana et al., 2025).

Engineering movement in architecture demands careful consideration of structural integrity, mechanical reliability, energy efficiency, and safety. The introduction of motion fundamentally alters structural behavior, as dynamic forces, friction, inertia, and fatigue must be accounted for alongside conventional static loads.

Key engineering considerations in kinetic architecture extend beyond conventional static structural analysis and require a comprehensive understanding of how movement affects performance over time. One of the primary concerns is structural load transfer during motion. When a structure moves, its load paths continuously shift, meaning that

Smart Civil and Structural Engineering. DOI: https://doi.org/10.1016/B978-0-443-30223-7.00014-0

equilibrium conditions are no longer purely static. Engineers must evaluate not only the stationary configuration but also the transitional states that occur throughout the movement cycle. Dynamic loading, resulting from acceleration, deceleration, wind interaction, or operational forces, can generate stresses that exceed those present under static conditions. Consequently, advanced finite element modeling (FEM) and parametric simulation tools are essential for predicting structural behavior, identifying critical stress points, and ensuring safety under both static and dynamic scenarios.

The choice and design of the actuation system, which permits movement within the structure, is another essential component. Electric motors, hydraulic pistons, pneumatic systems, linear drives, and smart material actuators like shape-memory alloys are just a few examples of the many different types of actuators. Depending on the size of the building, the amount of force needed, the degree of accuracy needed, energy efficiency objectives, and long-term maintenance concerns, each system has unique benefits and drawbacks. The Mercedes-Benz Stadium's retractable roof system (2017) is a modern, expansive model. Its eight-panel oculus top is powered by a synchronized motor-driven system that is intended to precisely and redundantly control substantial structural weight (Fig. 6.1). The project illustrates how factors like duration, force requirements, movement frequency, and long-term operating reliability affect the choice of actuator. To guarantee seamless operation, dependability, and alignment with architectural intent, the actuation mechanism's integration with the structural system needs to be properly designed.

Control and feedback mechanisms further define the performance of contemporary kinetic systems. Modern movable structures rely on networks of sensors and digital control platforms to monitor environmental and operational variables such as solar radiation, wind speed, occupancy patterns, temperature fluctuations, or structural strain. These data streams feed into control algorithms that determine the timing, extent, and mode of movement. Through this continuous feedback loop, kinetic structures function as intelligent systems capable of autonomous adaptation. In doing so, they closely align with the broader principles of smart structures, where responsiveness and performance optimization are embedded directly within the architectural system. The climate-responsive façade of One Ocean Pavilion (2012), designed for Expo 2012, exemplifies advanced feedback integration (Fig. 6.2). Its kinetic lamella system reacts to environmental inputs such as sunlight and temperature through a digital control system, transforming the façade into an adaptive environmental regulator rather than a static envelope.

Finally, durability and maintenance represent critical engineering priorities. Because kinetic systems involve repeated cycles of motion, their components are subject to mechanical wear, fatigue, and environmental degradation. Design strategies must therefore emphasize robust joint detailing, the use of corrosion-resistant materials, redundant safety mechanisms, and clearly accessible maintenance pathways. Reliability becomes especially crucial when large structural elements are in motion or when public safety is involved. Long-term performance depends not only on innovative design but also on careful anticipation of operational stresses and lifecycle demands (Nemati Aghdam and Dixit, 2026).

Figure 6.1 Retractable roof system of Mercedes-Benz Stadium, showing perspective views of the roof in closed and fully open configurations. The diagram illustrates the radial sliding motion of the eight-panel oculus system and the geometric transformation achieved through synchronized motorized actuation.

Kinetic structures can be classified according to the nature and mechanics of their movement (Choi et al., 2017). One primary category is rigid-body movement, in which entire components rotate, slide, or translate without undergoing deformation. In such systems, structural elements maintain their geometric integrity while changing position, as seen in retractable stadium roofs or rotating building segments. The engineering emphasis in rigid-body systems lies in hinge detailing, track systems, load redistribution during motion, and synchronization of large-scale components. Fig. 6.3 illustrates four basic motion types applied to kinetic façade systems: translation, rotation, angular deformation, and linear deformation. The schematic diagrams distinguish the geometric behavior of rigid plates and membranes, highlighting the mechanical differences between rigid-body motion and deformable systems. This classification supports a clearer understanding of how movement is generated and controlled in adaptive façade

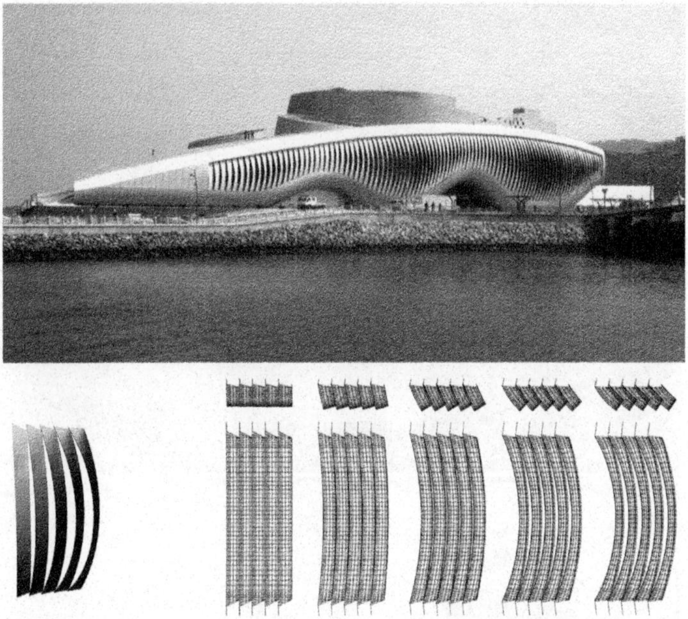

Figure 6.2 Climate-responsive façade of One Ocean Pavilion, illustrating the kinetic lamella system in varying configurations.

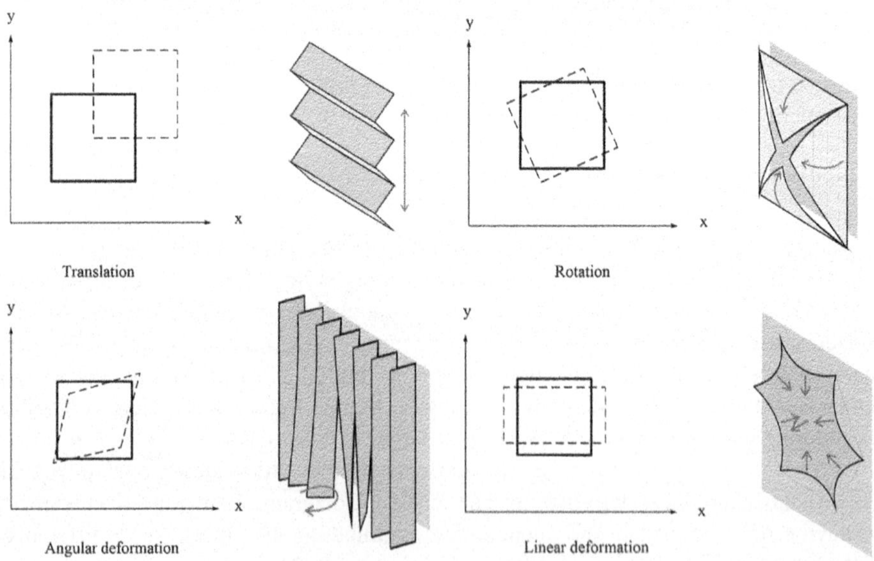

Figure 6.3 Schematic diagrams of fundamental motion types in kinetic façade systems.

technologies. Angular deformation occurs when rigid plates rotate around a hinge, pivot, or fixed axis, producing a change in angle (θ). The motion is rotational, and the transformation is governed by $\Delta\theta \neq 0$ while the component itself remains geometrically undeformed. Such systems typically rely on articulated joints and hinge mechanisms. Linear deformation occurs when rigid plates or membranes translate along a straight path (x, y, or z direction) without rotation. In this case, the motion is governed by Δx, Δy, or $\Delta z \neq 0$, and the displacement is achieved through sliding tracks, telescopic systems, or linear actuators. Unlike angular systems, no change in orientation occurs during movement.

Deployable structures represent an important typology of kinetic systems operating across multiple spatial scales (Wang et al., 2024). These systems expand, unfold, or extend from compact configurations into operational forms through articulated geometric transformation (Wang et al., 2025). At the building scale, scissor-hinge mechanisms enable rapid spatial expansion in pavilions and retractable roofs. At the envelope scale, foldable façade systems, often derived from origami geometries, achieve controlled angular deployment for environmental regulation. At the interior scale, articulated modular assemblies demonstrate deployability through joint-controlled reconfiguration. Across all scales, the design of deployable systems requires careful consideration of folding logic, joint articulation, locking mechanisms, structural stability in both collapsed and deployed states, and load redistribution during transformation.

As illustrated in Fig. 6.4, a scissor-classification system is presented through deployment-type diagrams and scale-model exercises of a deployable structure. The figure includes a foldable triangulated scissor grid with varying structural thickness,

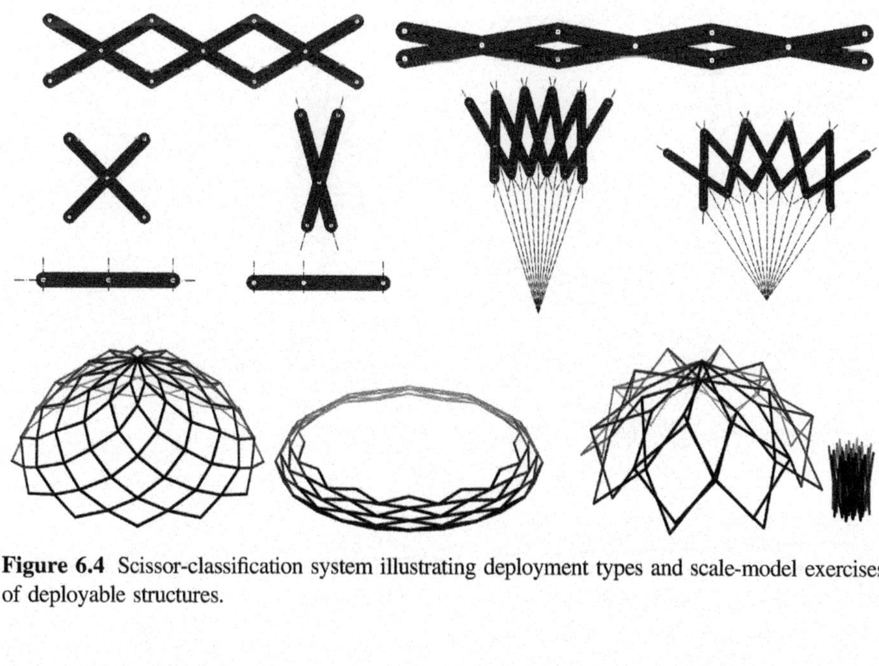

Figure 6.4 Scissor-classification system illustrating deployment types and scale-model exercises of deployable structures.

describing the generation of conical surfaces through coordinated joint rotation. The configuration studies demonstrate how articulated linear members produce geometric transformation, allowing the system to transition from a condensed state to expanded spatial arrangements while maintaining structural continuity and stability.

Transformable structures constitute a third major category of kinetic systems, typically associated with façades and building envelopes that adapt their geometry to regulate environmental performance (Brenner et al., 2026). Unlike large-scale deployable structures that transform overall spatial configuration, transformable structures operate through localized adjustments responding to solar radiation, wind, temperature, or occupancy patterns. Their primary objective is to optimize daylight penetration, natural ventilation, solar shading, and visual connectivity while enhancing indoor environmental quality. These structures commonly incorporate operable panels, rotating louvers, sliding screens, folding modules, or responsive membrane layers that adjust configuration in real time (Matheou et al., 2020). Although the movement is generally localized rather than global, its cumulative impact on thermal regulation, glare control, and energy performance is substantial.

As illustrated in Fig. 6.5, transformable structure principles are further explored through geometric studies, including the assembly of rhombi loops arranged along a circular framework and a transformable reciprocal frame structure. These configurations demonstrate how repeated modular elements can generate adaptive surface behavior through rotational articulation and interdependent structural relationships. The diagrams highlight how localized joint movement produces coordinated geometric transformation, enabling surface reconfiguration while maintaining structural continuity and load distribution.

Figure 6.5 Transformable structure studies illustrating (top) the assembly of rhombi loops arranged along a circular geometry and (bottom) a transformable reciprocal frame structure.

Elastic or morphing structures form a more advanced typology, in which materials or structural systems deform continuously rather than moving as discrete rigid components. Through controlled tension adjustments, variable stiffness mechanisms, or smart material actuation, these systems alter curvature, surface configuration, or structural behavior (Rodinò and Maletta, 2025). Although each typology requires distinct engineering strategies, they all depend on precise coordination between structural stability, mechanical systems, and control logic to ensure safe and reliable performance.

As illustrated in Fig. 6.6, morphing principles can operate through different deformation mechanisms. One example demonstrates a lamella-based structural system in which surface transformation is achieved through a combined bending and torsional buckling effect, generating adaptive curvature through controlled instability. Another example presents a self-forming wooden assembly that responds to environmental conditions, where programmed material anisotropy induces autonomous curvature without mechanical hinges or external actuation. Together, these illustrations emphasize how continuous deformation, whether mechanically induced or materially driven, enables adaptive spatial transformation through distributed strain rather than discrete joint rotation.

One of the major inspirations for the engineering movement in architecture is the pursuit of environmental feedback. Kinetic systems enable buildings to regulate solar gain, optimize daylight penetration, enhance natural ventilation, and reduce overall

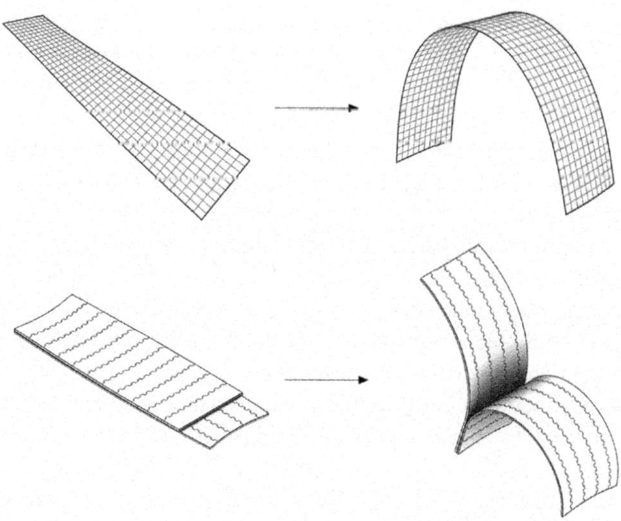

Figure 6.6 Illustrations of elastic and morphing structural systems. (Top) Lamella-based surface transformation achieved through combined bending and torsional buckling effects. (Bottom) Self-forming wooden assembly demonstrating environmentally induced curvature through programmed material behavior.

energy consumption. By dynamically adjusting their configuration, buildings can respond to changing climatic conditions rather than relying solely on static design strategies.

Environmental applications of kinetic systems include automated louvers that track the sun's trajectory, responsive shading screens that open or close according to radiation intensity, operable façades that adjust to wind direction, and roof systems that retract to facilitate natural cooling. Through such mechanisms, buildings evolve from passive environmental filters into active climate moderators. This transition enhances occupant comfort, improves indoor environmental quality, and reduces dependence on energy-intensive mechanical heating, ventilation, and air conditioning systems. Movement, in this context, becomes an instrument of performance optimization and sustainability. Beyond environmental performance, kinetic structures significantly expand the possibilities for spatial adaptability. Movable partitions, sliding walls, retractable seating systems, and transformable floors allow a single architectural footprint to accommodate diverse functions over time. Cultural venues, exhibition halls, educational facilities, and multipurpose civic buildings particularly benefit from such adaptability, as spatial configurations can be modified according to programmatic requirements.

In addition, user-responsive kinetic systems foster interaction between occupants and the built environment. Through sensors, programmable interfaces, and digital control platforms, interior configurations can adapt to occupancy density, event type, acoustic requirements, or even individual preferences. This interactive dimension reinforces the human-centered orientation of smart architecture, positioning users not merely as passive occupants but as active participants in shaping spatial conditions.

Movement introduces a powerful esthetic and experiential dimension to architecture. When buildings transform, they acquire a temporal quality; form is no longer fixed but evolves through sequences of motion. Shadows shift, surfaces open and close, volumes expand or contract, and spatial perceptions change accordingly. This dynamic character generates a layered architectural experience in which time becomes an integral component of design. Kinetic expression often conveys technological sophistication and innovation. However, esthetic effectiveness depends on the coherence between movement and architectural purpose. When motion emerges logically from environmental, structural, or functional requirements, it strengthens the unity of form and performance. Conversely, when deployed solely as spectacle, movement risks becoming superficial. Successful kinetic architecture therefore integrates expressive transformation with technical and environmental rationale (Barozzi et al., 2016).

Contemporary kinetic architecture increasingly operates in conjunction with smart structural technologies (Dąbrowska-Żółtak and Szalwa, 2025). Structural health monitoring systems track stress, displacement, and fatigue to ensure safe operation during repeated movement cycles. At the same time, advanced control systems and artificial intelligence algorithms optimize motion patterns based on predictive environmental data and usage scenarios. In such configurations, sensing, decision-making, and mechanical response form an integrated cyber-physical network.

This integration positions kinetic systems as a natural extension of smart architecture. Responsiveness, adaptability, and intelligence are embedded not only in digital data processing but also in the physical transformation of structural components. Movement thus becomes part of a broader performance ecosystem in which structural behavior, environmental conditions, and user interaction are dynamically coordinated.

Engineering movement in architecture represents a fundamental shift from static construction toward dynamic performance. By combining structural analysis, mechanical actuation, digital control, and material innovation, kinetic systems transform buildings into responsive entities capable of adaptation and interaction. As environmental challenges intensify and functional demands grow increasingly complex, the capacity of architecture to move intelligently, efficiently, and purposefully will play a critical role in shaping resilient, sustainable, and experientially rich built environments.

6.2 Creating dynamic and adaptable spaces

The accelerating pace of technological innovation, social transformation, and environmental uncertainty has rendered rigid architectural programming increasingly inadequate. Buildings are no longer expected to fulfill a singular, static function throughout their lifespan; instead, they must accommodate evolving uses, fluctuating occupancy densities, and shifting climatic conditions. In this context, the creation of dynamic and adaptable spaces has emerged as a fundamental objective of kinetic architecture. Contemporary design practice increasingly views architecture not as a fixed artifact but as a responsive framework capable of transformation.

An environment that can undergo physical transformation, spatial reconfiguration, or environmental modulation in response to contextual inputs or functional demands is referred to as dynamic space. The ability of a structure to change its environmental, structural, or spatial features over time without requiring major renovation is known as adaptability. All of these ideas work together to reframe architecture as a dynamic system as opposed to a fixed thing. The mechanical and structural tools necessary to facilitate such change are provided by kinetic systems, which permit volumes, interfaces, and spatial boundaries to move, bend, spin, slide, or expand (Johnson et al., 2019). As a result, the movement is not coincidental; rather, it becomes an essential component of the performance, immediately integrating flexibility into the logic of the architecture.

A clear illustration of the distinction between fixed infrastructure and adaptable enclosure can be observed in the architectural strategy of The Shed (arts center). The project embodies a dual logic: a permanent structural and service core that anchors the building spatially and mechanically, and a large-scale kinetic outer shell that redefines the usable volume through controlled movement. While the base building remains structurally stable and programmatically consistent, the deployable shell slides along exposed rails to create an expanded performance hall when required (Fig. 6.7). This separation between static framework and transformable envelope demonstrates how adaptability can be embedded within architectural systems without compromising structural clarity. The project thus exemplifies the transition from static programming toward a dynamic model in which spatial capacity evolves in response to event-based demands and temporal conditions.

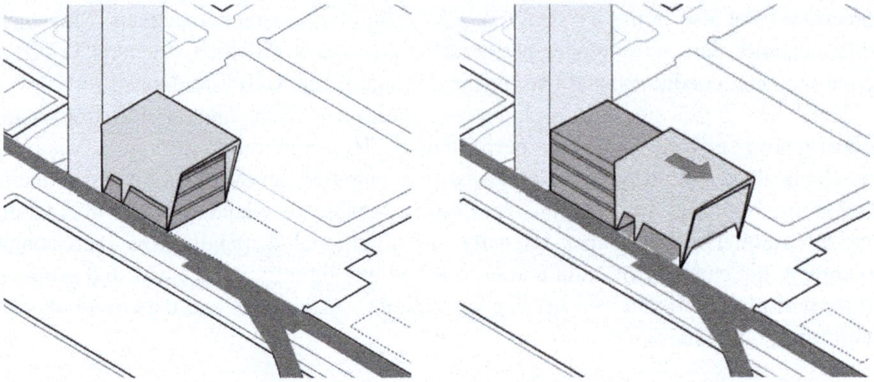

Figure 6.7 Conceptual diagram of The Shed illustrating the distinction between stable structural infrastructure and kinetic spatial enclosure.

One of the most direct strategies for achieving adaptability lies in the integration of movable architectural components. Sliding and folding walls, rotating partitions, retractable seating systems, deployable platforms, and operable ceilings or floors enable rapid spatial transformation. Through such mechanisms, a single volume may serve multiple programmatic purposes within short time intervals.

Educational and corporate environments also benefit from movable partitions that scale rooms according to occupancy density and activity type (Vuscan and Muntean, 2025). Engineering these systems requires close coordination between structural design and mechanical components. Load-bearing systems must accommodate transitional states during motion, ensuring stability in both open and closed configurations. Parametric modeling tools and dynamic structural simulations are typically employed to analyze variable states, evaluate stress redistribution, and test movement sequences before construction (Ramilo and Embi, 2014). While interior partitions represent one of the most immediate forms of adaptability, mechanical reconfiguration operates across multiple architectural scales, extending from lightweight interior elements to long-span structural systems and deployable building envelopes.

Mechanical spatial reconfiguration operates across multiple architectural scales, ranging from building envelopes to primary structural systems and interior partitions. Fig. 6.8 illustrates this multiscalar adaptability through three complementary mechanisms. At the structural scale, long-span systems create column-free interiors that permit flexible spatial arrangements without structural obstruction. At the largest scale, deployable or retractable envelope systems enable buildings to expand or contract their enclosed volume in response to climatic or programmatic demands. At the interior scale, lightweight mechanical partitions allow rapid subdivision or merging of spaces without altering the primary structure. These systems demonstrate that adaptability is not confined to a single architectural component. Instead, spatial flexibility emerges through coordinated mechanical strategies operating at envelope, structural, and interior levels.

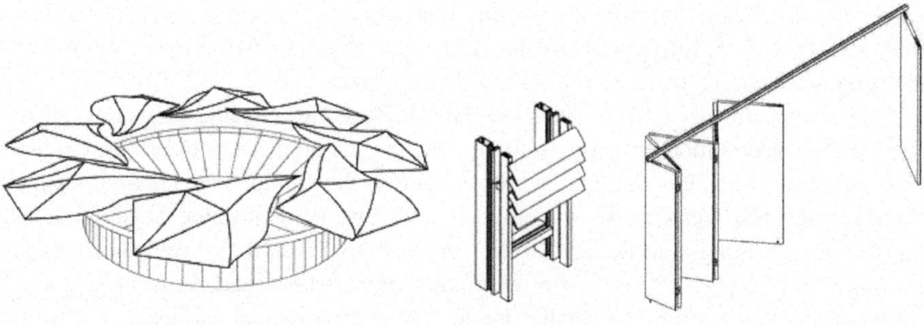

Figure 6.8 Illustrative diagram of mechanical spatial reconfiguration systems across architectural scales (structural span system, large-scale envelope, interior mechanical partition).

Dynamic spatial transformation extends beyond interior reconfiguration to include the building envelope. Adaptive façades and operable skins enable real-time environmental regulation, redefining the relationship between interior and exterior conditions (Zhang et al., 2022). Through controlled motion, kinetic envelope systems act as responsive environmental filters rather than static barriers (Han et al., 2025). Such systems may incorporate automated louvers that track solar position, rotating shading panels, sliding perforated screens, or tension-adjustable membranes (Uribe et al., 2019).

Increasingly, dynamic architecture integrates user interaction as a central design parameter. Sensor-based systems detect occupancy density, movement patterns, daylight levels, or thermal preferences, triggering spatial adjustments that enhance comfort and functionality. Through programmable control systems, spatial transformation becomes a coordinated response rather than manual intervention. Responsive environments employ sensor networks to optimize environmental performance and workspace conditions. Interactive installations further demonstrate how façade elements or interior components may respond to pedestrian proximity, reinforcing the concept of architecture as an interface between human behavior and environmental systems.

Dynamic architecture also engages with time as an active design variable (Posadas et al., 2025). Buildings may adopt different configurations throughout the day, season, or event cycle. Stadium roofs open during favorable weather and close during inclement conditions; cultural venues adjust acoustics and seating layouts depending on performance type; educational spaces transition between collaborative and lecture modes.

The realization of adaptable environments depends on digital integration. Sensors collect real-time data related to climate, occupancy, and structural behavior. Microcontrollers and centralized building management systems process this information, while actuators execute spatial transformations. Building information modeling and parametric platforms serve as the digital backbone, allowing designers to simulate movement sequences and evaluate performance implications prior to construction. In advanced applications, artificial intelligence algorithms analyze usage data over time, refining transformation patterns to optimize energy consumption and spatial efficiency. Thus adaptability becomes increasingly data-driven, merging architecture with computational intelligence.

Adaptability directly contributes to resilience. Buildings capable of spatial transformation can accommodate programmatic change without demolition or extensive reconstruction, reducing material waste and preserving embodied energy. Dynamic systems may also respond to extreme climatic events: adjustable shading devices mitigate overheating during heatwaves, and operable ventilation enhances passive cooling during power outages. The integration of environmental responsiveness with structural flexibility strengthens both operational and ecological resilience, positioning kinetic architecture as a strategy for sustainable longevity.

Creating dynamic and adaptable spaces represents a decisive evolution in contemporary architectural practice. Through kinetic systems, responsive envelopes, transformable structures, and digital control technologies, buildings transcend static limitations and actively engage with users and the environment. Adaptability enhances flexibility, environmental performance, resilience, and experiential richness. As cities confront climate volatility, resource constraints, and rapidly shifting programmatic demands, the capacity for spatial transformation will become essential rather than optional. In this context, kinetic architecture provides a framework for designing buildings engineered not merely to endure but to evolve.

6.3 Case studies in kinetic design

The realization of kinetic architecture in contemporary practice demonstrates that movement, whether physical, perceptual, or environmental, can be embedded within architectural systems to enhance performance, adaptability, and user experience. The following case studies illustrate diverse interpretations of kinetic design, ranging from media-driven dynamic façades to mechanically transformable building envelopes. Together, they reveal how motion in architecture can operate across multiple scales: from visual animation to full spatial reconfiguration.

6.3.1 Case study 1: Galleria Centercity, Cheonan—media façade and dynamic interior

The Galleria Centercity in Cheonan, South Korea, designed by UNStudio in collaboration with GANSAM Architects, represents a sophisticated reinterpretation of kinetic architecture through optical motion and media integration rather than mechanical movement (Fig. 6.9). The building challenges traditional definitions of kinetic design by creating the perception of motion through light, geometry, and digital animation, transforming the façade into a dynamic urban interface.

The building envelope consists of a double-layered façade system composed of vertical aluminum profiles arranged in a way that produces a moiré effect, generating a constantly shifting visual appearance depending on the viewer's position and lighting conditions.

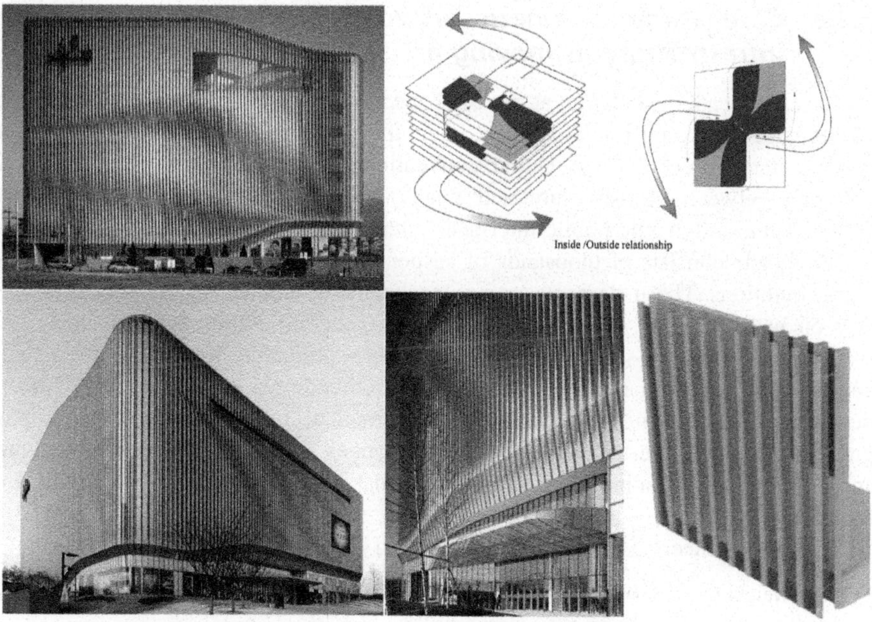

Figure 6.9 Galleria Centercity, Cheonan (UNStudio + GANSAM).

During the daytime, the façade appears as a reflective, monochromatic surface, while at night it transforms into a large-scale media screen animated by programmed light sequences. These animations create the illusion of fluid motion across the surface, effectively turning the building into an urban-scale kinetic installation.

Importantly, this project expands the definition of kinetic architecture beyond mechanical systems. As noted in research on kinetic façades, such designs can rely on light-based animation rather than physical transformation, producing movement through visual perception. In this sense, the building functions as a "media-kinetic" system, where digital content becomes the driver of architectural dynamism.

The interior further reinforces this concept of movement through spatial organization. The building is structured around a central atrium with stacked, rotating plateaus, creating a continuous upward flow of circulation. Curved geometries, layered platforms, and integrated lighting systems generate a sense of fluid motion, guiding users through the space in a nonlinear, exploratory manner.

From a performance perspective, the façade also contributes to environmental control. The outer layer reduces direct solar gain, while openings in the inner layer allow controlled daylight penetration, improving energy efficiency. Thus the project demonstrates how kinetic perception, environmental performance, and spatial experience can be integrated into a unified architectural system. Galleria Centercity exemplifies a shift toward nonmechanical kinetic architecture, where motion is achieved through optical effects, lighting, and digital media, expanding the conceptual boundaries of kinetic design.

6.3.2 Case study 2: kinetic artwork, Brisbane airport—urban infrastructure as moving art

The kinetic façade installation at the Brisbane Airport car park, developed by Urban Art Projects, represents a large-scale integration of art, engineering, and environmental interaction (Fig. 6.10). Unlike conventional buildings, this project operates as a hybrid between infrastructure and kinetic sculpture, transforming a utilitarian structure into a dynamic public interface.

The façade consists of thousands of suspended aluminum panels that respond to wind conditions. These elements are not mechanically actuated by motors but instead rely on natural forces, particularly airflow, to generate continuous movement. As wind passes across the surface, the panels oscillate, producing wave-like patterns that animate the façade in real time. This system embodies a form of passive kinetic design, where motion emerges from environmental interaction rather than programmed control. The façade continuously changes appearance depending on wind speed, direction, and atmospheric conditions, making it inherently unpredictable and temporally variable.

From an engineering perspective, the system required careful calibration of:

- panel weight and suspension systems;
- aerodynamic behavior;
- structural anchoring to withstand wind loads.

The resulting installation transforms the parking structure into a living surface, where movement is not imposed but emerges from the interaction between material and environment. Architecturally, the project redefines the role of infrastructure. Rather than remaining visually static and purely functional, the building becomes an experiential and

Figure 6.10 Kinetic Artwork, Brisbane Airport (Urban Art Projects).

performative element within the urban landscape. The façade engages travelers through motion, light reflection, and sound, enhancing the sensory quality of an otherwise utilitarian environment. This project demonstrates environmentally driven kinetic systems, where motion is generated passively through natural forces, emphasizing sustainability, minimal energy use, and continuous variability.

6.3.3 Case study 3: Ballet Mécanique housing, Zurich—transformable residential façade

The Ballet Mécanique housing project in Zurich represents a compelling example of mechanically transformable kinetic architecture applied at the residential scale (Fig. 6.11). Designed by Manuel Herz Architects and completed in 2017, the building integrates a façade system composed of foldable metal louvers that physically transform into balconies.

The building itself is a relatively simple cubic volume; however, its identity is defined by its kinetic envelope. The façade incorporates triangular and rectangular louver panels that can be opened outward through mechanical systems. When closed, these elements function as :

- sun-shading devices;
- privacy screens;
- protective outer layers.

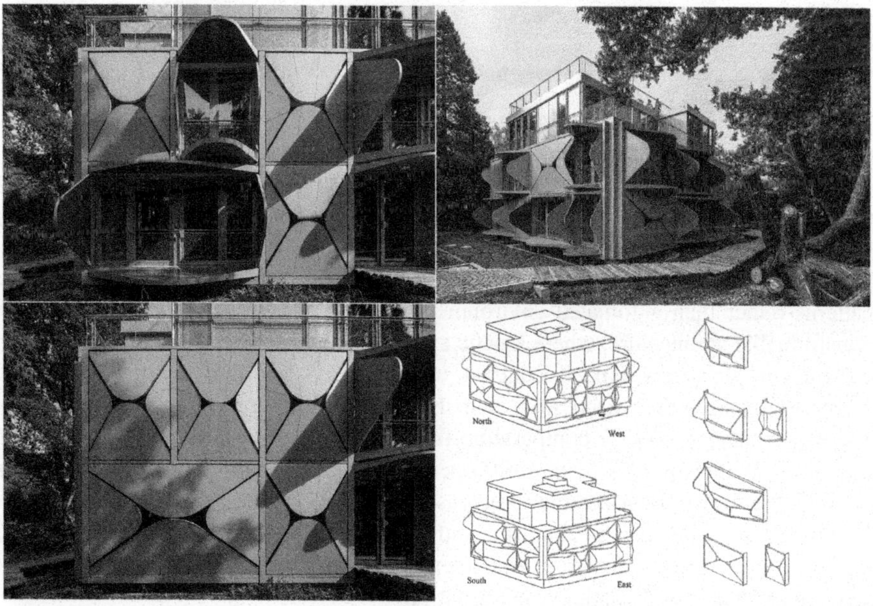

Figure 6.11 Ballet Mécanique Housing, Zurich (Manuel Herz Architects).

Table 6.1 Comparative analysis of kinetic design strategies.

Project	Type of kinetic system	Driving mechanism	Architectural impact
Galleria Centercity	Media/optical kinetic façade	Digital animation and lighting	Perceptual motion, urban identity
Brisbane Airport Artwork	Passive kinetic façade	Wind/environmental forces	Continuous natural movement
Ballet Mécanique	Mechanically transformable façade	User-controlled actuation	Spatial adaptability

When opened, they unfold into walkable balconies, effectively extending the interior living space into the exterior environment. This dual functionality allows the façade to operate simultaneously as an environmental regulator and spatial generator. The transformation process significantly alters the building's appearance. As panels open and close throughout the day, the façade shifts from a compact, monolithic surface to a fragmented, colorful composition. The interior surfaces of the panels, painted in vivid hues, are revealed only when opened, introducing a dynamic visual identity that evolves over time. From an engineering standpoint, the system integrates:

- hinged structural frames;
- mechanical or hydraulic actuation;
- fold-out safety railings;
- load-bearing support for temporary balcony use.

The façade is thus not merely an esthetic device but a structurally active system, capable of supporting human occupation when deployed. Importantly, the building operates in direct relation to user behavior. Residents control the opening and closing of panels, allowing the building to "respond" to daily routines—opening in the morning for light and outdoor access and closing at night for privacy and enclosure. This introduces a human-centered kinetic logic, where movement is driven by habitation patterns rather than automated environmental control. Ballet Mécanique exemplifies transformable architecture, where kinetic systems redefine spatial boundaries, enabling buildings to physically expand, contract, and adapt to user needs.

To synthesize the discussion and highlight their conceptual and technical distinctions, the following comparative framework categorizes the selected projects according to their modes of motion, operational drivers, and architectural implications. Table 6.1 summarizes the three case studies according to their type of kinetic system, primary driving mechanism, and resulting architectural impact, illustrating the spectrum of kinetic design from perceptual and environmentally driven motion to mechanically actuated spatial transformation.

Together, they demonstrate that kinetic architecture operates along a spectrum:

- perceptual motion (light, media, optical effects);
- environmental motion (wind-driven, passive systems);
- mechanical motion (actuated, transformable structures).

This spectrum reflects the expanding definition of kinetic architecture within contemporary practice. The examined projects confirm that kinetic design is not a singular approach but a multidimensional strategy encompassing physical movement, environmental responsiveness, and perceptual dynamism. Whether through digitally animated façades, wind-activated installations, or transformable building envelopes, kinetic architecture redefines buildings as active participants in their environment. In the context of smart structures, these case studies illustrate how motion, whether real or perceived, can enhance performance, adaptability, and user engagement. As technologies continue to evolve, the integration of kinetic systems will play an increasingly critical role in shaping responsive, intelligent, and experientially rich architectural environments.

References

Askar, R., Bragança, L., Gervásio, H., 2021. Adaptability of buildings: a critical review on the concept evolution. Applied Sciences 11 (10), 4483. Available at https://doi.org/10.3390/app11104483.

Barozzi, M., Lienhard, J., Zanelli, A., Monticelli, C., 2016. The sustainability of adaptive envelopes: developments of kinetic architecture. Procedia Engineering 155, 275–284. Available at https://doi.org/10.1016/j.proeng.2016.08.029.

Brenner, M., Silvestru, V.-A., Studer, P., Denoréaz, T., Bambach, M., Taras, A., Langenberg, S., 2026. Structural behavior of a wire arc additively manufactured steel eye bar for the repair of a historic high-tech facade. Structures 84, 111073. Available at https://doi.org/10.1016/j.istruc.2026.111073.

Choi, S.-J., Lee, D.-S., Jo, J.-H., 2017. Method of deriving shaded fraction according to shading movements of kinetic façade. Sustainability 9 (8), 1449. Available at https://doi.org/10.3390/su9081449.

Dąbrowska-Żółtak, K., Szalwa, A., 2025. Urban canvas in motion: the role of kinetic and media facades in urban space design. Arts 14 (1), 6. Available at https://doi.org/10.3390/arts14010006.

Han, S., Seo, J., Lee, H., 2025. Development of a solar-tracking movable Louver with a PV module for building energy reduction. Buildings 15 (12). Available at https://doi.org/10.3390/buildings15122100.

Johnson, A., Zheng, S., Nakano, A., Schierle, G., Choi, J.H., 2019. Adaptive kinetic architecture and collective behavior: a dynamic analysis for emergency evacuation. Buildings 9 (2). Available at https://doi.org/10.3390/buildings9020044.

Lee, J.H., Ostwald, M.J., Kim, A.M.J., 2021. Characterizing smart environments as interactive and collective platforms: a review of the key behaviors of responsive architecture. Sensors 21 (10). Available at https://doi.org/10.3390/s21103417.

Matheou, M., Phocas, M.C., Christoforou, E.G., Müller, A., 2018. On the kinetics of reconfigurable hybrid structures. Journal of Building Engineering 17, 32–42. Available at https://doi.org/10.1016/j.jobe.2018.01.013.

Matheou, M., Couvelas, A., Phocas, M.C., 2020. Transformable building envelope design in architectural education. Procedia Manufacturing 44, 116–123. Available at https://doi.org/10.1016/j.promfg.2020.02.212.

Nemati Aghdam, M., Dixit, M.K., 2026. A systematic review of methods and frameworks integrating environmental sustainability and resilience in building design and construction. Building and Environment 293, 114381. Available at https://doi.org/10.1016/j.buildenv.2026.114381.

Posadas, S.A., Morales-Luna, G., Fernández-Anaya, G., 2025. Exo-structures and domotic algorithms for dynamic architecture. Ain Shams Engineering Journal 16 (10). Available at https://doi.org/10.1016/j.asej.2025.103583.

Ramilo, R., Embi, M.R.B., 2014. Critical analysis of key determinants and barriers to digital innovation adoption among architectural organizations. Frontiers of Architectural Research 3 (4), 431–451. Available at https://doi.org/10.1016/j.foar.2014.06.005.

Rodinò, S., Maletta, C., 2025. Design development and testing of a morphable vehicle underbody shield actuated by shape memory alloys. Smart Materials and Structures 34 (6), 065001. Available at https://doi.org/10.1088/1361-665x/addeb1.

Soana, V., Minooee Sabery, S., Bosi, F., Wurdemann, H., 2025. Elastic robotic structures: a multidisciplinary framework for the design and control of shape-morphing elastic system for architectural and design applications. Construction Robotics 9 (1). Available at https://doi.org/10.1007/s41693-024-00128-8.

Uribe, D., Vera, S., Bustamante, W., McNeil, A., Flamant, G., 2019. Impact of different control strategies of perforated curved louvers on the visual comfort and energy consumption of office buildings in different climates. Solar Energy 190, 495–510. Available at https://doi.org/10.1016/j.solener.2019.07.027.

Vuscan, S., Muntean, R., 2025. Building conversion: enhancing sustainability through multifunctionality and movable interior systems. Sustainability 17 (7), 3182. Available at https://doi.org/10.3390/su17073182.

Wang, B., Zhu, J., Zhong, S., Liang, W., Guan, C., 2024. Space deployable mechanics: a review of structures and smart driving. Materials & Design 237, 112557. Available at https://doi.org/10.1016/j.matdes.2023.112557.

Wang, Y., Wang, X., Qu, H., Hu, B., Kuo, C.H., Guo, S., 2025. Origami-inspired deployable spatial structures with terrain adaptability. Engineering Structures 345. Available at https://doi.org/10.1016/j.engstruct.2025.121412.

Zhang, X., Zhang, H., Wang, Y., Shi, X., 2022. Adaptive façades: review of designs, performance evaluation, and control systems. Buildings 12 (12). Available at https://doi.org/10.3390/buildings12122112.

Adaptive building design: shaping environments

7

7.1 What is adaptive architecture?

Architecture has traditionally been conceived as a static discipline, producing buildings designed to resist environmental forces and accommodate predetermined functions over extended periods. This paradigm, rooted in durability, permanence, and structural stability, reflects an understanding of buildings as inert objects operating within relatively stable environmental and social contexts. However, contemporary challenges—including climate change, rapid technological advancement, resource scarcity, and evolving patterns of human behavior—have fundamentally challenged this assumption (Pelsmakers and Brown, 2025). The increasing variability of climatic conditions, coupled with the demand for higher environmental performance and user-centered design, necessitates a shift from static resistance toward dynamic responsiveness (Sánchez-Silva and Baker, 2024).

Within this evolving context, adaptive architecture has emerged as a transformative design approach that reconceptualizes buildings as dynamic, interactive, and responsive systems. Rather than functioning as passive enclosures, adaptive buildings actively engage with their environment and users, continuously adjusting their spatial, environmental, and operational characteristics. This paradigm aligns architecture more closely with ecological systems, where continuous feedback, adaptation, and optimization define performance over time (Barbhuiya et al., 2025).

Adaptive architecture can therefore be understood as a design framework in which buildings possess the capacity to sense, interpret, and respond to changing internal and external conditions. These responses may occur through physical transformation, material behavior, environmental regulation, or intelligent control systems. At its core, adaptive architecture is defined by three interrelated attributes:

- responsiveness, referring to the ability to detect environmental and user-driven stimuli, such as temperature fluctuations, solar radiation, occupancy patterns, and airflow, and react accordingly;
- flexibility, which enables buildings to accommodate multiple spatial configurations and evolving functional requirements without extensive structural intervention;
- transformability, denoting the capacity to physically alter form, geometry, or performance through mechanical systems, responsive materials, or digitally controlled mechanisms.

These attributes distinguish adaptive architecture from conventional static design by emphasizing performance variability, system integration, and temporal evolution. In this sense, architecture is no longer conceived as a finished object but as an ongoing process of interaction between built form, environment, and users.

The operation of adaptive architecture can be understood across multiple levels of technological and systemic complexity (Staehr et al., 2025). At the most fundamental

Smart Civil and Structural Engineering. DOI: https://doi.org/10.1016/B978-0-443-30223-7.00013-9

level, passive adaptive systems rely on intrinsic material properties or geometric configurations to respond to environmental conditions without external energy input. Examples include thermally responsive materials, hygroscopic surfaces, phase-change materials, and self-shading geometries. These systems are governed by physical principles such as heat transfer, expansion and contraction, and pressure differentials, enabling low-energy responsiveness embedded directly within the material or form.

At a more advanced level, active adaptive systems incorporate mechanical components and control mechanisms. These include motorized shading devices, operable façades, adjustable structural elements, and dynamic ventilation systems. Active systems depend on real-time data collected through sensors and processed by control units, enabling buildings to respond dynamically to changing environmental conditions. While more energy-intensive than passive systems, they offer greater precision and responsiveness.

The most sophisticated level is represented by intelligent adaptive systems, which integrate digital technologies such as artificial intelligence, machine learning, and predictive algorithms. These systems enable buildings to learn from historical data, anticipate future conditions, and optimize performance proactively rather than reactively. Through this integration, buildings evolve into cyber-physical systems, where digital intelligence and physical infrastructure are tightly interconnected (Ghansah and Lu, 2023).

The functionality of adaptive architecture relies on the integration of several interconnected components:

1. Sensing systems: Sensors continuously collect real-time data on environmental variables (solar radiation, temperature, humidity, wind) and user activity (occupancy, movement patterns).
2. Data processing and control: Building management systems, parametric design tools, or AI-based platforms analyze incoming data and determine appropriate responses.
3. Actuation mechanisms: Actuators translate digital decisions into physical transformations, enabling movement, adjustment, or modulation of building components.
4. Material and structural integration: Adaptive behavior is embedded within materials and structural systems, ensuring that responsiveness is intrinsic rather than applied superficially.

This feedback-driven loop establishes a continuous cycle of sensing, analysis, and response, allowing buildings to maintain optimal environmental performance while minimizing resource consumption (Emedo et al., 2025). Adaptive architecture intersects with related domains such as smart structures and kinetic architecture, but extends beyond them in scope and integration. Smart structures emphasize sensing, data processing, and intelligent control, while kinetic architecture focuses on movement and mechanical transformation. Adaptive architecture synthesizes these dimensions into a holistic framework that combines intelligence, responsiveness, and physical transformation within a unified system.

A key driver of adaptive architecture is the need to improve environmental performance (Aigwi et al., 2023). Buildings account for a significant proportion of global energy consumption and greenhouse gas emissions, and static design strategies often fail to respond effectively to dynamic climatic conditions. Adaptive systems address this limitation by enabling buildings to regulate solar gain, optimize daylighting, enhance natural ventilation, and reduce dependence on mechanical heating, ventilation, and air conditioning (HVAC) systems (Chen et al., 2025). As a result, adaptive buildings can achieve higher levels of energy efficiency, thermal comfort, and environmental sustainability.

At the same time, adaptive architecture fundamentally redefines the relationship between occupants and space. Users are no longer passive recipients of predefined environmental conditions but active participants in shaping their surroundings. Through responsive systems, occupants can influence lighting, temperature, airflow, and spatial configurations, resulting in increased comfort, productivity, and well-being. This shift reflects a broader transition toward human-centered design, where adaptability enhances both performance and user experience.

Despite its advantages, adaptive architecture introduces several challenges. The integration of multiple systems increases technical complexity and requires advanced coordination among disciplines. Initial costs may be higher due to the incorporation of sensors, actuators, and intelligent control systems. Maintenance requirements also increase, particularly for moving components and digital infrastructures. Furthermore, successful implementation demands interdisciplinary collaboration among architects, engineers, material scientists, and data specialists.

Addressing these challenges requires a holistic and integrated design approach in which adaptability is embedded from the earliest conceptual stages. Rather than being treated as an add-on technology, adaptive strategies must inform spatial organization, structural systems, material selection, and environmental design from the outset.

Recent advancements in façade technology have significantly improved the performance, efficiency, and functionality of building envelopes. These developments reflect a shift toward smarter and more adaptive systems that enhance both sustainability and occupant comfort in the built environment. Among these innovations, adaptive façades have emerged as a key strategy for achieving responsive and energy-efficient building design (Loonen et al., 2013). Adaptive façades represent a cutting-edge development in contemporary architecture, characterized by their ability to respond dynamically to changing environmental conditions. These systems incorporate advanced materials and control technologies that allow façade components to modify their configuration or properties in response to external stimuli such as solar radiation, temperature, and wind. This dynamic behavior enables buildings to continuously adjust to their surroundings, thereby improving indoor environmental quality while reducing overall energy demand (Machard et al., 2023). A schematic representation of the operational logic and components of adaptive façade systems is illustrated in Fig. 7.1, highlighting the interaction between environmental inputs, control systems, and responsive façade elements.

Key features and performance benefits include:

- Dynamic shading: Adaptive façades integrate movable shading elements that adjust according to solar position, optimizing daylight penetration while minimizing unwanted

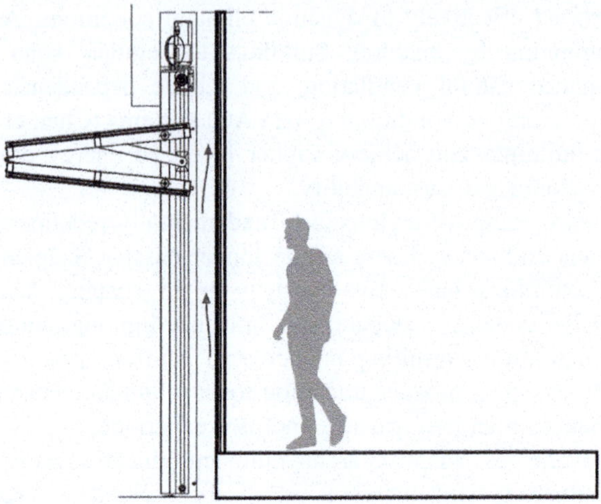

Figure 7.1 Schematic illustration of an adaptive façade system.

heat gain. This contributes to reduced cooling loads and improved visual comfort (Lee et al., 2021).

- Ventilation control: Operable façade components, such as louvers or vents, regulate airflow in response to indoor and outdoor conditions, enhancing natural ventilation and indoor air quality (Alkhatib et al., 2021).
- Thermal regulation: The use of smart or responsive materials allows façades to alter their thermal properties, reducing heat transfer and improving overall energy performance (Karanafti and Theodosiou, 2024).

Adaptive architecture is not merely an extension of traditional building design but a fundamental redefinition of architectural practice. By integrating sensing, intelligence, and transformation, buildings evolve into responsive systems capable of adapting to environmental, social, and functional change. As the built environment faces increasing pressures from climate change and urbanization, adaptive architecture will play a critical role in shaping resilient, efficient, and human-centered spaces. In this emerging paradigm, architecture is no longer static; it is dynamic, interactive, and continuously evolving.

7.2 Responding to changing conditions

Adaptive architecture derives much of its value from its ability to respond meaningfully to change. Unlike conventional buildings that are typically designed around fixed assumptions of climate, occupancy, and use, adaptive buildings are conceived as responsive systems capable of adjusting their behavior in relation to dynamic environmental, social, and operational conditions. This responsiveness enables buildings not only to maintain performance over time but also to improve comfort, efficiency, resilience, and long-term usability. In this context, responding to changing conditions is not simply a technical feature; it is a core design principle that redefines the relationship between architecture, users, and environment.

Environmental unpredictability is one of the main forces behind adaptive design. Buildings are always subject to shifting weather conditions, such as variations in solar radiation, temperature, humidity, wind speed, precipitation, and seasonal changes. Conventional static envelopes frequently find it difficult to effectively maintain indoor comfort in such unpredictable circumstances, which results in higher energy consumption and worse environmental performance. In order to overcome this difficulty, adaptive buildings use systems that react to climate stimuli either actively or passively. For instance, dynamic shading devices reduce glare and solar heat gain while permitting beneficial daylight penetration by adjusting their position based on solar angle and intensity. While responsive insulation systems can change heat resistance in accordance with seasonal requirements, operable façades can adjust ventilation channels based on wind direction and indoor air quality. The building envelope becomes an intelligent mediator between the inner and outdoor surroundings as a result of adaptive design.

Thermal comfort is one of the primary domains in which adaptive buildings demonstrate measurable benefits (Mohan et al., 2026). Conventional heating and cooling systems often operate on preset schedules or broad assumptions about occupant needs. Adaptive systems, by contrast, can respond in real time to occupancy patterns, indoor temperatures, and user preferences. Through integrated sensors, predictive control algorithms, and zoned environmental systems, buildings can optimize thermal conditions while minimizing energy use. Smart glazing technologies, such as electrochromic glass, can change tint in response to solar exposure, reducing overheating and cooling loads (Wen et al., 2025). Similarly, phase-change materials integrated into walls or ceilings can absorb and release heat depending on internal temperature fluctuations, enhancing passive thermal regulation. These strategies support both occupant well-being and operational sustainability.

Adaptive architecture also responds to changing patterns of occupancy and human activity. Contemporary buildings increasingly need to accommodate diverse users, variable schedules, and evolving functional requirements. Offices may shift between individual work, collaborative meetings, and public events; educational buildings may support lectures, group learning, and digital instruction; residential buildings may need to accommodate work-from-home arrangements, family growth, or aging occupants. Adaptive design addresses this through spatial flexibility and programmable environments. Movable partitions, retractable walls, modular furnishings, and transformable floor plans enable spaces to be reconfigured quickly and efficiently. Sensor-driven lighting, acoustics, and ventilation systems can also adjust according to occupancy density and activity type, ensuring that environmental conditions remain aligned with user needs.

In addition to daily functional adaptation, buildings must also respond to long-term changes in use and societal needs. One of the most important dimensions of adaptability is the capacity for buildings to remain relevant over time without requiring major reconstruction. This concept, often referred to as “long-life loose-fit” design, emphasizes structural and spatial systems that can accommodate future change. Examples include generous floor-to-floor heights, modular service cores, accessible mechanical systems, and open structural grids that permit reconfiguration. By anticipating change rather than resisting it, adaptive buildings extend their useful lifespan, reduce material waste, and support circular economy principles (van Laar et al., 2025).

Another important component of adaptive response is resilience to harsh events and disturbances. The frequency and intensity of heatwaves, floods, storms, and other environmental pressures have increased due to climate change. By implementing mechanisms that react to such occurrences, adaptive buildings can improve resilience. Architecture may reduce risk through features flood-responsive foundations, retractable storm barriers, automatic smoke control ventilation systems, and façades that adapt to intense sun exposure. Adaptive buildings may incorporate battery storage, smart microgrids, and renewable energy controllers into their energy systems to enable continuous operation in the event of grid failures. These tactics guarantee that structures are not only effective in everyday situations but also competent to continue essential operations in emergency situations.

Adaptive reactions are made possible by the integration of digital technology. The building is equipped with sensors that gather information on occupancy patterns, energy consumption, structural performance, and environmental variables. Building management systems, digital twins, or AI-based platforms that examine patterns and enhance building operations are used to process this data. Predictive rather than reactive adaptation is made possible by these systems' ability to foresee future circumstances (Ibrahim et al., 2026). For example, occupancy trends might guide ventilation schedules ahead of time, or weather forecasts can be utilized to precool a building before a heatwave. A significant development in adaptive design, this move toward predictive intelligence turns buildings into learning systems that gradually enhance performance.

Importantly, adaptive response must also be considered from a human-centered perspective. Buildings should not only react to environmental variables but also support the psychological, social, and experiential needs of occupants. Access to daylight, acoustic comfort, personal environmental control, and spatial diversity all contribute to user well-being. Adaptive systems that let users manage their surroundings, such as movable windows, customized heat controls, or adjustable shade, can improve user satisfaction and sense of agency. Designers must, however, steer clear of unduly complicated systems that diminish usability or foster technology dependence. An effective adaptive architecture strikes a balance between automation and user-friendliness.

From a design and engineering perspective, responding to changing conditions requires an integrated and anticipatory approach. Adaptability must be embedded from the earliest conceptual stages rather than added as a technological afterthought. Architects, engineers, environmental consultants, and digital specialists must collaborate to define performance goals, identify relevant variables, and design coordinated response strategies. Computational simulation tools, parametric workflows, and performance modeling are essential in testing adaptive scenarios before construction, ensuring that systems are both effective and buildable. Ultimately, responding to changing conditions is central to the future of architecture. As environmental uncertainty, urban density, and user expectations continue to evolve, buildings must become more responsive, resilient, and capable of transformation. Adaptive architecture offers a framework through which buildings can actively engage with change rather than passively endure it. By integrating responsive envelopes, flexible spaces, intelligent controls, and resilient systems, adaptive buildings redefine architecture as a living, evolving interface between people, technology, and environment. In this sense, adaptive architecture does not merely improve building performance—it fundamentally reshapes how built environments are conceived, experienced, and sustained over time. It positions architecture as an active participant in ecological and social systems, capable of learning from change, responding to uncertainty, and supporting more sustainable and humane futures.

This adaptive logic is particularly evident in building envelope systems, where the façade serves as the primary interface between indoor and outdoor environments. Among the most effective strategies for responding to changing climatic conditions is the use of ventilated façade systems, which are designed to regulate heat transfer, airflow, moisture, and acoustic performance through a layered and responsive envelope configuration. Unlike conventional static façades, ventilated façades provide an intermediate cavity that facilitates controlled air movement, allowing the building skin to respond more efficiently to fluctuations in solar radiation, outdoor temperature, wind pressure, and humidity.

To enhance the adaptive potential of such systems, an integrated lifecycle approach is increasingly used that combines advanced sensing technologies, digital simulation tools, and continuous monitoring strategies from design through construction and field operation (Fig. 7.2). This approach enables the façade to be understood not merely as a passive enclosure but as an active environmental regulator whose performance can be measured, evaluated, and optimized over time. By incorporating performance feedback at multiple stages of the building lifecycle, ventilated façades can be continuously refined to improve energy efficiency, indoor comfort, and long-term resilience (Es-sakali et al., 2024).

A key component of this methodology is the systematic identification, selection, and deployment of sensors to monitor the façade's thermal, environmental, ventilation, and acoustic behavior. Thermal sensors are used to track surface and cavity temperatures, while environmental sensors measure parameters such as humidity, airflow velocity, and solar exposure. Acoustic sensors may also be integrated to evaluate sound attenuation and occupant comfort. Each sensor is assigned a specific identification code and incorporated into a wireless monitoring network to facilitate real-time data collection, storage, and analysis (Fig. 7.3). Standardized measurement protocols are established to ensure consistency and reliability, while simulation tools are used in parallel to support calibration, validate performance trends, and predict system behavior under varying environmental scenarios.

This feedback-based framework allows continuous assessment of how ventilated façades perform under dynamic conditions, including seasonal temperature changes, daily solar exposure cycles, wind-driven ventilation effects, and occupancy-related indoor loads. The resulting data supports targeted improvements in cavity airflow

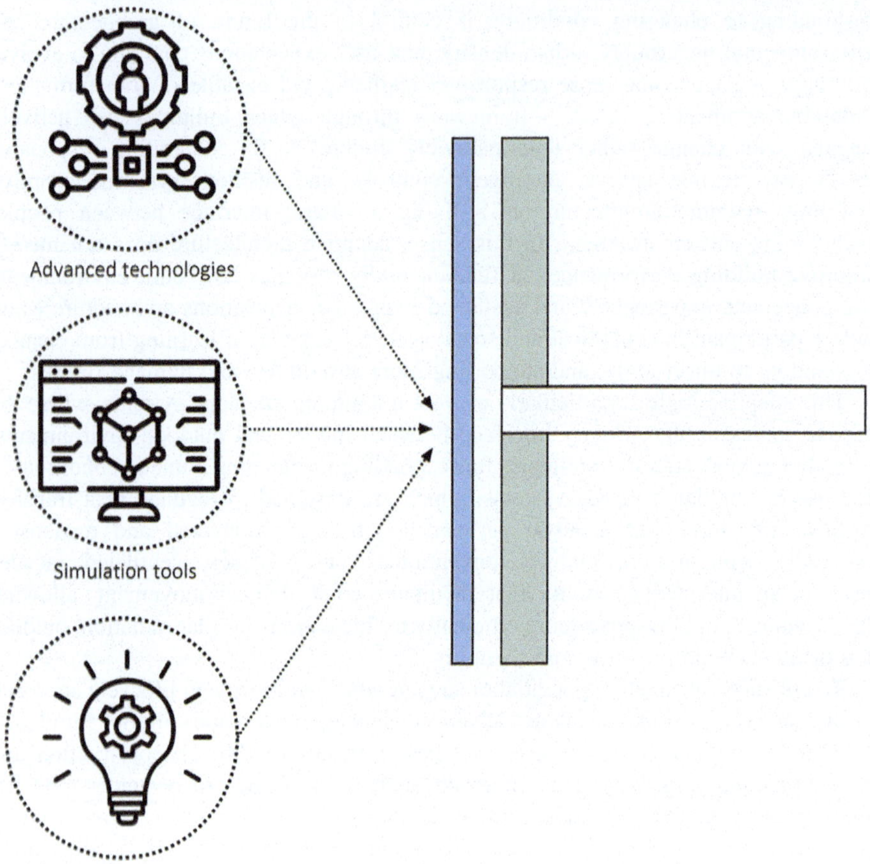

Figure 7.2 Integrated lifecycle methodology for adaptive ventilated façade performance.

behavior, thermal buffering, moisture control, and acoustic insulation. In this way, the ventilated façade evolves from a static protective layer into an intelligent and adaptive building component capable of dynamically regulating environmental exchanges. This example clearly demonstrates how adaptive building design can transform the building envelope into a responsive system that enhances long-term sustainability, resilience, and occupant well-being.

7.3 Case studies of adaptive building designs

The practical value of adaptive architecture is most clearly demonstrated through built projects that successfully respond to environmental variability, user needs, and long-term operational demands. While the principles of adaptive building design are rooted in flexibility, responsiveness, and performance optimization, their true

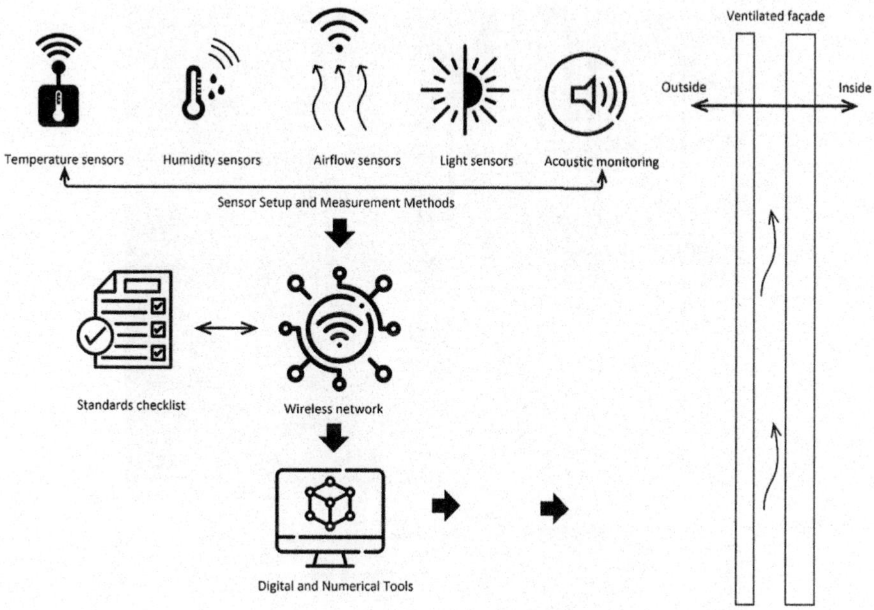

Figure 7.3 Sensor configuration and measurement strategy for adaptive ventilated façade monitoring.

significance emerges when these principles are translated into real-world applications. Contemporary adaptive buildings increasingly integrate responsive façades, intelligent environmental systems, flexible spatial configurations, and digital control platforms to create environments that can evolve in relation to climate, occupancy, and functional change. The following case studies illustrate different dimensions of adaptive building design. Together, they demonstrate how architecture can transition from static shelter to responsive environmental infrastructure, capable of sensing, adjusting, and learning over time.

The following case studies illustrate different adaptive design approaches, ranging from ventilated double-skin façades and daylight-responsive shading systems to hybrid ventilation strategies and monitored performance-based envelopes. Together, they demonstrate how architecture can evolve from static enclosure to responsive environmental infrastructure.

7.3.1 Case study 1: Alto Tower—ventilated double-skin façade and user-controlled comfort

The Alto Tower represents a significant advancement in adaptive high-rise architecture, where the building envelope is reconceptualized as an active environmental mediator rather than a static protective layer (Fig. 7.4). In line with the broader theoretical framework of adaptive architecture outlined in this chapter, the

Figure 7.4 Alto Tower: Adaptive double-skin façade and ventilation strategy.

project demonstrates how façade systems can operate as dynamic interfaces that regulate environmental exchange, enhance occupant comfort, and optimize building performance over time. The design reflects a shift from conventional sealed high-rise envelopes toward permeable, responsive systems that integrate passive strategies with user interaction and controlled environmental modulation.

At the core of the building's adaptive strategy is its ventilated double-skin façade, a multilayered envelope system composed of an outer glazed skin and an inner thermally insulated façade separated by a continuous air cavity. This cavity functions as a transitional environmental zone, mediating heat transfer, airflow, and acoustic performance. The spatial depth of the façade allows it to operate not merely as a barrier but as a microclimatic buffer, capable of adapting its behavior in response to external climatic conditions and internal comfort requirements.

From a thermodynamic perspective, the cavity plays a crucial role in regulating heat exchange. During periods of high solar radiation, shortwave solar energy penetrates the outer glazing and is partially absorbed within the cavity. This leads to air temperature stratification, which activates stack-driven ventilation within the cavity. As warmer air rises and exits through upper openings, cooler air is drawn in from lower openings, establishing a continuous airflow cycle. This process effectively removes excess heat before it reaches the inner façade, thereby reducing solar heat gain and lowering cooling loads. The result is a significant improvement in energy performance, particularly in high-rise buildings where façade exposure is extensive and solar gains can be substantial.

In colder conditions, the façade operates differently. The cavity acts as an insulating layer, reducing convective heat loss and enhancing the overall thermal resistance of the building envelope. By trapping a layer of relatively stable air between the two façade skins, the system minimizes heat transfer from the interior to the exterior. This dual-mode operation—cooling through ventilation in summer and insulating in winter—illustrates the adaptive capacity of the system to respond to seasonal variability.

A defining characteristic of the Alto Tower is the integration of user-controlled operable windows and ventilation openings within the inner façade layer. Unlike fully sealed high-rise buildings that rely exclusively on mechanical systems, this design introduces a hybrid environmental strategy that combines passive climatic regulation with occupant agency. Users are able to adjust airflow, ventilation rates, and local environmental conditions according to their preferences, thereby enhancing perceived comfort and satisfaction.

This user-centered adaptability aligns with contemporary theories of adaptive thermal comfort, which emphasize the role of occupant control and behavioral adjustment in achieving comfort conditions. By enabling direct interaction with the façade, the building supports a more personalized indoor environment, reducing reliance on centralized HVAC systems and allowing localized environmental tuning. Importantly, this approach also contributes to psychological comfort, as occupants gain a sense of control over their surroundings, an aspect often overlooked in fully automated smart building systems.

Beyond thermal performance, the double-skin façade contributes significantly to acoustic insulation, particularly in dense urban environments where external noise levels can be high. The cavity acts as a sound buffer, attenuating noise transmission and improving indoor acoustic quality (Hu et al., 2021). This is especially important for high-rise office or residential buildings located in urban centers, where façade design must simultaneously address environmental performance and user well-being.

Daylighting performance is another critical dimension of the system. The transparency of the outer glazing layer allows natural light to penetrate deep into the interior, reducing the need for artificial lighting. At the same time, the cavity and inner façade layer help modulate light intensity, preventing glare and excessive brightness. This balanced daylighting strategy enhances visual comfort while contributing to overall energy efficiency.

From an architectural perspective, the double-skin façade introduces a layered spatial and visual quality that reflects the building's environmental logic. The depth of the façade, combined with the visibility of cavity airflow and operable elements, expresses the dynamic nature of the building envelope. Rather than concealing its performance, the façade communicates its adaptive behavior, reinforcing the integration of environmental function and architectural expression.

The Alto Tower also illustrates the importance of system integration in adaptive architecture. The effectiveness of the double-skin façade depends on the coordination between passive airflow mechanisms, user-controlled elements, and supporting mechanical systems where necessary. This integration ensures that the building can operate efficiently across varying climatic conditions while maintaining occupant comfort.

However, the implementation of such systems is not without challenges. Double-skin façades require careful consideration of issues such as overheating within the cavity, maintenance access, fire safety, and construction cost. The Alto Tower demonstrates how these challenges can be addressed through precise engineering, appropriate material selection, and the integration of control strategies that regulate airflow and thermal performance.

In a broader context, this case study exemplifies how adaptive façade systems can redefine the role of the building envelope in high-rise architecture. By combining passive environmental regulation, user interaction, and layered design strategies, the Alto Tower achieves a balance between energy efficiency, occupant comfort, and architectural quality. It highlights the importance of designing façades not as static components but as responsive environmental systems that actively contribute to building performance.

Ultimately, the Alto Tower demonstrates that adaptive architecture does not rely solely on complex digital technologies or fully automated systems. Instead, it shows the effectiveness of integrating passive climatic principles, simple mechanical strategies, and user engagement to create buildings that are both efficient and human-centered. In doing so, it provides a valuable reference for future high-rise developments seeking to address the challenges of energy consumption, environmental variability, and occupant well-being in increasingly complex urban environments.

7.3.2 Case Study 2: University of Southern Denmark Kolding Campus—responsive daylight and solar control system

The Kolding Campus of the University of Southern Denmark (SDU), designed by Henning Larsen Architects and completed in 2014, represents a benchmark in adaptive educational architecture, where environmental performance, user comfort, and architectural expression are seamlessly integrated through a responsive façade system (Fig. 7.5). The project exemplifies a shift from static building envelopes toward dynamic systems that actively mediate between external climatic conditions and internal spatial requirements. Within the broader framework of adaptive architecture discussed in this chapter, the campus demonstrates how façade systems can function as intelligent environmental regulators while simultaneously contributing to the identity and experiential quality of the building.

At the core of the building's adaptive strategy is its kinetic perforated shutter façade, consisting of approximately 1600 triangular steel panels distributed across the

Figure 7.5 SDU Kolding Campus: Responsive perforated shutter façade (*SDU*, University of Southern Denmark).

building envelope. These panels are mounted on a digitally controlled mechanism that allows continuous and precise adjustment in response to environmental inputs. Each panel operates as part of an integrated network, responding to real-time data such as solar radiation intensity, sun angle, façade orientation, and interior daylight demands. This system transforms the façade into a responsive layer capable of dynamically controlling solar exposure and daylight penetration.

The operation of the façade is based on a sensor-driven control system integrated within the building management infrastructure. Sensors continuously monitor external and internal conditions, and the collected data is processed through algorithms that determine the optimal configuration of the shading panels. As environmental conditions change throughout the day, the panels adjust incrementally, ensuring that interior spaces maintain balanced lighting and thermal conditions.

This adaptive mechanism allows the façade to regulate multiple environmental parameters simultaneously:

- daylight penetration, ensuring sufficient natural illumination for learning and working activities while reducing reliance on artificial lighting systems;
- solar heat gain, minimizing overheating and lowering cooling energy demand;

- glare control, enhancing visual comfort in classrooms, lecture halls, and study areas;
- visual connection, maintaining outward views while filtering incoming light.

Unlike conventional fixed shading devices, which are designed for static or average conditions, the SDU façade operates as a continuously responsive environmental filter. This enables the building to adapt to short-term fluctuations such as cloud cover as well as long-term seasonal variations, maintaining stable indoor environmental quality. A defining architectural feature of the system is the perforation pattern of the triangular panels, which introduces a sophisticated approach to daylight modulation. Instead of simply blocking sunlight, the perforated surfaces diffuse incoming light, producing soft and evenly distributed illumination within interior spaces. This creates a dynamic interplay of light and shadow that evolves throughout the day, enriching spatial perception and enhancing the overall learning environment. The façade thus performs both environmental and experiential functions, demonstrating the integration of performance-driven design with architectural quality.

From a thermal performance perspective, the adaptive shading system plays a critical role in reducing solar heat gains, particularly on highly exposed façades. By limiting direct solar radiation during peak hours, the system significantly reduces cooling loads and contributes to the building's overall energy efficiency. At the same time, the ability of the panels to open during periods of lower solar intensity allows beneficial daylight and passive heat gains when appropriate, supporting seasonal performance optimization.

The façade system is further integrated with natural ventilation strategies, enabling mixed-mode environmental operation. Operable windows and ventilation pathways allow fresh air to circulate through interior spaces, while the shading system helps regulate thermal conditions to support comfortable airflow. This integration enhances indoor air quality and reduces dependence on fully mechanical ventilation systems.

In addition to passive strategies, the building incorporates energy-efficient mechanical systems that operate in coordination with the adaptive façade. This layered approach ensures that environmental control is achieved through a hierarchy of strategies, prioritizing passive and low-energy solutions before mechanical intervention is required. Such integration reflects a comprehensive approach to sustainable design, where multiple systems work together to optimize building performance.

The orientation and form of the building are also carefully considered to support the effectiveness of the adaptive façade. The geometry of the campus maximizes daylight access while minimizing excessive solar exposure. The responsive shading system complements this strategy by providing fine-tuned control at the façade level, ensuring that each elevation responds appropriately to its specific environmental conditions.

As an educational facility, the SDU Kolding Campus accommodates diverse and variable occupancy patterns, including lectures, group work, individual study, and social interaction. The adaptive façade supports this diversity by maintaining consistent environmental conditions across different spaces and usage scenarios. Spaces with higher occupancy densities benefit from increased shading and ventilation, while areas requiring greater daylight access can receive higher levels of natural illumination through adjusted panel configurations.

From a user experience perspective, the kinetic façade enhances environmental awareness and engagement. The visible movement of the panels and the changing quality of light allow occupants to perceive the building as a responsive system. This dynamic interaction strengthens the relationship between users and their environment, reinforcing the principles of adaptive architecture (Sommese et al., 2024).

Architecturally, the façade establishes a strong and recognizable identity for the campus. The continuously changing configuration of the triangular panels creates a building envelope that is visually dynamic and contextually responsive. This evolving appearance reflects real-time environmental conditions, linking performance directly to architectural expression. The building thus becomes a physical manifestation of adaptability, where form and function are closely interconnected.

Despite its advantages, the implementation of such a system involves challenges related to mechanical complexity, control precision, and maintenance requirements. The large number of moving components necessitates robust engineering solutions and reliable control systems to ensure long-term functionality. Regular maintenance and calibration are required to sustain performance over time. Nevertheless, the successful operation of the SDU Kolding Campus demonstrates that these challenges can be effectively addressed through integrated design and careful coordination.

In a broader context, this project illustrates how adaptive façade systems can redefine daylight and solar control in contemporary architecture. By moving beyond static shading strategies, the building achieves high environmental performance while enhancing spatial quality and user experience. It demonstrates that adaptive architecture can operate simultaneously at technical, environmental, and experiential levels.

Ultimately, the SDU Kolding Campus serves as a model for future educational buildings, showing how responsive design strategies can support sustainability, improve occupant comfort, and enrich architectural expression. Through the integration of kinetic systems, digital control technologies, and passive design principles, the project exemplifies the potential of adaptive architecture to create environments that are efficient, responsive, and engaging.

7.3.3 Case study 3: Q1 Building, ThyssenKrupp Quarter, Essen—hybrid ventilation and intelligent façade design

The Q1 Building within the ThyssenKrupp Quarter in Essen, designed by JSWD Architekten in collaboration with Ingenhoven Architects and completed in 2010, represents a highly developed example of adaptive workplace architecture (Fig. 7.6). The project demonstrates how integrated façade systems, natural ventilation strategies, and solar control mechanisms can be combined to produce a responsive, energy-efficient, and user-centered office environment. Within the broader discourse of adaptive architecture, the building illustrates how environmental performance can be achieved through the careful coordination of passive design principles and intelligent system integration.

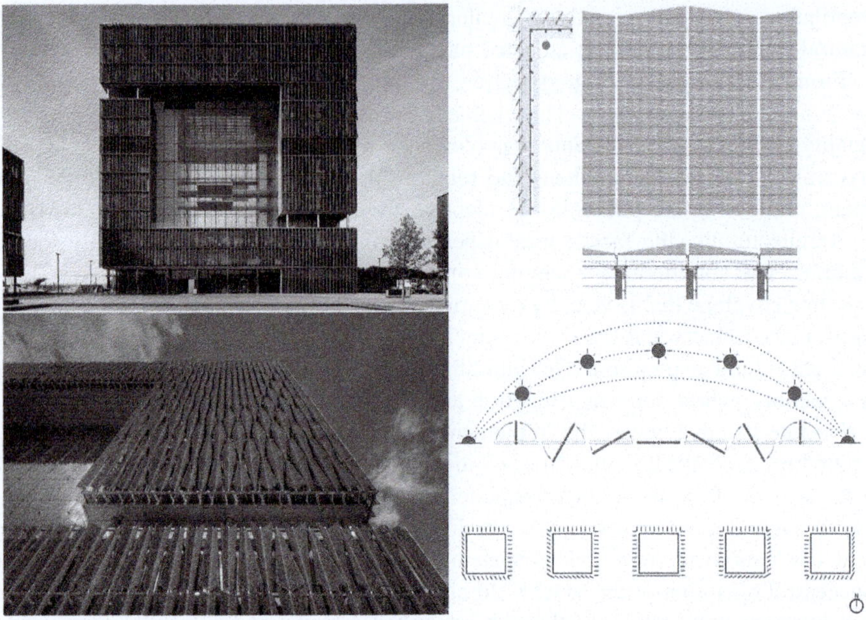

Figure 7.6 Q1 Building, ThyssenKrupp Quarter: Hybrid ventilation and solar control system.

A central component of the building's adaptive strategy is its hybrid ventilation system, which combines natural ventilation with mechanical support. This approach allows the building to operate under varying climatic conditions by prioritizing passive airflow when external conditions are favorable and supplementing it with mechanical systems when necessary. The hybrid system is closely integrated with a high-performance double-skin façade and a centrally organized atrium, forming a comprehensive environmental network that regulates airflow, thermal conditions, and indoor air quality.

The double-skin façade consists of an external glazed layer and an internal insulated façade separated by a ventilated cavity. This cavity functions as an environmental buffer, reducing heat transfer while enabling controlled airflow. Through strategically positioned openings, fresh air enters the cavity and is distributed into interior spaces. At the same time, warm air is exhausted through upper openings, creating a continuous airflow cycle that enhances natural ventilation and reduces cooling demand.

The environmental strategy of the Q1 Building is structured around several interrelated adaptive elements:

- operable façade panels, allowing controlled intake of fresh air and enabling user interaction with the indoor environment;
- cavity-assisted ventilation, where the façade cavity supports airflow and thermal regulation;
- central atrium air circulation, facilitating vertical airflow and enhancing stack-driven ventilation;

- solar-responsive shading systems, regulating solar radiation and improving visual and thermal comfort.

These components operate together as an integrated system, enabling the building to respond dynamically to environmental changes. The façade acts as a responsive environmental filter, balancing daylight access, solar control, and ventilation performance. This coordinated operation ensures that indoor conditions remain stable while minimizing energy consumption.

A key adaptive feature of the building is the use of external sun-shading devices, which are designed according to façade orientation and solar exposure. These shading systems reduce direct solar radiation during peak periods while allowing controlled daylight penetration at other times. As a result, they minimize overheating, reduce glare, and contribute to overall energy efficiency. This adaptive shading strategy plays a critical role in maintaining visual comfort in office environments, where excessive brightness or glare can negatively affect productivity.

The central atrium is a defining spatial and environmental element of the building. It functions as a vertical ventilation shaft, supporting stack-driven airflow by allowing warm air to rise and exit at higher levels. This process is complemented by horizontal airflow through the façade, creating a continuous natural ventilation system. The integration of the atrium with the façade demonstrates how spatial organization and environmental performance can be closely linked in adaptive design.

From a thermal perspective, the hybrid system enables the building to operate in multiple modes. During mild weather conditions, natural ventilation is maximized, reducing reliance on mechanical systems. In more extreme conditions, mechanical cooling or heating systems are activated to maintain comfort. This mixed-mode operation ensures that energy use is optimized while maintaining consistent indoor environmental quality.

Daylighting is another important aspect of the building's performance. The façade design allows for deep penetration of natural light into interior spaces, reducing the need for artificial lighting. At the same time, the combination of shading devices and façade layering prevents glare and excessive brightness. The atrium further enhances daylight distribution by bringing natural light into the central areas of the building, improving both energy performance and spatial quality.

From the perspective of occupant experience, the building supports adaptive comfort by combining environmental stability with opportunities for user interaction. Operable façade elements allow occupants to influence local airflow conditions, while the overall system maintains consistent thermal and visual comfort. This balance between automation and user control is essential in creating a responsive and user-friendly workplace environment.

The double-skin façade also contributes to acoustic performance, acting as a buffer against external noise. This allows the building to utilize natural ventilation without compromising indoor acoustic comfort, which is particularly important in urban office settings.

Architecturally, the Q1 Building expresses its environmental strategy through a transparent and layered façade composition. The visibility of the double-skin system and shading devices reflects the building's adaptive behavior, linking performance with architectural expression. The design emphasizes clarity and openness, reinforcing the concept of the building as a responsive system rather than a sealed enclosure.

Despite its advantages, the implementation of such an integrated system requires careful consideration of technical complexity, system coordination, and maintenance. The interaction between natural and mechanical systems must be precisely managed to ensure efficient operation. Additionally, the long-term performance of operable façade components and shading devices depends on regular maintenance and monitoring. The Q1 Building demonstrates that these challenges can be successfully addressed through integrated design processes and interdisciplinary collaboration.

In a broader context, this project illustrates how adaptive architecture can transform office buildings into high-performance environmental systems. By integrating façade design, ventilation strategies, and spatial organization, the building achieves a holistic approach to environmental control. It moves beyond reliance on conventional mechanical systems, instead utilizing natural processes and responsive design to create a sustainable and efficient workplace.

From a workplace performance perspective, the adaptive strategies implemented in the Q1 Building contribute to:

- improved access to natural daylight, supporting visual comfort and productivity;
- enhanced indoor air quality, promoting health and well-being;
- reduced temperature fluctuations, ensuring stable and comfortable indoor conditions.

Ultimately, the Q1 Building demonstrates how adaptive design principles can redefine the office building typology. Through the integration of passive strategies, intelligent systems, and user interaction, the project exemplifies a responsive architectural approach that aligns environmental performance with occupant needs. It stands as a key reference for future developments seeking to achieve sustainability, efficiency, and adaptability in the built environment.

7.3.4 Case study 4: Kuggen Building—solar-responsive kinetic façade system

The Kuggen, designed by Wingårdhs and completed in 2011, represents a significant exploration of solar-responsive architecture in which kinetic façade elements are employed to regulate environmental performance while reinforcing architectural identity (Fig. 7.7). The project is situated within a university campus environment and is designed to accommodate office and research functions, where visual comfort, thermal stability, and daylight quality are essential performance criteria.

A defining characteristic of the building is its solar-responsive kinetic shading system, integrated primarily on the upper floors where solar exposure is most intense. The façade incorporates a series of movable external shading panels that adjust their configuration based on the position and intensity of solar radiation. These elements operate through a mechanically actuated system governed by solar-tracking logic, allowing the façade to respond dynamically to diurnal and seasonal variations in sunlight.

Figure 7.7 Kuggen Building: Solar-responsive kinetic façade system.

The environmental strategy of the Kuggen Building is rooted in the principle of adaptive solar modulation, where the façade continuously balances the need for daylight with the requirement to limit solar heat gain. During periods of high solar exposure, the shading elements extend outward, reducing direct radiation on the glazed surfaces and minimizing heat transfer into the building. Conversely, during periods of lower solar intensity, the panels retract or adjust to allow increased daylight penetration, supporting passive illumination strategies.

This dynamic response contributes to several key environmental performance outcomes:

- reduction of peak solar loads, which decreases the demand for mechanical cooling systems;
- improved daylight autonomy, enabling interior spaces to maintain adequate illumination levels with reduced artificial lighting;
- enhanced visual comfort through the mitigation of glare and excessive brightness.

Unlike fully sensor-driven intelligent systems, the Kuggen façade operates within a predefined parametric control framework, where solar geometry informs the movement of shading elements. This places the system within the category of active kinetic façades, characterized by mechanical adaptability and environmental responsiveness without extensive reliance on real-time data processing. The system demonstrates how relatively simple control logics can produce effective adaptive performance when aligned with environmental principles.

The building's cylindrical geometry plays an important role in supporting its adaptive behavior. The rounded form reduces uniform solar exposure across the façade, distributing radiation more evenly and allowing the shading system to operate more efficiently. This geometric strategy illustrates the integration of form and performance, where architectural design enhances the effectiveness of adaptive systems.

From a material and construction perspective, the façade system combines lightweight structural elements with durable external cladding, ensuring that the movable components can operate reliably over time. The design addresses critical considerations such as wind loads, mechanical stress, and maintenance access, which are essential for the long-term performance of kinetic façade systems.

Architecturally, the Kuggen Building expresses its adaptive logic through a dynamic and evolving façade appearance. The movement of shading panels creates a constantly changing visual pattern, reflecting environmental conditions and reinforcing the concept of architecture as a responsive system. This dynamic quality enhances the building's identity and contributes to its recognition as an example of environmentally driven design.

From a user perspective, the adaptive façade supports a stable and comfortable indoor environment, particularly in terms of visual and thermal conditions. By reducing glare and controlling daylight levels, the system enhances the usability of interior spaces, which is critical in office and research settings. Although direct user control is limited compared to other adaptive systems, the automated response ensures consistent environmental performance.

Despite its advantages, the implementation of kinetic façade systems introduces challenges related to mechanical complexity, system durability, and maintenance requirements. The reliability of moving components, the need for periodic inspection, and the integration of control systems must be carefully addressed during design and operation. The Kuggen Building demonstrates that these challenges can be effectively managed through integrated engineering and performance-driven design.

In summary, the Kuggen Building illustrates how solar-responsive kinetic systems can serve as an effective adaptive strategy in temperate climates. It highlights the potential of combining geometric design, mechanical adaptability, and environmental logic to achieve high-performance building envelopes. The project contributes to the broader discourse on adaptive architecture by demonstrating that responsiveness can be achieved through both technological systems and architectural form.

7.3.5 Case study 5: Surry Hills Library and Community Centre—intelligent adaptive façade and environmental control system

The Surry Hills Library and Community Centre, designed by Francis-Jones Morehen Thorp (FJMT) Studio and completed in 2009, represents an advanced example of intelligent adaptive architecture, in which digital technologies, environmental sensing, and responsive façade systems are integrated into a cohesive operational framework (Fig. 7.8). The project is a mixed-use public building that includes library spaces, community facilities, and educational areas, each with distinct environmental and occupancy requirements.

At the core of the building's adaptive strategy is a computerized sun-tracking control system, which governs the operation of external shading devices, operable

Figure 7.8 Surry Hills Library and Community Centre: Intelligent adaptive façade system.

louvers, and ventilation components. This system continuously monitors environmental conditions and adjusts façade elements in real time to maintain optimal indoor environmental quality. The integration of sensing, computation, and actuation positions the building within the category of cyber-physical architectural systems, where digital intelligence directly influences physical performance.

The adaptive façade operates through a multilayered control framework, consisting of:

- environmental sensors, measuring solar radiation, temperature, daylight levels, and occupancy conditions;
- centralized control algorithms, processing data, and determining optimal system responses based on predefined performance criteria;
- actuated façade elements, including adjustable louvers and ventilation openings that respond to control signals.

This integrated system enables the building to regulate multiple environmental parameters simultaneously, including:

- daylight control, ensuring consistent illumination levels across different interior zones;
- solar heat gain reduction, minimizing cooling loads, and improving energy efficiency;
- natural ventilation optimization, enhancing airflow and indoor air quality;
- glare reduction, improving visual comfort in reading and working areas.

A key strength of the system is its ability to coordinate multiple environmental functions, rather than treating them as independent systems. For example, the positioning of shading louvers is synchronized with ventilation strategies, ensuring that solar control does not restrict airflow. This level of integration reflects a higher degree of adaptive intelligence, where the building operates as a unified environmental system. From a performance perspective, the adaptive façade contributes to:

- reduced energy consumption through decreased reliance on mechanical cooling and artificial lighting;
- improved indoor environmental quality, supporting thermal comfort, air quality, and visual conditions;
- operational flexibility, allowing the building to respond to changing occupancy patterns and external conditions.

The building also incorporates passive design strategies, including optimized orientation, thermal mass, and natural ventilation pathways, which work in conjunction with the adaptive façade. This combination of passive and active systems reflects a hierarchical environmental approach, where low-energy strategies are prioritized before mechanical intervention.

Architecturally, the façade expresses its adaptive behavior through the visible modulation of louvers and shading elements. The changing configuration of the façade reflects environmental conditions and system responses, reinforcing the concept of architecture as a responsive and evolving system. This dynamic expression contributes to the building's identity as a contemporary public facility.

From a user experience perspective, the adaptive system enhances comfort and usability by maintaining stable environmental conditions across diverse spaces, including reading rooms, study areas, and community spaces. The system operates largely in the background, minimizing the need for user intervention while ensuring consistent performance.

However, the implementation of intelligent adaptive systems introduces challenges related to system complexity, integration, and maintenance. The coordination of sensors, control algorithms, and actuators requires careful design and calibration. Long-term performance depends on the reliability of digital infrastructure and the ability to update or adjust control strategies as conditions change. In summary, the Surry Hills Library and Community Center demonstrates how computational control and environmental responsiveness can be integrated to create high-performance, user-centered buildings. It represents a significant step toward intelligent adaptive architecture, where buildings are capable of sensing, learning, and responding to their environment in real time.

To clarify the distinct adaptive strategies employed in the selected projects, a comparative synthesis is necessary to highlight their primary environmental response mechanisms, operational logic, and architectural contributions. Table 7.1 summarizes their primary environmental response mechanisms, operational logic, and architectural contributions.

While each case study, the Alto Tower, the SDU Kolding Campus, the Q1 Building, the Kuggen, and the Surry Hills Library and Community Center, operates within distinct climatic, functional, and technological contexts, all five projects collectively demonstrate how adaptive architecture can be realized through the integration of environmental intelligence, responsive building envelopes, and user-oriented design strategies.

Table 7.1 Comparative analysis of adaptive building design strategies.

Project	Adaptive strategy	Response type	Main benefits
Alto Tower	Ventilated double-skin façade	Passive + user-controlled	Thermal comfort, natural ventilation, acoustic control
SDU Kolding Campus	Kinetic perforated shutters	Active automated	Daylight optimization, glare control, and reduced cooling
Q1 Building	Hybrid ventilation + shading	Passive + active	Energy efficiency, air quality, and solar control
Kuggen	Solar-tracking movable sunscreen	Active kinetic	Solar control, reduced overheating, improved daylight balance
Surry Hills Library and Community Center	Computational adaptive façade system	Intelligent (sensor-driven)	Integrated control of light, heat, and ventilation

From a comparative perspective, the Alto Tower emphasizes a passive and user-interactive adaptive approach, where the ventilated double-skin façade operates as a climatic buffer. The system relies primarily on physical environmental processes such as stack-driven ventilation and thermal insulation, complemented by user-controlled openings that enable occupants to regulate airflow and comfort conditions. This approach highlights the effectiveness of combining low-energy passive strategies with human agency, particularly in high-rise environments where façade performance is critical. In contrast, the SDU Kolding Campus represents a digitally controlled active adaptive system, in which kinetic façade elements respond continuously to real-time environmental data. The automated perforated shutters regulate daylight, glare, and solar heat gain with a high degree of precision. This project demonstrates how sensor integration and computational control can transform the façade into an active environmental interface capable of fine-grained performance optimization.

The Q1 Building illustrates a hybrid adaptive strategy, integrating passive and active systems within a coordinated environmental framework. Its double-skin façade, operable components, atrium-based airflow organization, and solar shading devices operate collectively to enable mixed-mode ventilation. This systems-oriented approach emphasizes synergy between multiple environmental mechanisms, balancing energy efficiency, indoor air quality, and user comfort.

The Kuggen introduces a solar-responsive kinetic façade strategy, where movable shading elements follow the sun's trajectory to regulate solar exposure. Unlike fully sensor-driven intelligent systems, this approach relies on solar geometry and preprogrammed control logic. It represents an intermediate level of adaptation, bridging passive environmental design and fully intelligent systems. The project demonstrates how mechanical responsiveness and geometric design can effectively control daylight and heat gain while maintaining architectural expressiveness.

Finally, the Surry Hills Library and Community Center exemplifies an intelligent adaptive system, characterized by the integration of sensors, control algorithms, and actuated façade elements. The computerized sun-tracking system regulates shading, ventilation, daylight, and thermal performance simultaneously. This multivariable control capability positions the building at the highest level of adaptive complexity, where the façade operates as part of a cyber-physical system capable of real-time environmental optimization.

Taken together, these projects reveal that adaptive architecture operates across multiple scales of intervention:

- material and component scale, where individual façade elements such as louvers, glazing systems, and shading devices respond to environmental stimuli;
- building envelope scale, where systems such as double-skin façades and ventilated layers regulate heat transfer, airflow, and daylight;
- spatial scale, where atria, ventilation paths, and flexible interior layouts support environmental performance and user interaction;
- system scale, where sensing, data processing, and actuation mechanisms coordinate building behavior dynamically.

This multiscalar framework demonstrates that adaptability is not confined to a single technology but emerges from the integration of diverse strategies across architectural, mechanical, and digital systems. A key insight from this expanded comparison is that adaptive architecture is fundamentally defined by its operational logic rather than specific technologies. The five case studies collectively illustrate a spectrum of responsiveness:

- passive environmental modulation (Alto Tower);
- mechanically driven kinetic adaptation (Kuggen);
- automated façade control (SDU Kolding);
- hybrid integrated systems (Q1 Building);
- intelligent, sensor-driven optimization (Surry Hills Library).

This spectrum reflects an evolution from reactive and rule-based systems toward predictive and intelligent systems, where buildings increasingly incorporate data-driven decision-making processes.

Another critical dimension is the role of performance feedback and continuous optimization. In the more advanced cases, particularly the Surry Hills Library, the integration of sensing and control systems enables ongoing adjustment based on real-time conditions. This feedback-driven approach represents a shift from static performance assumptions toward dynamic operational models, which are essential in the context of climate variability and changing user demands.

From an environmental standpoint, the integration of adaptive strategies leads to substantial improvements in energy performance, particularly through reductions in cooling loads, enhanced daylight utilization, and optimized natural ventilation. These strategies contribute not only to energy efficiency but also to broader sustainability goals, including reduced carbon emissions and improved resource management.

Equally important is the contribution of adaptive architecture to occupant experience and well-being. Systems that regulate thermal comfort, daylight, and air quality create healthier and more productive indoor environments. In cases where user interaction is incorporated, such as the Alto Tower, occupants gain a greater sense of control over their environment, further enhancing satisfaction.

In addition to performance and comfort, adaptive systems influence architectural expression. The dynamic behavior of façades—whether through moving panels, changing shading patterns, or responsive surfaces—introduces a temporal dimension to architecture. Buildings are no longer static compositions but evolving entities that reflect environmental conditions and operational processes.

As climate change intensifies and urban environments become increasingly complex, the importance of adaptive architecture will continue to grow. Buildings must respond not only to immediate environmental conditions but also to long-term changes in climate, energy systems, and patterns of use. Adaptive strategies provide a robust framework for addressing these challenges by enabling buildings to remain efficient, resilient, and responsive over time.

The expanded set of case studies confirms that adaptive architecture is best understood as an integrated and interdisciplinary design approach. By combining responsive façades, flexible environmental systems, and continuous performance feedback, adaptive buildings redefine architecture as an active participant in ecological and technological systems. In this emerging paradigm, architecture is no longer conceived as static and fixed but as a dynamic, evolving system capable of learning from and responding to its environment.

References

Aigwi, I.E., Duberia, A., Nwadike, A.N., 2023. Adaptive reuse of existing buildings as a sustainable tool for climate change mitigation within the built environment. Sustainable Energy Technologies and Assessments 56. Available at https://doi.org/10.1016/j.seta.2022.102945.

Alkhatib, H., Lemarchand, P., Norton, B., O'Sullivan, D.T.J., 2021. Deployment and control of adaptive building facades for energy generation, thermal insulation, ventilation and daylighting: a review. Applied Thermal Engineering 185, 116331. Available at https://doi.org/10.1016/j.applthermaleng.2020.116331.

Barbhuiya, S., Das, B.B., Adak, D., Rajput, A.S., 2025. Next-generation building envelopes: smart materials, energy efficiency and environmental impact. Next Materials 9, 101226. Available at https://doi.org/10.1016/j.nxmate.2025.101226.

Chen, Y., Ozaki, A., Yang, X., Arima, Y., Li, Z., Choi, Y., 2025. Adaptive model-based advanced natural ventilation control strategy for mixed-mode residential buildings in Japan. Building and Environment 273. Available at https://doi.org/10.1016/j.buildenv.2025.112718.

Emedo, C., Wada, O.Z., Clement David-Olawade, A., Ling, J., Esan, D.T., Ijiwade, J., et al., 2025. AI-driven transformations in smart buildings: a review of energy efficiency and sustainable operations. Digital Engineering 7, 100068. Available at https://doi.org/10.1016/j.dte.2025.100068.

Es-sakali, N., Idrissi Kaitouni, S., Ait Laasri, I., Oualid Mghazli, M., Cherkaoui, M., Pfafferott, J., 2024. Static and dynamic glazing integration for enhanced building efficiency and indoor comfort with thermochromic and electrochromic windows. Thermal Science and Engineering Progress 52. Available at https://doi.org/10.1016/j.tsep.2024.102681.

Ghansah, F.A., Lu, W., 2023. Cyber-physical systems and digital twins for "cognitive building" in the construction industry. Construction Innovation 25 (3), 787–818. Available at https://doi.org/10.1108/ci-07-2022-0164.

Hu, Z., Maxit, L., Cheng, L., 2021. Acoustic design and analyses of a double skin façade system. Applied Acoustics 173. Available at https://doi.org/10.1016/j.apacoust.2020.107727.

Ibrahim, N.I., Vering, C., Giannetti, N., Kim, M., Sholahudin, Storek, T., et al., 2026. State-of-the-art of the digital twin concept in HVAC+R systems for thermal management of buildings. Renewable and Sustainable Energy Reviews 229. Available at https://doi.org/10.1016/j.rser.2025.116601.

Karanafti, A., Theodosiou, T., 2024. Summer thermal performance analysis of an opaque ventilated façade operating under the dynamic insulation principle. Energy and Buildings 312, 114193. Available at https://doi.org/10.1016/j.enbuild.2024.114193.

Lee, D., Cho, Y.H., Jo, J.H., 2021. Assessment of control strategy of adaptive façades for heating, cooling, lighting energy conservation and glare prevention. Energy and Buildings 235. Available at https://doi.org/10.1016/j.enbuild.2021.110739.

Loonen, R.C.G.M., Trčka, M., Cóstola, D., Hensen, J.L.M., 2013. Climate adaptive building shells: state-of-the-art and future challenges. Renewable and Sustainable Energy Reviews 25, 483–493. Available at https://doi.org/10.1016/j.rser.2013.04.016.

Machard, A., Inard, C., Alessandrini, J.M., Devys-Peyre, F., Martinez, S., Ribéron, J., et al., 2023. Climate change influence on buildings dynamic thermal behavior during summer overheating periods: an in-depth sensitivity analysis. Energy and Buildings 284, 112758. Available at https://doi.org/10.1016/j.enbuild.2022.112758.

Mohan, M., Dewangan, S.K., Soni, S.K., Kalbande, V.P., Choudhari, M.S., Jain, R., et al., 2026. Building thermal comfort and energy savings: a review of performance, challenges, opportunities, and recommendations in phase change material integration within building elements. Applied Thermal Engineering 300, 131105. Available at https://doi.org/10.1016/j.applthermaleng.2026.131105.

Pelsmakers, S., Brown, J.B., 2025. Architectural Thinking in a Climate Emergency. Routledge, London. Available at https://doi.org/10.4324/9781003293903.

Sommese, F., Hosseini, S.M., Badarnah, L., Capozzi, F., Giordano, S., Ambrogi, V., et al., 2024. Light-responsive kinetic façade system inspired by the Gazania flower: a biomimetic approach in parametric design for daylighting. Building and Environment 247. Available at https://doi.org/10.1016/j.buildenv.2023.111052.

Staehr, E.R., Stevik, T.K., Houck, L.D., 2025. Adaptability in the building process: a multifaceted perspective across the life cycle of a building. Buildings 15 (7), 1119. Available at https://doi.org/10.3390/buildings15071119.

Sánchez-Silva, M., Baker, J.W., 2024. Dynamic infrastructure systems: advancing sustainable urbanization and climate change. Environment Systems and Decisions 44 (3), 489–499. Available at https://doi.org/10.1007/s10669-024-09969-8.

van Laar, B., Greco, A., Remøy, H., Gruis, V., Hamida, M.B., 2025. Towards desirable futures for the circular adaptive reuse of buildings: a participatory approach. Sustainable Cities and Society 122. Available at https://doi.org/10.1016/j.scs.2025.106259.

Wen, J., Sun, J., Meng, X., 2025. Effect of window orientation on the thermal performance of electrochromic glass. Case Studies in Thermal Engineering 69, 105990. Available at https://doi.org/10.1016/j.csite.2025.105990.

Collaboration and synergy: architects and engineers working together

8

8.1 The importance of collaboration

Collaboration between architects and engineers has always been fundamental to the realization of successful built environments (Rinke and Hvejsel, 2025). In contemporary architectural practice, however, this relationship has evolved from a supportive interaction into a central driver of design innovation and performance. As buildings are expected to address increasingly complex and interrelated challenges—ranging from sustainability and climate resilience to digital integration, material efficiency, and user well-being—collaboration is no longer optional; it is essential. The growing sophistication of structural systems, building envelopes, environmental technologies, and computational workflows has transformed architecture into a highly interconnected discipline that depends on shared knowledge, continuous communication, and collective problem-solving.

Historically, architectural and engineering processes were often organized in a linear and sequential manner. Architects would develop the spatial concept, formal language, and esthetic direction during the early design phases, while engineers were introduced later to assess structural feasibility and technical performance (Napps et al., 2026). Although this approach could produce functional buildings, it frequently resulted in inefficiencies, design conflicts, and missed opportunities for innovation. Structural systems were often imposed as constraints rather than integrated as generative elements, and environmental strategies were frequently applied reactively instead of being embedded from the outset (Chew et al., 2024). In contrast, contemporary architectural challenges demand an integrated and simultaneous design approach. Modern buildings must respond to multiple performance criteria at once, including the following:

- structural safety and stability;
- energy efficiency and carbon reduction;
- climate responsiveness and resilience;
- indoor environmental quality and occupant well-being;
- construction efficiency and material optimization;
- lifecycle adaptability and long-term performance;
- digital integration and operational intelligence.

These requirements are deeply interconnected and cannot be effectively addressed within isolated disciplinary frameworks. Instead, they require a collaborative design culture in which architects, structural engineers, environmental consultants, fabricators, and digital specialists work together from the earliest conceptual stages (Singh et al., 2011). In such a

Smart Civil and Structural Engineering. DOI: https://doi.org/10.1016/B978-0-443-30223-7.00006-1

model, collaboration is not merely a process of coordination; it becomes a creative and strategic partnership that shapes the conceptual, technical, and experiential dimensions of a project.

At its core, collaboration enables a shift from compromise to synthesis. Rather than negotiating conflicting priorities, interdisciplinary teams can develop solutions in which architectural vision and engineering performance reinforce one another. Architects contribute expertise in spatial organization, human-centered design, cultural context, and experiential quality, while engineers provide knowledge of structural behavior, material systems, environmental performance, and technical feasibility. When these perspectives are integrated early, the resulting design achieves a higher level of coherence, efficiency, and innovation (Mahdavi et al., 2025).

One of the most significant outcomes of effective collaboration is the alignment of form and performance. In high-quality contemporary projects, structural systems are not concealed but become integral to architectural identity. Similarly, environmental systems are embedded within spatial design, contributing to both performance and user experience. Through collaboration:

- structural logic can inform spatial organization and form-making;
- material selection can support both performance goals and esthetic intent;
- environmental systems can enhance comfort, atmosphere, and energy efficiency;
- construction strategies can improve both feasibility and design quality.

This integrated approach enables buildings to function as coherent systems, where each component contributes to multiple objectives simultaneously.

The importance of collaboration becomes particularly evident in projects involving advanced systems such as long-span structures, lightweight envelopes, kinetic façades, and adaptive building technologies (Wang et al., 2024). These systems require careful coordination between disciplines to balance the following:

- geometric ambition and architectural expression;
- structural stability and load distribution;
- fabrication constraints and construction methods;
- environmental and operational performance.

Without close collaboration, such systems risk becoming either technically unfeasible or architecturally compromised.

The emergence of smart structures and adaptive architecture has further intensified the need for interdisciplinary collaboration (Bao et al., 2024). Contemporary buildings increasingly incorporate responsive systems, sensor networks, intelligent materials, and digitally controlled environments. Designing these systems requires coordinated decision-making across multiple domains, including:

- sensor integration and data collection strategies;
- control algorithms and system responsiveness;
- material behavior and performance characteristics;
- user interaction and comfort preferences;
- maintenance, adaptability, and lifecycle performance.

This level of complexity cannot be managed through isolated expertise. Instead, it requires a holistic and integrated design process in which knowledge is shared, and decisions are made collectively.

Digital technologies have played a transformative role in enabling this shift. Tools such as parametric modeling, building information modeling (BIM), digital twins, and performance simulation platforms allow architects and engineers to collaborate within shared digital environments (Semjén and Szép, 2025). These tools support:

- real-time structural and environmental analysis;
- integrated performance evaluation;
- clash detection and coordination;
- cost estimation and resource optimization;
- lifecycle assessment and predictive modeling.

By working within a common digital framework, interdisciplinary teams can test, evaluate, and refine design solutions collaboratively. This approach reduces errors, improves efficiency, and enables more informed decision-making throughout the design and construction process (Nemati Aghdam and Dixit, 2026).

Beyond design development, collaboration also contributes to long-term building performance and resilience. Projects developed through integrated teamwork tend to perform better over time because operational considerations, maintenance strategies, and lifecycle impacts are addressed early. Collaborative design can significantly reduce:

- material waste during construction;
- delays caused by coordination conflicts;
- operational inefficiencies in building systems;
- long-term maintenance and repair costs.

These benefits are particularly important in the context of climate change, where buildings must be capable of adapting to evolving environmental conditions and patterns of use. Collaboration also enhances the qualitative and experiential dimensions of architecture. When architects and engineers work together as equal contributors, structural and environmental systems can become expressive elements of design. Buildings can achieve not only technical excellence but also spatial richness, visual clarity, and cultural significance. This integration transforms engineering systems from hidden infrastructure into visible and meaningful components of architectural identity.

A well-known example of such collaborative synergy is the Center Pompidou, designed by Renzo Piano and Richard Rogers with engineering contributions from Peter Rice (Mallgrave, 2017). In this project, structural and mechanical systems were intentionally exposed and integrated into the architectural expression. The external structural frame, service ducts, and circulation systems were designed as visible elements, demonstrating how technical systems can shape architectural identity when collaboration is fully integrated.

In contemporary practice, this approach has evolved further with the development of smart and adaptive buildings, where structure, systems, and user experience are designed simultaneously. This represents a shift from simple coordination to true coauthorship, where architects and engineers jointly contribute to the conceptual and technical development of a project (Olsen and Mac Namara, 2025).

Ultimately, collaboration is not merely a professional requirement; it is the foundation of meaningful innovation in the built environment. As architecture continues to intersect with digital technologies, sustainability imperatives, and adaptive systems, the ability to work across disciplines will become increasingly critical. The importance of collaboration lies not only in improving efficiency or resolving technical challenges but also in expanding the boundaries of what architecture can achieve.

Through shared knowledge, mutual respect, and integrated thinking, architects and engineers can create buildings that are technically robust, environmentally responsible, and experientially compelling. In this sense, collaboration is not only a method of working—it is a transformative force that shapes the future of architecture, enabling the creation of built environments that are resilient, intelligent, and deeply responsive to human and environmental needs.

8.2 Effective communication between disciplines

Effective communication between architects and engineers is a foundational condition for successful collaboration and integrated design. While both disciplines contribute essential expertise to the built environment, they often operate with different priorities, vocabularies, methodologies, and representational tools. Architects typically emphasize spatial quality, user experience, form-making, and cultural expression, whereas engineers focus on structural integrity, material efficiency, system reliability, and measurable performance outcomes. These differences are not limitations; rather, they represent complementary perspectives that, when properly aligned, can generate highly innovative and high-performing architectural solutions. However, without a structured and intentional communication framework, these differences can lead to misunderstandings, fragmented workflows, inefficiencies, and compromised design quality.

In contemporary architectural practice, particularly in the design of smart, adaptive, and high-performance buildings, communication extends far beyond the exchange of drawings, reports, or technical specifications. It is an iterative, multidirectional process that evolves throughout the entire project lifecycle, from early conceptual design to construction, operation, and postoccupancy evaluation. Effective communication supports shared understanding, coordinated decision-making, and collective authorship. It ensures that architectural intent and engineering logic are not developed in isolation but are continuously negotiated and refined through dialog and feedback.

One of the most significant challenges in interdisciplinary collaboration lies in the translation of ideas across disciplinary boundaries (Klein, 2023). Architects often communicate through sketches, conceptual diagrams, and spatial narratives that express qualitative aspects of design, such as atmosphere, movement, and user experience. Engineers, by contrast, rely on quantitative tools such as calculations, simulations, analytical models, and performance metrics. Bridging this gap requires the development of hybrid representational methods that integrate qualitative and quantitative information into a shared communicative framework. Such hybrid tools may include:

- Performance-informed diagrams, where spatial configurations are directly linked to environmental or structural data;
- Integrated section drawings, combining architectural geometry with structural systems and environmental flows;
- Parametric visualizations, where changes in form are directly associated with performance outputs such as daylight levels, energy use, or structural stress.

These forms of representation enable both disciplines to engage with the same design information from their respective perspectives, fostering a deeper and more productive dialog. The emergence of digital design environments has significantly transformed interdisciplinary communication. Platforms such as BIM, parametric modeling tools (e.g., Grasshopper), and integrated simulation software allow architects and engineers to operate within a shared digital ecosystem (Berlato et al., 2025). These tools do not merely facilitate documentation; they function as interactive collaborative environments in which geometry, material properties, structural behavior, and environmental performance are interconnected. Within these environments, communication becomes embedded within the model itself. A shared digital model enables:

- real-time updates and synchronization across disciplines;
- immediate feedback on design decisions, including structural feasibility and environmental performance;
- reduction of coordination errors and inconsistencies between drawings and specifications.

This integrated workflow supports iterative design loops, where architects and engineers continuously refine design solutions through mutual input. Rather than following a linear process in which design is developed first and then evaluated, interdisciplinary teams engage in a simultaneous process of design and analysis, leading to more informed and optimized outcomes.

Despite the transformative potential of digital tools, effective communication remains fundamentally dependent on human interaction and organizational structure. Regular interdisciplinary meetings, design workshops, and coordination sessions are essential for aligning expectations, resolving conflicts, and maintaining a coherent project vision. These interactions provide opportunities for clarifying assumptions, discussing alternatives, and negotiating trade-offs between competing priorities such as esthetics, cost, performance, and constructability.

Early-stage collaboration is particularly critical. Decisions made during the conceptual design phase have a disproportionate impact on structural systems, environmental performance, and construction feasibility (Kanyilmaz et al., 2023). When engineers are involved from the outset, they can contribute to shaping the design rather than merely responding to it. This approach, often referred to as integrated design, ensures that performance considerations are embedded within the architectural concept rather than added as constraints at later stages. Another essential aspect of effective communication is the definition of roles and responsibilities. While interdisciplinary collaboration encourages shared authorship, clarity in decision-making authority is necessary to avoid ambiguity and delays. Each discipline must understand:

- its own scope of work and deliverables;
- how its contributions interface with those of other disciplines;
- the decision-making hierarchy within the project team.

Clear communication protocols, standardized documentation practices, and robust version control systems further support coordination and accountability. These mechanisms ensure that information is consistent, up-to-date, and accessible to all team members throughout the project lifecycle. Language plays a critical role in shaping interdisciplinary understanding. Technical terminology, if used without consideration of its audience, can create barriers rather than facilitate communication. Developing a shared vocabulary—or at least a mutual understanding of key concepts—helps ensure that ideas are interpreted consistently. In many cases, visual communication proves more effective than verbal or textual descriptions, as diagrams, models, and simulations can convey complex relationships in a more intuitive and accessible manner.

The increasing complexity of contemporary building systems further amplifies the importance of communication (Subramaniam et al., 2022). Projects that incorporate smart technologies, adaptive façades, sensor networks, and intelligent control systems require collaboration not only between architects and structural engineers but also with specialists in mechanical engineering, data science, material science, and software development. In such multidisciplinary contexts, communication must operate across multiple layers, integrating physical design, digital infrastructure, and operational logic.

An illustrative example of effective interdisciplinary communication can be observed in the development of the 30 St Mary Axe, commonly known as "The Gherkin," designed by Norman Foster and engineered by Arup. The project's innovative diagrid structural system and natural ventilation strategy were the result of continuous and iterative collaboration between architectural and engineering teams. The building's aerodynamic form, structural efficiency, and environmental performance were developed simultaneously through advanced modeling and simulation tools. This integrated communication process enabled the realization of a building that achieves both technical excellence and strong architectural identity.

Beyond design and construction, effective communication also plays a crucial role in the operational phase of buildings. As buildings become more complex and technologically integrated, communication must extend to facility managers, operators, and users. The transfer of knowledge from design teams to operational stakeholders is essential for ensuring that building systems function as intended. Postoccupancy evaluation and feedback loops further enhance this process by allowing designers and engineers to learn from real-world performance and user experience (Al Mughairi et al., 2023).

Ultimately, effective communication between disciplines is not simply about the transmission of information; it is about the construction of shared understanding. It requires openness to different perspectives, adaptability in thinking, and a willingness to engage in continuous dialog. Successful interdisciplinary communication enables architects and engineers to transcend disciplinary boundaries and operate as a cohesive design team.

In the context of contemporary architecture, where performance, adaptability, and sustainability are central concerns, effective communication becomes a critical driver of innovation. It allows complex systems to be integrated seamlessly, ensures that design intent is preserved throughout the project lifecycle, and supports the creation of built environments that are both technically robust and experientially rich. As the demands placed on architecture continue to evolve, the ability to communicate effectively across disciplines will remain one of the most important competencies in shaping resilient, high-performance, and human-centered design solutions (Cai et al., 2026).

8.3 Success stories in collaboration

The transformative potential of collaboration between architects and engineers is most clearly demonstrated through built projects in which design ambition and technical innovation are developed as a unified process. In such projects, collaboration moves beyond coordination toward genuine coauthorship, where architectural form, structural logic, environmental performance, and construction strategies are conceived simultaneously. These success stories illustrate how interdisciplinary synergy enables solutions that would be unattainable within isolated disciplinary frameworks.

A defining characteristic of successful collaboration is the early integration of engineering intelligence into the design process (Zhang et al., 2025). Rather than treating structure and systems as constraints, these projects position them as generative drivers of architectural expression. This integrated approach allows complex geometries, advanced material systems, and high-performance environmental strategies to emerge from a shared design logic. The following case studies highlight how collaborative practices have redefined the relationship between architecture and engineering, producing buildings that are innovative, efficient, and experientially rich.

8.3.1 Case study 1: Heydar Aliyev Center—seamless form through integrated structural design

The Heydar Aliyev Center, designed by Zaha Hadid in collaboration with Arup, stands as a seminal example of how effective interdisciplinary communication enables the realization of highly complex architectural geometries (Fig. 8.1). The project is distinguished by its continuous, flowing surfaces that dissolve conventional distinctions between roof, wall, and ground, resulting in a unified spatial and formal expression (Bhooshan, 2017).

At the conceptual level, the design is based on a principle of surface continuity, where architectural elements are not treated as discrete components but as part of an

Figure 8.1 Heydar Aliyev Center—Integrated structural system and continuous envelope geometry.

uninterrupted geometric system. This approach challenges traditional construction logic, which typically relies on orthogonal grids and clearly separated building elements. Achieving such a fluid and continuous form required the development of an advanced structural strategy capable of accommodating complex double-curved geometries while maintaining stability, efficiency, and constructability.

The structural system consists of a reinforced concrete core combined with a three-dimensional space frame structure that supports the building envelope. The concrete core provides primary structural stability and houses vertical circulation and service systems, while the space frame forms a flexible structural network that adapts to the continuously varying geometry of the building. This hybrid system allows for efficient load distribution while preserving the architectural intent of seamless form.

One of the most critical challenges in the project was the translation of complex geometry into buildable components. The free-form surfaces could not be constructed directly without rationalization. To address this, the design team employed advanced parametric modeling tools, which enabled the continuous geometry to be discretized into a system of coordinated structural elements and façade panels. These tools allowed architects and engineers to work within a shared digital environment, where geometric, structural, and fabrication constraints could be evaluated simultaneously. Through parametric modeling, the team was able to:

- analyze surface curvature and geometric continuity;
- optimize the structural configuration of the space frame;
- coordinate the alignment between structural members and façade systems;
- ensure precision in fabrication and assembly processes.

This approach facilitated an iterative and collaborative design process, in which architectural intent and engineering performance were continuously refined through feedback loops. Rather than following a linear sequence of design and evaluation, the project evolved through ongoing dialog between disciplines, supported by shared digital models.

The façade system represents another key aspect of integration. It consists of thousands of uniquely shaped panels, each designed to conform precisely to the building's complex geometry. These panels are mounted onto the underlying space frame, forming a continuous outer skin that reinforces the perception of fluidity. The development of this façade required extensive coordination between design, engineering, and fabrication teams to ensure that each panel could be manufactured and installed accurately.

Material selection was critical in achieving both performance and visual continuity. The use of glass-fiber-reinforced concrete and glass-fiber-reinforced polyester enabled the production of lightweight yet durable panels capable of accommodating complex curvature. These materials provided the necessary flexibility to achieve smooth surface transitions while maintaining structural performance and long-term durability.

The integration of digital technologies extended beyond design into fabrication and construction processes. Digital models were directly linked to computer-controlled manufacturing systems, allowing for precise production of components. This integration minimized errors, improved construction efficiency, and ensured that the built structure accurately reflected the design intent.

From a performance perspective, the building demonstrates how complex geometry can be aligned with environmental and structural considerations. The continuous form contributes to aerodynamic efficiency, reducing wind loads on the structure. The envelope system also supports daylight modulation and thermal performance, contributing to interior comfort conditions while maintaining the architectural vision.

The success of the project is fundamentally rooted in effective communication between architects and engineers. Collaboration began at the earliest conceptual stages and continued throughout design development, fabrication, and construction. Shared digital platforms enabled real-time coordination, allowing both disciplines to contribute to decision-making processes and resolve challenges collaboratively. Importantly, the project exemplifies a shift toward integrated design methodology, where structure is not treated as a secondary system applied to an architectural form but as an intrinsic component of the design itself. The structural system does not merely support the architecture; it actively shapes and enables it.

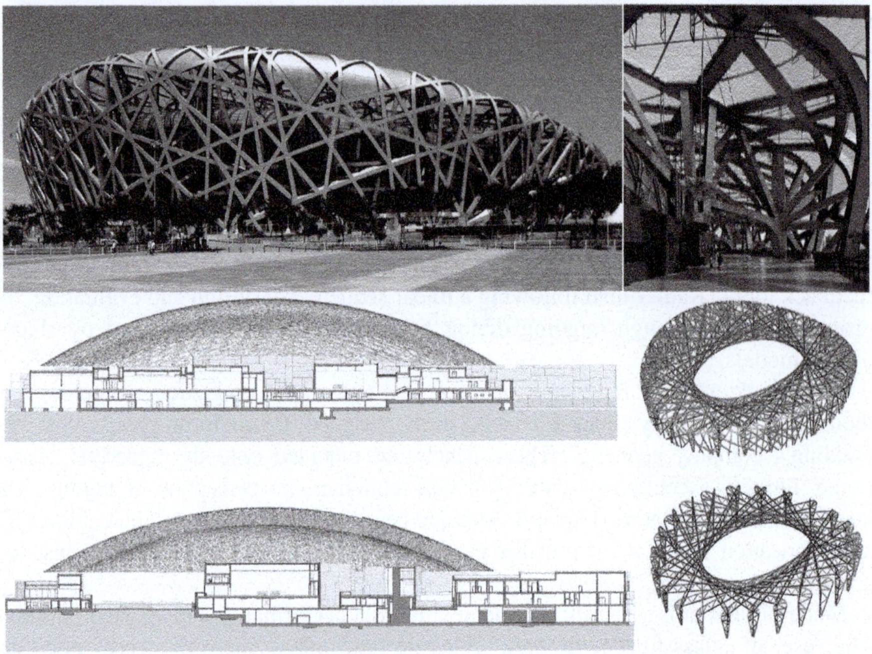

Figure 8.2 Beijing National Stadium (Bird's Nest)—Structural lattice as architectural expression.

From a broader perspective, the Heydar Aliyev Center demonstrates the potential of combining parametric design, advanced materials, and digital fabrication within a collaborative framework. It shows how complex architectural concepts can be realized when interdisciplinary communication is effectively managed and supported by appropriate technological tools.

In conclusion, the Heydar Aliyev Center illustrates how seamless architectural form can be achieved through integrated structural design and coordinated interdisciplinary collaboration. The project stands as a benchmark for contemporary practice, demonstrating that when communication between disciplines is clear, continuous, and supported by shared methodologies, it is possible to transform complex conceptual ideas into coherent, buildable, and high-performance architectural systems.

8.3.2 Case study 2: Beijing National Stadium—structural expression as architectural identity

The Beijing National Stadium, widely known as the "Bird's Nest," designed by Herzog and de Meuron in collaboration with Ai Weiwei and engineered by Arup, stands as a defining example of how structural systems can become the primary driver of architectural identity (Fig. 8.2). The project demonstrates a high level of integration between architecture and engineering, where structural performance, material efficiency, and visual expression are unified within a single design framework.

At the core of the stadium's concept is its irregular steel lattice structure, which serves simultaneously as the main load-bearing system and the building's architectural expression. Unlike conventional stadium designs, where structure is often concealed behind cladding or secondary elements, the Bird's Nest exposes its structural framework as the dominant visual and spatial feature. This approach transforms structure from a purely technical necessity into a central element of architectural meaning and identity.

The structural system consists of a complex network of interwoven steel members forming a seemingly random configuration. However, this apparent randomness is carefully controlled and engineered. The geometry is precisely calibrated to ensure efficient load distribution, structural stability, and redundancy. The lattice acts as a three-dimensional structural shell, supporting the roof, resisting lateral forces, and transferring loads to the foundation system. This dual function—structural and architectural—required an exceptionally high level of coordination between architects and engineers.

The realization of this system depended on an intensive collaborative design process. Architects and engineers worked closely from the earliest conceptual stages to ensure that the structural system could satisfy multiple performance criteria while maintaining the intended visual complexity. Key design challenges included:

- structural stability under complex and variable loading conditions;
- optimization of material use in a large-span structure;
- resistance to seismic forces in a seismically active region;
- accommodation of dynamic loads such as wind and crowd movement.

To address these challenges, the project relied heavily on advanced computational modeling and simulation tools. These tools enabled the analysis of structural behavior within the irregular lattice system and supported an iterative process of refinement. Through digital simulation, the design team was able to:

- map and evaluate load paths throughout the structure;
- optimize the size, orientation, and distribution of steel members;
- test the system under multiple scenarios, including extreme environmental conditions;
- improve structural efficiency while maintaining architectural intent.

This process reflects a performance-driven design methodology, in which form is not predefined but evolves in response to structural and environmental requirements. The result is a structure that appears organic and informal yet is highly rational and engineered with precision.

Seismic performance was a critical aspect of the design. The stadium incorporates a robust structural strategy that combines the steel lattice with a reinforced concrete substructure. This integrated system enhances stability and allows the building to withstand seismic forces while maintaining its structural integrity. The coordination between these systems required careful planning and continuous communication between disciplines.

Material efficiency was another key objective. Although the structure appears dense and visually complex, the distribution of steel members is optimized to use material only where necessary. The lattice operates as a network-based structural system, where loads are distributed across interconnected elements rather than concentrated in a few primary members. This approach increases redundancy and resilience while reducing unnecessary material use.

The construction process presented significant challenges due to the unique geometry of the structure. Each steel component required precise fabrication and careful assembly. Digital design models were directly linked to fabrication processes, ensuring high accuracy and minimizing construction errors. The sequencing of assembly was also carefully planned to maintain structural stability throughout the construction phase.

From an architectural perspective, the stadium demonstrates how structural logic can define both form and experience. The exposed lattice creates a dynamic visual identity that changes with light and perspective. Internally, the structure contributes to a rich spatial experience, where the interplay of light, shadow, and structural depth enhances the perception of scale and complexity.

The project also represents a broader shift in contemporary architecture toward the integration of engineering innovation and architectural expression. By eliminating the traditional separation between structure and form, the Bird's Nest exemplifies an approach in which technical systems are fully integrated into the architectural concept. Structure is not applied to the design; it generates and defines it.

The success of the Beijing National Stadium is fundamentally rooted in effective interdisciplinary communication. Continuous collaboration between architects, engineers, and specialists enabled the alignment of creative vision with technical feasibility. Shared digital tools and iterative workflows ensured that design decisions were informed by both esthetic and performance considerations.

In conclusion, the Beijing National Stadium illustrates how structural engineering can move beyond its conventional role to become a primary generator of architectural identity. The project demonstrates that when communication between disciplines is strong and design processes are fully integrated, it is possible to create buildings in which structure, form, and performance are inseparable. This case stands as a benchmark for integrated design, highlighting the potential of collaboration to produce innovative, efficient, and iconic architectural solutions.

8.3.3 Case study 3: The Louvre Abu Dhabi–light, structure and cultural expression

The Louvre Abu Dhabi, designed by Jean Nouvel in collaboration with Buro Happold, represents a significant example of how interdisciplinary collaboration can integrate structural complexity with environmental performance and cultural expression (Fig. 8.3). The project demonstrates how architecture and engineering can work together to produce a building that is not only technically advanced but also deeply responsive to its context and experiential qualities.

The defining feature of the museum is its large-span domed roof, approximately 180 meters in diameter, which appears to float above a composition of gallery volumes. Rather than functioning as a solid enclosure, the dome is conceived as a multilayered

Figure 8.3 Louvre Abu Dhabi–Layered dome structure and filtered daylight strategy.

geometric structure, composed of interlocking patterns derived from abstracted regional design principles. This layered configuration creates a permeable canopy that filters sunlight, producing the characteristic "rain of light" effect within the spaces below.

This lighting phenomenon is both an environmental strategy and an architectural expression. The dome moderates solar radiation by diffusing direct sunlight, thereby reducing heat gain and improving thermal comfort in semioutdoor spaces. At the same time, the filtered light creates a dynamic spatial atmosphere, enhancing the visual and sensory experience of visitors. This dual function illustrates how environmental performance and architectural identity can be closely integrated through design.

From a structural perspective, the dome is an advanced engineering system that balances large-span performance with visual lightness. It is composed of multiple overlapping layers of steel elements arranged in a complex spatial configuration. Each layer contributes to load distribution and structural stability, while the combined system allows the dome to span a vast area with minimal visible support. The structure is supported at a limited number of points, reinforcing the perception that the dome is hovering above the building. The realization of this system required extensive use of advanced computational modeling and digital simulation tools. These tools enabled the design team to:

- analyze the structural behavior of the multilayered dome system;
- optimize the geometric configuration for both stability and material efficiency;
- coordinate the relationship between structural elements and the light-filtering pattern;
- evaluate environmental performance, including daylight distribution and thermal effects.

Through this process, the dome was developed as a performance-driven design, where structural, environmental, and architectural considerations were addressed simultaneously. The design evolved through iterative collaboration, with continuous feedback between architects and engineers.

Material selection and fabrication played a crucial role in achieving the required precision and performance. The dome is constructed from steel components arranged in a layered framework, with each element fabricated to exact specifications. Digital models were directly linked to manufacturing processes, enabling accurate production and efficient assembly. This integration of design and fabrication ensured that the complex geometry could be realized at full scale.

In addition to its structural and visual qualities, the dome contributes to the creation of a microclimate within the museum complex. The shaded areas beneath the canopy benefit from reduced solar exposure and improved thermal comfort, allowing for semiopen circulation spaces that are less dependent on mechanical conditioning. This approach reflects a broader commitment to sustainable design principles, where passive environmental strategies are integrated with advanced engineering systems.

The success of the Louvre Abu Dhabi is fundamentally rooted in effective interdisciplinary communication. The complexity of the project required close coordination between architects, structural engineers, environmental consultants, and fabrication specialists. Shared digital platforms facilitated real-time collaboration, enabling the team to align design intent with technical performance throughout the project lifecycle.

Beyond its technical achievements, the project demonstrates how collaboration can produce architecture that is both innovative and culturally meaningful. The dome is not only a structural solution but also a central element of the building's identity, shaping both its appearance and the experience of its spaces. It illustrates how engineering systems can actively contribute to architectural expression rather than simply supporting it.

Building on the preceding case studies—the Heydar Aliyev Center, the Beijing National Stadium, and the Louvre Abu Dhabi—it becomes evident that collaboration between architects and engineers enables solutions that extend beyond conventional disciplinary limits. In each project, the integration of design intent with structural logic and environmental performance resulted in buildings that are not only technically sophisticated but also architecturally expressive and experientially rich. These outcomes were not achieved through isolated efforts but through continuous dialog, shared methodologies, and a commitment to coauthorship.

At the Heydar Aliyev Center, collaboration allowed a highly complex, continuous form to be translated into a buildable structural system through parametric modeling and iterative refinement. The project demonstrates how geometric ambition can be aligned with structural feasibility when communication is embedded within a shared digital framework. Similarly, the Beijing National Stadium illustrates how structural systems can evolve into primary architectural expressions. The exposed steel lattice is not an applied feature but the defining element of the building's identity, made possible through performance-driven design and close interdisciplinary coordination. In the case of the Louvre Abu Dhabi, collaboration enabled the integration of structural complexity with environmental and cultural narratives, resulting in a dome that simultaneously performs as a climatic moderator, structural system, and experiential device.

These examples collectively demonstrate that collaboration is not merely a supportive component of the design and construction process; it is a primary driver of architectural innovation. When architects and engineers engage as equal partners from the earliest stages of design, they are able to synthesize diverse forms of knowledge—spatial, structural, environmental, and technological—into cohesive and high-performing solutions. This synthesis allows for the emergence of architectures in which form, structure, and performance are inseparable, each reinforcing and enhancing the others.

A key outcome of such collaboration is the ability to operate across multiple scales simultaneously. At the material and component level, advanced fabrication techniques and engineered materials enable precision and adaptability. At the structural level, innovative systems such as space frames, lattice shells, and large-span domes support complex geometries and long-span performance. At the environmental level, integrated strategies for daylight modulation, thermal control, and passive cooling contribute to energy efficiency and user comfort. At the system level, digital tools and simulation platforms enable real-time coordination and performance optimization. These multiscalar integrations are only possible through effective interdisciplinary communication and shared design processes.

In the context of smart structures and adaptive systems, the importance of collaboration becomes even more pronounced. Contemporary buildings increasingly incorporate responsive technologies, sensor networks, and data-driven control systems that require coordination across architecture, structural engineering, environmental engineering, and digital disciplines. The ability to integrate these systems into a coherent design depends on continuous information exchange and collaborative problem-solving. Without such integration, the complexity of modern building systems can lead to fragmentation rather than innovation.

Furthermore, collaboration plays a critical role in advancing sustainability-driven design. The projects discussed demonstrate how environmental performance can be embedded within architectural form rather than treated as an afterthought. Whether through aerodynamic shaping, daylight filtering, or passive climate control, these strategies emerge from the intersection of architectural creativity and engineering analysis. This integrated approach not only improves energy efficiency but also enhances the experiential quality of spaces, contributing to occupant well-being and long-term building performance.

Another important dimension is the role of collaboration in fostering resilience and adaptability. Buildings designed through integrated processes are better equipped to respond to changing environmental conditions, evolving user needs, and technological advancements. By aligning structural robustness with environmental responsiveness and operational flexibility, interdisciplinary teams can create built environments that remain relevant and effective over time.

Ultimately, these case studies affirm that the future of architecture lies in the continued strengthening of interdisciplinary partnerships. As design challenges become more complex and multifaceted, the separation between disciplines becomes increasingly impractical. Instead, architecture must be understood as a collaborative endeavor in which creativity, engineering expertise, and technological innovation converge (Hong et al., 2025).

In this emerging paradigm, the role of the architect and the engineer evolves from that of independent contributors to collaborative coauthors, jointly responsible for shaping the built environment. This shift not only enhances the quality and performance of individual projects but also expands the possibilities of architectural practice as a whole.

In conclusion, the examined projects demonstrate that when collaboration is effectively structured and sustained, it becomes a powerful catalyst for innovation. It enables the creation of buildings that are not only technically advanced but also responsive, resilient, and meaningful. As the demands placed on architecture continue to grow—driven by climate change, technological development, and societal transformation—interdisciplinary collaboration will remain indispensable in shaping the next generation of built environments.

References

Al Mughairi, M., Beach, T., Rezgui, Y., 2023. Post-occupancy evaluation for enhancing building performance and automation deployment. Journal of Building Engineering 77, 107388. Available at https://doi.org/10.1016/j.jobe.2023.107388.

Bao, D.W., Burry, J., Xie, Y.M., 2024. Transforming architecture: the role of interdisciplinary collaboration in design and fabrication. Architectural Intelligence 3, 34. Available at https://doi.org/10.1007/s44223-024-00077-w.

Berlato, M., Binni, L., Durmus, D., Gatto, C., Giusti, L., Massari, A., et al., 2025. Digital platforms for the built environment: a systematic review across sectors and scales. Buildings 15 (14). Available at https://doi.org/10.3390/buildings15142432.

Bhooshan, S., 2017. Parametric design thinking: a case-study of practice-embedded architectural research. Design Studies 52, 115–143. Available at https://doi.org/10.1016/j.destud.2017.05.003.

Cai, S., Li, L., Gou, Z., 2026. Optimizing open-plan office passive retrofits: a human-centric, climate-adaptive approach to lighting, thermal comfort, and energy savings. Building and Environment 287. Available at https://doi.org/10.1016/j.buildenv.2025.113894.

Chew, Z.X., Wong, J.Y., Tang, Y.H., Yip, C.C., Maul, T., 2024. Generative design in the built environment. Automation in Construction 166, 105638. Available at https://doi.org/10.1016/j.autcon.2024.105638.

Hong, J., Wu, H., Wu, H., Li, K., Wang, D., Shrestha, A., 2025. How technology convergence drives the invention of carbon-neutral building technologies? Investigation through the lens of binary innovation theory. Building and Environment 271. Available at https://doi.org/10.1016/j.buildenv.2025.112623.

Kanyilmaz, A., Dang, V.H., Kondratenko, A., 2023. How does conceptual design impact the cost and carbon footprint of structures? Structures 58, 105102. Available at https://doi.org/10.1016/j.istruc.2023.105102.

Klein, J.T., 2023. Boundary discourse of crossdisciplinary and cross-sector research: refiguring the landscape of science. Minerva 61 (1), 31–52. Available at https://doi.org/10.1007/s11024-022-09474-9.

Mahdavi, A., Berger, C., Bandurski, K., Hoes, P.J., Azar, E., Lu, Y., 2025. Beyond technology: meeting complex multi-stakeholder challenges in the built environment. Building and Environment 281. Available at https://doi.org/10.1016/j.buildenv.2025.113228.

Mallgrave, H.F., 2017. Centre Pompidou: Renzo Piano and Richard Rogers: Twentieth-Century Architecture. In: Mallgrave, H.F. (Ed.), Companion to the History of Architecture. Wiley, pp. 1–10. Available at https://doi.org/10.1002/9781118887226.wbcha155.

Napps, D., Staudt, J., Saluz, U., Chen, X., Steiner, D., Zong, C., et al., 2026. Evaluation of building design variants in early phases on the basis of adaptive detailing strategies. Buildings 16 (4). Available at https://doi.org/10.3390/buildings16040685, www.mdpi.com/journal/buildings.

Nemati Aghdam, M., Dixit, M.K., 2026. A systematic review of methods and frameworks integrating environmental sustainability and resilience in building design and construction. Building and Environment 293. Available at https://doi.org/10.1016/j.buildenv.2026.114381, https://www.sciencedirect.com/science/journal/03601323.

Olsen, C.J., Mac Namara, S.C., 2025. Restructuring Collaborations between Architects and Engineers. Informa UK Limited, pp. 172–179. Available at 10.1201/9781003658641-21.

Rinke, M., Hvejsel, M.F., 2025. Structures and Architecture: REstructure REmaterialize REthink REuse. CRC Press. Available at 10.1201/9781003658641.

Semjén, Á.Á., Szép, J., 2025. Integrating generative and parametric design with BIM: a literature review of challenges and research gaps in construction design. Applications in Engineering Science 23, 100253. Available at https://doi.org/10.1016/j.apples.2025.100253.

Singh, V., Gu, N., Wang, X., 2011. A theoretical framework of a BIM-based multi-disciplinary collaboration platform. Automation in Construction 20 (2), 134–144. Available at https://doi.org/10.1016/j.autcon.2010.09.011.

Subramaniam, C., Ismail, S., Rani, W.N.M.W.M., Mahdiyar, A., 2022. Improving project communications management practices in the construction sector during the COVID-19 pandemic: a Malaysian scenario. Buildings 12 (9). Available at https://doi.org/10.3390/buildings12091291, www.mdpi.com/journal/buildings.

Wang, G., Fang, J., Yan, C., Huang, D., Hu, K., Zhou, K., 2024. Advancements in smart building envelopes: a comprehensive review. Energy and Buildings 312. Available at https://doi.org/10.1016/j.enbuild.2024.114190, https://www.sciencedirect.com/science/journal/03787788.

Zhang, B., Mo, Y., Li, B., 2025. Reimagining synergy between architects and computational systems: Insights from web-based generative design applications. Frontiers of Architectural Research. Available at https://doi.org/10.1016/j.foar.2025.07.009, https://www.sciencedirect.com/science/journal/20952635.

Smart structures in practice: case studies

9

9.1 Innovative projects and their design principles

The translation of smart structure theory into built reality is most effectively understood through contemporary architectural projects that successfully integrate advanced technologies, adaptive systems, and interdisciplinary design approaches (Schwegmann et al., 2024). These projects do not treat intelligence as an additional layer applied after design completion; instead, they embed responsiveness, sensing, and performance optimization directly into the architectural, structural, and environmental logic of the building. In this context, "smartness" evolves from a collection of discrete technological components into a holistic design paradigm that fundamentally reshapes how buildings are conceived, constructed, and operated (Campodonico Avendano et al., 2024).

Rather than viewing buildings as static objects, contemporary smart architecture positions them as dynamic, data-driven systems capable of sensing, learning, and adapting over time. This shift reflects broader transformations in digital technology, environmental awareness, and user expectations (Schmidt et al., 2017). Buildings are increasingly expected to operate as cyber-physical systems, where physical components, such as structure, envelope, and mechanical systems, are continuously informed and optimized through digital data flows and intelligent control mechanisms.

This section critically examines pioneering projects that demonstrate the maturity of smart structures as an integrated design approach. These case studies highlight the convergence of the following:

- data-driven operational systems;
- computational and parametric design methodologies;
- intelligent materials and responsive components;
- user-centered interaction and control strategies.

Collectively, they illustrate a transition from conventional form-making toward adaptive, performance-driven, and responsive environments.

9.1.1 *Case study 1: The Edge—data-driven intelligent building design*

The Edge, designed by PLP Architecture, is widely recognized as one of the most advanced examples of a fully integrated smart office building (Al Dakheel et al., 2020). It represents a paradigm shift in which architecture, digital infrastructure, and user interaction are combined into a continuous feedback-driven system (Fig. 9.1). Rather than functioning as a passive container for activity, the building operates as an intelligent platform that actively responds to both environmental conditions and occupant behavior (Verma and Gopalakrishnan, 2025).

Smart Civil and Structural Engineering. DOI: https://doi.org/10.1016/B978-0-443-30223-7.00002-4

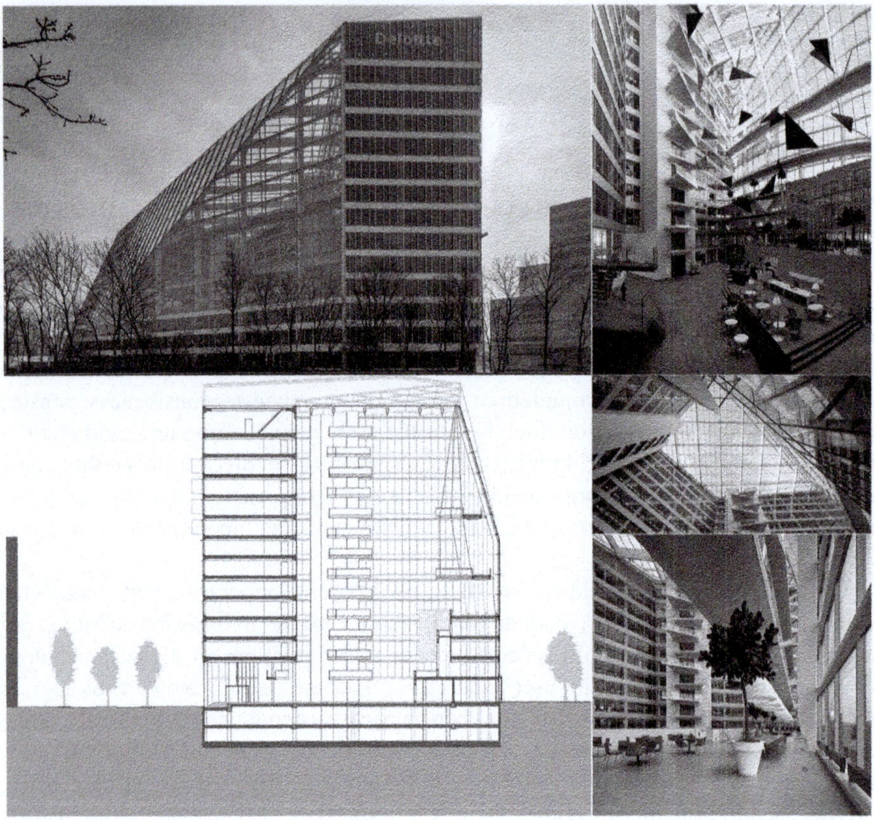

Figure 9.1 The Edge—Elevation, section, and multiperspective views of a data-driven intelligent building.

At the core of its intelligence is a dense network of internet of things sensors distributed throughout the building (Martínez et al., 2025). These sensors collect real-time data on a wide range of parameters, including:

- occupancy and movement patterns;
- indoor temperature and humidity levels;
- daylight availability and artificial lighting demand;
- energy consumption and system performance.

This data is transmitted to a centralized building management system, where it is processed and translated into operational decisions. The system dynamically regulates building functions such as lighting, heating, cooling, and ventilation in real time. Unlike conventional buildings that rely on fixed schedules or predefined zoning, the Edge continuously adapts its performance based on actual usage patterns, significantly improving efficiency and reducing energy waste.

A fundamental design principle in this project is data-driven adaptability. The building does not operate according to static assumptions; instead, it learns from user behavior and environmental conditions. For example, lighting systems adjust automatically based on occupancy and daylight availability, while heating, ventilation, and air conditioning systems respond to localized thermal demands rather than treating entire zones uniformly. This fine-grained responsiveness allows the building to achieve high levels of energy efficiency while maintaining optimal indoor comfort.

Another defining aspect of the Edge is its emphasis on user-centered interaction (Escamilla et al., 2026). Occupants are not passive recipients of environmental conditions; they actively participate in shaping their surroundings. Through a dedicated mobile application, users can:

- adjust lighting and temperature settings;
- locate available workspaces;
- personalize their working environment;
- interact with building services in real time.

This integration of user input into building operation represents a significant shift toward personalized environmental control, where architecture responds to individual preferences rather than imposing uniform conditions. The building effectively becomes a responsive interface between people and space.

From an architectural perspective, the design also incorporates environmental intelligence at the spatial and formal level. The building's orientation, atrium configuration, and façade design are optimized to maximize daylight penetration while minimizing glare and heat gain. A large central atrium acts as both a social hub and an environmental moderator, enhancing natural light distribution and supporting passive climate strategies (Verma et al., 2026). These architectural decisions reduce reliance on mechanical systems and complement the building's digital intelligence.

Energy performance is further enhanced through the integration of renewable energy systems and efficient infrastructure. The building is designed to operate with low energy demand, supported by intelligent energy distribution and monitoring systems. The combination of passive design strategies and active digital control results in a highly efficient and sustainable building operation.

Importantly, the Edge demonstrates that smart architecture is not defined solely by the presence of advanced technology but by the integration and coordination of systems. The success of the project lies in how data, digital platforms, physical infrastructure, and user interaction are seamlessly interconnected. This integrated approach transforms the building into a living system that continuously evolves in response to internal and external inputs.

From a broader perspective, the Edge exemplifies several key principles of smart building design:

- Continuous feedback loops, where data informs real-time decision-making;
- Adaptive performance, where systems respond dynamically to changing conditions;
- User engagement, where occupants actively interact with the building environment;
- Integrated design, where architecture and technology are developed simultaneously;
- Operational intelligence, where building performance is continuously optimized.

These principles reflect a fundamental transformation in architectural thinking. Buildings are no longer static compositions but interactive and learning systems capable of improving performance over time. The Edge illustrates how smart structures can redefine the relationship between architecture, technology, and users. It demonstrates that intelligence in architecture is not merely a technological attribute but a systemic quality emerging from the integration of data, design, and human interaction. As such, it provides a model for future buildings that are not only energy-efficient and technologically advanced but also adaptive, responsive, and deeply aligned with user needs and environmental conditions (Ma and Liu, 2026).

9.1.2 Case study 2: Jewel Changi Airport—environmental simulation and experiential intelligence

The Jewel Changi Airport, designed by Safdie Architects in collaboration with Buro Happold and completed in 2019, represents a landmark project in the evolution of smart structures at the intersection of environmental simulation, architectural form, and user experience (Fig. 9.2). Conceived as both an infrastructural hub and a public destination, the project demonstrates how large-scale architecture can integrate advanced environmental intelligence with spatial and experiential design.

At the core of the project is a large-span glass-and-steel dome that encloses a carefully controlled indoor microclimate. This enclosed environment accommodates dense vegetation, public circulation spaces, and the iconic indoor waterfall, creating a hybrid condition between building and landscape. The complexity of maintaining stable environmental conditions within such a vast enclosed volume required the use of advanced environmental simulation tools throughout the design process. These tools informed critical decisions related to solar control, daylight distribution, thermal regulation, humidity management, and airflow dynamics.

Figure 9.2 Jewel Changi Airport—Sectional organization and experiential interior–exterior integration.

A defining design principle in this project is climate-responsive design through simulation (Mahlan et al., 2024). Unlike conventional workflows in which simulation is used only to verify performance after design decisions are made, Jewel Changi Airport employs simulation as a generative design driver. Computational models were used iteratively to shape the geometry of the dome, optimize the performance of the glazing system, and define the integration of mechanical and passive environmental strategies. This approach ensured that environmental performance was embedded within the architectural concept from the earliest stages, rather than applied as a secondary layer.

The building envelope plays a critical role in achieving environmental control (Lozinsky et al., 2025). The glass dome incorporates high-performance glazing systems designed to regulate solar radiation while maximizing daylight penetration. The geometry of the dome itself contributes to environmental performance by facilitating controlled airflow and thermal stratification, allowing warm air to rise and be managed efficiently within the upper zones of the structure. These passive principles are complemented by active systems that maintain stable temperature and humidity levels, particularly important for supporting the extensive indoor vegetation.

Another key principle demonstrated in this project is integrated ecosystem design. Architecture, landscape, and environmental engineering are not treated as separate components but as parts of a unified system. The interior landscape is not merely decorative; it actively contributes to the environmental performance of the building by influencing humidity levels, air quality, and thermal comfort (Ageli and Chassiakos, 2026). Water features, planting systems, and airflow strategies are carefully coordinated to create a balanced and sustainable indoor ecosystem.

This integration results in a building that functions as a hybrid environment, blurring the distinction between interior and exterior space. Visitors experience conditions that resemble an outdoor garden while remaining within a controlled architectural enclosure. This approach reflects a broader shift in contemporary architecture toward biophilic design, where natural elements are incorporated into built environments to enhance well-being and environmental quality (Jones et al., 2025).

A particularly important dimension of Jewel Changi Airport is its emphasis on experiential intelligence. Environmental systems are not optimized solely for efficiency or technical performance; they are calibrated to enhance sensory and spatial experience. The project carefully orchestrates the following:

- light diffusion through the dome structure, creating dynamic visual conditions;
- airflow patterns that contribute to thermal comfort without discomfort;
- temperature gradients that support both human comfort and plant growth;
- acoustic conditions shaped by water features and spatial configuration.

The central waterfall, for example, is not only a visual focal point but also plays a role in cooling and humidifying the surrounding air. The interaction between water, light, and air creates a continuously changing environment that engages users on multiple sensory levels. This demonstrates how environmental performance can be transformed into a design asset, contributing to the identity and experience of the space.

From a technological perspective, the project relies on a combination of digital modeling, real-time monitoring, and integrated control systems. Environmental data is continuously assessed to ensure that indoor conditions remain within optimal ranges. This feedback-driven operation allows the building to maintain stability despite external climatic variations and fluctuating occupancy levels.

The scale of the project further emphasizes the importance of interdisciplinary collaboration. The successful realization of Jewel Changi Airport required close coordination between architects, structural engineers, environmental engineers, landscape designers, and digital specialists. Shared digital platforms and simulation tools enabled these teams to work within a common framework, ensuring that design decisions were informed by both performance data and experiential considerations.

In broader terms, Jewel Changi Airport exemplifies several key principles of smart and adaptive architecture:

- simulation-driven design, where environmental modeling informs form and system integration;
- hybrid environmental systems, combining passive and active strategies;
- ecosystem integration, where natural and built elements function together;
- experiential performance, where environmental conditions enhance user experience;
- continuous environmental control, supported by monitoring and feedback systems.

These principles illustrate a shift from purely functional infrastructure toward experience-driven environmental architecture, where performance, technology, and human perception are closely interconnected. In fact, Jewel Changi Airport demonstrates how smart structures can operate at multiple levels simultaneously—technical, environmental, and experiential. It redefines the role of large-scale public architecture by integrating digital simulation, environmental intelligence, and user-centered design into a cohesive system. The project stands as a powerful example of how contemporary architecture can move beyond static enclosure to create dynamic, responsive, and immersive environments that support both performance efficiency and human experience.

9.1.3 Case study 3: Morpheus Hotel—computational design and structural intelligence

The Morpheus Hotel, designed by Zaha Hadid Architects and completed in 2018, represents a major advancement in the integration of computational design, structural optimization, and architectural expression (Fig. 9.3). The project exemplifies how digital design methodologies can transform structural systems into primary generators of architectural form, resulting in a building where geometry, performance, and identity are inseparably linked.

The most distinctive feature of the Morpheus Hotel is its free-form steel exoskeleton, which functions simultaneously as the building's primary structural system and its defining architectural element. Unlike conventional high-rise structures that rely on

Figure 9.3 Morpheus Hotel—Structural exoskeleton and spatial expression.

internal columns and repetitive floor grids, the exoskeleton carries the majority of structural loads along the exterior envelope. This strategy eliminates the need for internal columns across large areas, enabling highly flexible and open interior spaces that can accommodate diverse programmatic requirements.

From a structural perspective, the exoskeleton operates as a continuous load-bearing network, distributing forces efficiently across the building's irregular geometry. The structural system is particularly notable for its integration of large-scale voids carved through the building mass. These voids introduce significant geometric complexity, requiring careful redistribution of loads around openings while maintaining overall stability. The resulting structure is both structurally efficient and visually expressive, demonstrating how engineering constraints can drive innovative architectural solutions. A detailed understanding of the structural logic and its integration with architectural space is further clarified in Fig. 9.4, which presents a focused technical section of the Morpheus Hotel. The illustration highlights how the exoskeleton system is implemented as the primary load-bearing framework, transferring forces efficiently through its continuous geometric network. It also reveals the relationship between the external structural grid, floor plates, and interior spatial organization, demonstrating how structural elements are coordinated with programmatic requirements. The figure provides insight into connection details, load paths, and the interface between the façade and internal structure, emphasizing the role of computational design and digital fabrication in achieving precision and constructability. Through this technical representation, the integration of structural performance, architectural expression, and interior functionality becomes evident, reinforcing the concept of structure as an active and visible component of the overall design system.

The development of this geometry was made possible through advanced parametric and algorithmic design processes. Computational tools were used to

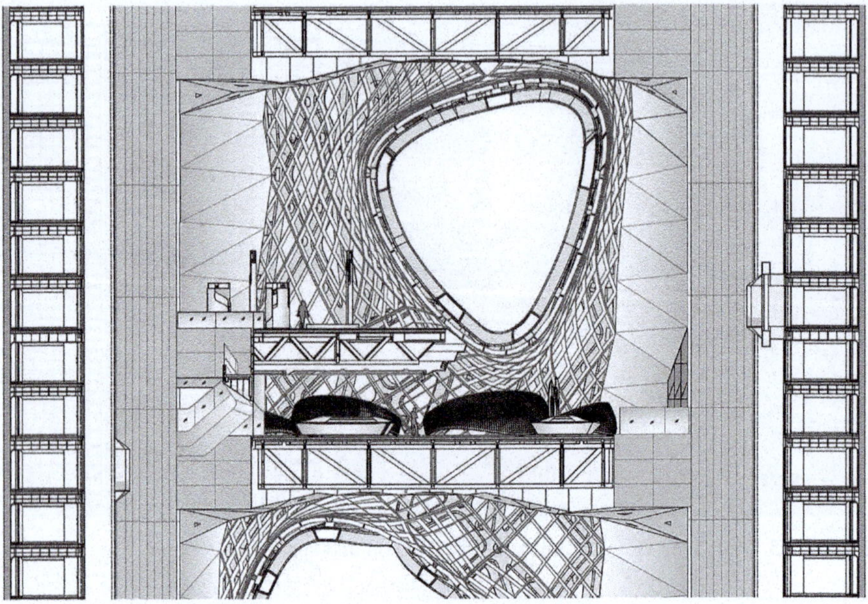

Figure 9.4 Morpheus Hotel—Technical section and structural implementation detail.

generate, test, and refine the building form through iterative cycles of analysis and optimization. These tools enabled the design team to:

- analyze structural load paths across complex geometries;
- optimize material distribution to reduce weight while maintaining strength;
- adjust geometric parameters in response to performance requirements;
- coordinate structural logic with architectural intent.

This process reflects the principle of structure as architecture, where structural behavior is not hidden behind cladding or secondary systems but is directly expressed in the building's form. The exoskeleton becomes both a functional system and a visual language, defining the building's identity through its structural logic.

Another central principle demonstrated in this project is computational optimization. The design process required the reconciliation of multiple constraints, including:

- structural performance under vertical and lateral loads;
- material efficiency and cost considerations;
- fabrication limitations and construction sequencing;
- architectural continuity and spatial quality.

Digital tools enabled these constraints to be addressed simultaneously within a unified workflow. Rather than treating structure, form, and construction as separate problems, the design process integrated them into a single computational framework, allowing for continuous refinement and coordination. The realization of the exoskeleton also depended on advanced digital fabrication and construction technologies. Each structural component was uniquely shaped and required precise fabrication to ensure proper alignment within the overall system. Digital models were directly linked to manufacturing processes, enabling high levels of accuracy and reducing construction errors. This integration of design and fabrication demonstrates how computational workflows can bridge the gap between complex digital geometry and physical construction.

In addition to its structural innovation, the Morpheus Hotel incorporates environmental responsiveness within its design. The building envelope includes façade elements that help regulate solar exposure and daylight penetration. The exoskeleton itself contributes to shading by creating varying depths across the façade, reducing direct solar gain in certain areas. These strategies support improved energy performance while maintaining visual transparency and spatial quality.

The large voids within the building also play a role in environmental performance and spatial experience. They introduce natural light into deeper parts of the building and create visual connections between interior spaces. At the same time, they enhance airflow patterns and contribute to the building's overall environmental strategy. This integration of structural form with environmental performance further reinforces the concept of multifunctional design, where a single system serves multiple roles.

From an experiential perspective, the Morpheus Hotel offers a highly dynamic spatial environment. The interplay between structure, light, and void creates constantly changing visual conditions, enhancing the user experience. Circulation spaces, atria, and interior volumes are shaped by the structural system, demonstrating how computational design can influence not only performance but also perception and movement.

The project also highlights the importance of interdisciplinary collaboration in realizing complex smart structures. Architects, structural engineers, façade specialists, and fabrication experts worked closely throughout the design and construction process. Shared digital platforms enabled real-time coordination, ensuring that geometric complexity, structural performance, and construction feasibility were aligned at every stage.

In a broader context, the Morpheus Hotel exemplifies several key principles of contemporary smart and computational architecture:

- Structural integration, where load-bearing systems define architectural form;
- Parametric design, enabling the exploration and optimization of complex geometries;
- Computational workflows, integrating design, analysis, and fabrication;
- Material efficiency, achieved through performance-driven optimization;
- Environmental responsiveness, embedded within the building envelope and form;
- Experiential complexity, where spatial quality emerges from structural logic.

These principles illustrate a shift from traditional design approaches toward data-informed and performance-driven architecture, where digital tools enable new forms of integration between disciplines. In conclusion, the Morpheus Hotel demonstrates how smart structures can emerge from the convergence of computational design and structural intelligence. It challenges conventional distinctions between structure and architecture, showing that engineering systems can serve as the primary drivers of form, space, and experience. By integrating digital design, structural optimization, and environmental responsiveness, the project expands the possibilities of contemporary architecture, offering a model for future buildings that are both technically advanced and architecturally expressive.

9.1.4 Synthesis of contemporary design principles

Across the examined case studies—The Edge, Jewel Changi Airport, and Morpheus Hotel—a coherent and advanced set of design principles emerges, reflecting the evolution of smart structures into a mature, integrated, and performance-driven discipline. These projects demonstrate that contemporary architecture is no longer defined by isolated systems or singular innovations but by the interconnected relationships between digital technologies, physical systems, environmental forces, and human experience.

Rather than functioning as static objects, buildings are increasingly conceived as dynamic and adaptive systems capable of sensing, processing, and responding to changing conditions. The following principles synthesize this transformation and provide a conceptual framework for understanding the next generation of architectural design.

9.1.5 Digital–physical integration

A defining characteristic of contemporary smart structures is the seamless integration between digital systems and physical building components (Bibri and Huang, 2025b). Buildings operate as cyber-physical systems, where real-time data continuously informs physical responses and operational decisions. In projects such as The Edge, sensor networks, data platforms, and control systems are fully embedded within the architectural framework, enabling continuous monitoring and optimization of environmental conditions, energy use, and occupancy patterns. This integration transforms buildings into interactive systems, where digital intelligence is not external but intrinsic to their operation. The boundary between design and operation becomes increasingly blurred, as buildings continue to evolve after construction through data-driven feedback loops.

9.1.6 Real-time adaptation

Another key principle is the capacity for real-time adaptation (Stefanopoulou et al., 2025). Contemporary buildings are designed to respond dynamically to fluctuations in climate, occupancy, and user behavior. Environmental systems such as lighting, ventilation, and thermal control are no longer static but adjust continuously to maintain optimal performance. For example, in Jewel Changi Airport, environmental conditions are carefully regulated through a combination of passive design strategies and active control systems informed by simulation and real-time monitoring. This responsiveness enhances both energy efficiency and user comfort, demonstrating how adaptability can operate at multiple scales, from localized environmental control to large-scale climatic regulation.

9.1.7 Computational design integration

The integration of computational design methodologies is central to the development of smart structures (Sun et al., 2026). Parametric modeling, algorithmic design, and simulation tools are not merely used for representation but function as generative and analytical instruments that shape architectural form and performance. In the Morpheus Hotel, computational design enabled the creation of a complex exoskeleton structure that optimizes load distribution while defining the building's architectural identity. Through iterative digital workflows, designers can explore multiple design scenarios, evaluate performance outcomes, and refine solutions in real time. This approach allows for a high level of integration between geometry, structure, and environmental performance, leading to more efficient and innovative design outcomes.

9.1.8 Material and structural intelligence

Smart structures increasingly rely on material and structural intelligence, where building components actively contribute to performance rather than serving as passive elements. Materials and structural systems are designed to respond to environmental conditions, optimize resource use, and enhance adaptability. In contemporary practice, structural systems are not hidden but become active participants in architectural expression and performance. The exoskeleton of the Morpheus Hotel, for instance, demonstrates how structural logic can simultaneously support loads, shape spatial experience, and contribute to environmental modulation. This principle reflects a broader shift toward multifunctional design, where each component performs multiple roles, increasing overall efficiency and coherence.

9.1.9 User-centered interaction

A critical dimension of smart architecture is the integration of user-centered interaction. Occupants are no longer passive recipients of predetermined conditions but active participants who influence building behavior through direct and indirect feedback. In The Edge, users interact with the building through digital interfaces that allow personalized control over environmental settings and workspace allocation. This level of engagement enhances comfort, satisfaction, and productivity while also providing valuable data that informs building operation. This shift toward human-centered design recognizes that building performance is not only a technical issue but also a matter of user experience and well-being.

9.1.10 Holistic sustainability

Sustainability in smart structures is approached as a holistic and integrated objective, rather than a set of isolated strategies (Bibri and Huang, 2025a). Environmental performance is embedded across all layers of design, including form, envelope systems, material selection, and operational strategies. Projects such as Jewel Changi Airport demonstrate how passive and active systems can be combined to create efficient and comfortable environments. The integration of daylight control, thermal regulation, and natural elements contributes to reduced energy consumption while enhancing spatial quality. This approach reflects a shift toward systems thinking, where environmental performance is considered as part of a broader ecological and technological network.

9.1.11 Transformability and flexibility

Contemporary buildings are increasingly designed for transformability and long-term adaptability. Rather than being fixed in function and configuration, they are capable of accommodating changing needs over time. Flexibility is achieved through both physical adaptability, such as open floor plans and modular systems, and operational adaptability, where building systems can be reconfigured through digital control. This ensures that buildings remain relevant and functional in the face of evolving user requirements and environmental conditions.

9.1.12 Toward a new architectural paradigm

Taken together, these principles demonstrate that smart structures are not defined by individual technologies but by the dynamic relationships established between systems, users, and environmental conditions. The examined projects reveal a profound transformation in architectural thinking, marking a shift from static, deterministic design toward dynamic, adaptive, and performance-driven systems.

In contemporary practice, buildings are increasingly conceived as living systems—capable of sensing, learning, and responding to both environmental stimuli and human needs. This evolution is driven by the convergence of the following:

- sensing and data technologies;
- computational design and simulation;
- advanced materials and structural systems;
- interdisciplinary collaboration

As a result, architecture is no longer a passive container but an active participant within ecological and socio-technical systems, continuously interacting with its context and users.

9.1.13 Implications for future design

In the face of global challenges such as climate change, resource scarcity, and rapid urbanization, the lessons derived from these innovative projects are particularly significant. They demonstrate that smart structures offer not only technological advancement but also a pathway toward more sustainable, resilient, and human-centered built environments. Looking forward, the emphasis will increasingly shift from isolated innovations to systemic integration, where intelligence is embedded across all scales—from material components to building systems and urban networks. The role of the architect and engineer will continue to evolve, requiring new forms of collaboration, digital literacy, and systems thinking. Ultimately, these contemporary case studies provide a foundational framework for the future of architecture—one defined by adaptability, responsiveness, and holistic performance. In this emerging paradigm, buildings are not static end-products but evolving systems, capable of continuous improvement and meaningful interaction with their environment and users.

9.2 Functionality and technological innovations

The ability of smart structures to transform conventional building functionality into dynamic, responsive, and continuously optimized systems lies at the core of their architectural and operational performance. In traditional buildings, functionality is typically predefined during the design phase and remains largely static throughout the building's lifecycle. Systems operate within fixed parameters, responding only in limited and often reactive ways to environmental variation or user demand. In contrast, smart structures introduce a fundamentally different paradigm in which functionality is fluid, adaptive, and data-driven, emerging from the continuous interaction between design intent, technological systems, and real-time operational conditions (Fang et al., 2025).

Within this context, functionality can no longer be understood as a singular or fixed attribute. Instead, it becomes a multilayered and evolving capacity, shaped by the integration of digital technologies, environmental intelligence, and human interaction. Technological innovation does not simply enhance existing building systems; it redefines how buildings operate, how they are experienced, and how they perform over time. The focus shifts from isolated components—such as sensors or façade systems—to the orchestration of technologies into cohesive operational frameworks, where buildings behave as intelligent, interconnected systems.

9.2.1 From controlled systems to semiautonomous operation

A defining shift in smart structures is the transition from manually controlled or preprogrammed systems to semiautonomous building operations. Contemporary buildings are increasingly capable of making low-level decisions independently, based on continuously updated data streams and predefined performance objectives. These decisions may include adjusting environmental conditions, reallocating energy resources, or modifying spatial usage patterns in response to real-time inputs. In advanced workplace and institutional environments, integrated management platforms coordinate multiple systems simultaneously. Rather than relying on static schedules, these platforms dynamically reorganize operational priorities throughout the day. For instance, spaces can be activated or deactivated based on occupancy, environmental systems can be modulated according to usage intensity, and energy distribution can be optimized to reduce peak loads. This introduces the concept of operational intelligence, where the building functions as an active agent managing its own performance.

Importantly, this autonomy does not eliminate human control. Instead, it redefines the role of users and operators, shifting their involvement from direct system control to strategic oversight. Users define goals—such as comfort levels, energy targets, or spatial preferences—while the building system determines how best to achieve them.

9.2.2 Interoperability and system integration

A major technological advancement in smart structures is the integration of traditionally separate building systems into a unified, interoperable network. Historically, structural, mechanical, electrical, and architectural systems were designed and operated independently, often resulting in inefficiencies and limited coordination. In smart buildings, these systems are connected through shared digital platforms and communication protocols, enabling continuous data exchange and coordinated operation. This integration supports the principle of interoperability, where functionality emerges from the interaction of multiple subsystems rather than from isolated components. For example, environmental data can influence spatial zoning strategies, while occupancy patterns can inform energy distribution and ventilation control. Structural monitoring systems may also contribute to operational decisions, particularly in adaptive or high-performance buildings. The result is a building that behaves as a coherent and synchronized system, capable of optimizing performance across multiple domains simultaneously. This reduces redundancy, enhances efficiency, and enables more sophisticated responses to complex conditions.

9.2.3 Temporal functionality and long-term adaptability

Technological innovation introduces a critical new dimension to architectural functionality: time-based adaptability. Buildings are no longer designed solely for immediate use but are increasingly conceived to accommodate changes over time, both in daily operation and long-term programmatic evolution. Digital infrastructure enables buildings to transition between different modes of use without requiring physical reconstruction. Through scheduling systems, occupancy analytics, and flexible service networks, spaces can be reconfigured to support a variety of functions—such as work, collaboration, events, or public engagement. This reflects the principle of temporal functionality, where the purpose of a building evolves in response to changing social, economic, and organizational needs. Such adaptability extends the useful life of buildings, reduces the need for resource-intensive renovations, and supports more sustainable patterns of use. Technology acts as the enabling layer that ensures continuity of performance while allowing for transformation.

9.2.4 Behavioral responsiveness and user integration

One of the most transformative aspects of smart structures is the integration of user behavior into building functionality. Buildings are no longer passive environments designed solely to accommodate users; they actively respond to and learn from user interactions. Occupancy patterns, movement flows, and environmental preferences are continuously analyzed to inform operational decisions. Through digital interfaces—often in the form of mobile applications or interactive control systems—users can directly influence environmental conditions such as lighting, temperature, and spatial allocation. This introduces the concept of behavioral responsiveness, where human activity becomes a key input in the building's operational logic. The result is a more personalized and adaptive environment in which functionality is coproduced by both the system and its users. This enhances comfort, satisfaction, and productivity while also improving system efficiency through more accurate demand-based operation.

9.2.5 Lifecycle continuity and performance integration

Another critical dimension of technological innovation is the extension of functionality across the entire building lifecycle. Traditional practice often treats design, construction, and operation as separate and disconnected phases. In smart structures, these stages are integrated through continuous data exchange and performance monitoring. Digital platforms enable the transfer of design intelligence into operational systems, ensuring that performance targets established during the design phase are maintained and refined over time. Building performance is continuously evaluated, and systems are adjusted accordingly to maintain efficiency and comfort. Maintenance strategies are also transformed through condition-based monitoring, where interventions are triggered by real-time performance data rather than predetermined schedules. This reduces unnecessary maintenance, extends system lifespan, and improves reliability. This approach reflects the principle of lifecycle continuity, where buildings are understood as evolving systems rather than static products.

9.2.6 Adaptive resilience and operational stability

In an era of increasing environmental uncertainty and operational complexity, technological innovation plays a key role in enhancing functional resilience. Smart structures incorporate redundancy, adaptability, and real-time monitoring to maintain performance under changing or adverse conditions. Integrated systems can respond to disruptions by redistributing loads, adjusting environmental controls, or isolating affected zones. For example, in the event of system failure or extreme environmental conditions, the building can reconfigure its operations to maintain essential functions. This capacity reflects the principle of adaptive resilience, where functionality is defined not by resistance to change but by the ability to respond, recover, and continue operating effectively.

9.2.7 Informational functionality and data-driven decision making

Beyond real-time control, smart buildings function as decision-support environments. Data generated by building systems is not only used for immediate adjustments but also for long-term analysis and strategic planning. In large-scale facilities, data analytics platforms provide insights into energy consumption, spatial utilization, and system performance. These insights inform decisions related to operational strategies, renovation planning, and resource allocation. In this way, buildings contribute to broader organizational and urban systems, supporting smarter infrastructure and more efficient resource management. This introduces the concept of informational functionality, where buildings act as sources of knowledge within a larger data ecosystem.

9.2.8 Redefining functionality in smart architecture

The technological innovations discussed above collectively redefine functionality as:

- operational rather than static;
- integrated rather than segmented;
- adaptive rather than fixed;
- user-influenced rather than predetermined;
- continuous rather than phase-based.

Smart structures operate as intelligent, evolving systems, where functionality emerges from the interaction between technology, users, and environmental conditions. The emphasis is not on individual technologies but on how these technologies are coordinated into a unified and responsive framework.

Technological innovation in smart structures is not simply about incorporating advanced components; it is about reconfiguring the fundamental logic of building operation. By integrating operational intelligence, system interoperability, temporal adaptability, behavioral responsiveness, and lifecycle continuity, smart structures move beyond traditional definitions of performance. Functionality is no longer a static attribute but a dynamic capability, continuously shaped by data, interaction, and adaptation. Buildings become responsive frameworks that evolve alongside their users and environments, offering new possibilities for efficiency, resilience, and human-centered design. In this emerging paradigm, the true value of technology lies not in its visibility but in its ability to seamlessly and intelligently enhance building performance—operating quietly in the background while enabling richer, more adaptable, and more sustainable built environments (Munaro et al., 2020).

9.3 Lessons from real-world applications

The implementation of smart structure concepts in built projects reveals a fundamental and often underestimated insight: innovation is not defined by the sophistication of individual technologies but by the degree to which these technologies are meaningfully integrated, operationally coherent, and aligned with human and environmental contexts. While theoretical frameworks often emphasize the potential of advanced systems, real-world applications provide a more rigorous and grounded perspective. They expose the complexities of execution, the constraints of construction and operation, and the critical importance of long-term usability and adaptability (Vandamme et al., 2026).

Rather than focusing on isolated technologies or individual project outcomes, this section synthesizes cross-cutting lessons derived from practice, offering a comprehensive understanding of how smart structures perform in reality. These lessons address key dimensions, including system integration, usability, lifecycle performance, data management, resilience, and interdisciplinary collaboration. Together, they establish a mature framework for guiding future smart architectural design.

9.3.1 Integration over technological accumulation

One of the most consistent findings across built projects is that successful smart structures are defined by integration rather than accumulation. Projects that introduce smart technologies as late-stage additions often encounter system incompatibility, fragmented control logic, and increased operational complexity. In contrast, high-performing buildings demonstrate that intelligence must be embedded from the earliest stages of design. Effective implementations are characterized by:

- early definition of system logic and performance targets;
- co-development of architectural, structural, and digital strategies;
- design decisions informed by measurable performance criteria.

This reinforces a critical principle: innovation is effective only when it is structurally embedded within the design process. Technologies must be aligned with spatial organization, material systems, and environmental strategies, functioning as integral components rather than external layers.

9.3.2 Managing complexity through intelligent simplicity

While smart structures inherently involve complex systems, real-world experience shows that uncontrolled complexity can undermine performance. Overly intricate automation systems, fragmented interfaces, and poorly coordinated subsystems often lead to maintenance difficulties, system failures, and user frustration. Successful projects instead demonstrate the value of intelligent simplicity, achieved through:

- clear system hierarchies and streamlined control frameworks;
- selection of robust, reliable, and interoperable technologies;
- user interfaces designed for clarity and accessibility.

In this approach, complexity is managed internally within the system architecture, while the user experience remains intuitive and legible. The effectiveness of a smart building is therefore measured not by the quantity of technology deployed, but by the coherence, reliability, and usability of its systems.

9.3.3 Human-centered functionality and user engagement

A recurring challenge in real-world applications is the disconnect between technological capability and actual user experience. Even highly advanced systems can fail when occupants do not understand, trust, or engage with them. This highlights the importance of human-centered functionality, where technology enhances rather than complicates the user experience. Lessons from practice emphasize that:

- interfaces must be intuitive, transparent, and responsive;
- occupants should retain meaningful control over their environment;
- behavioral patterns must inform system design and operation.

Buildings must not only respond to data but also communicate effectively with users, enabling interaction, feedback, and adaptability. In this context, functionality is co-produced through the interaction between systems and occupants.

9.3.4 Performance as a continuous process

Smart buildings are often designed to meet specific performance targets, such as energy efficiency or indoor environmental quality. However, real-world applications consistently reveal a gap between predicted and actual performance. This discrepancy arises from factors such as:

- variability in occupant behavior;
- changing environmental conditions;
- system degradation or miscalibration over time.

As a result, high-performing smart structures adopt a model of continuous performance management, incorporating:

- real-time monitoring and data collection;
- feedback loops and performance analytics;
- ongoing recalibration and system optimization.

This leads to a fundamental shift in perspective: performance is not a fixed outcome achieved at completion but an evolving process maintained throughout the building's lifecycle.

9.3.5 Interdisciplinary collaboration as a core requirement

The complexity of smart structures necessitates deep and sustained collaboration among architects, engineers, data specialists, and facility managers. Real-world projects demonstrate that fragmented workflows lead directly to fragmented systems. Successful implementations are defined by:

- early and continuous interdisciplinary engagement;
- shared digital platforms and integrated modeling environments;
- transparent communication and collective decision-making.

This reinforces the principle of collective authorship, where building performance emerges from coordinated expertise. Collaboration is not merely supportive—it is foundational to system coherence and innovation.

9.3.6 Lifecycle thinking and long-term value

A critical lesson from practice is that the value of smart technologies extends far beyond initial construction. Systems must be designed with a clear understanding of their long-term implications, including maintenance, upgrades, and operational costs. Projects that neglect lifecycle considerations often face:

- rapid technological obsolescence;
- increased maintenance complexity;
- declining performance over time.

In contrast, successful smart structures adopt a lifecycle-oriented approach, characterized by:

- modular and upgradable system architectures;
- accessible and well-documented maintenance strategies;
- integration with long-term facility management plans.

This ensures that buildings remain functional, efficient, and adaptable in the face of technological and environmental change.

9.3.7 From data collection to actionable intelligence

Smart buildings generate vast quantities of data, but real-world applications demonstrate that data alone does not guarantee improved performance. The key challenge lies in transforming raw data into meaningful and actionable insights. Effective systems:

- prioritize relevant and high-impact data streams;
- present information through clear and interpretable interfaces;
- support informed decision-making for users and operators.

This reflects the principle of actionable intelligence, where data serves as a tool for optimization rather than an end in itself. Without clear strategies for interpretation, data risks becoming either overwhelming or underutilized.

9.3.8 Resilient adaptability and flexibility

While optimization is a central goal in smart design, real-world experience highlights the importance of flexibility and resilience. Buildings designed for narrowly defined conditions often struggle to adapt to changing requirements. Successful projects incorporate:

- adaptable spatial configurations;
- scalable and interoperable technological systems;
- flexible operational frameworks.

This reflects the principle of resilient adaptability, ensuring that buildings can accommodate uncertainty and change without compromising performance.

9.3.9 Contextual intelligence and local adaptation

Smart technologies are not universally transferable solutions. Their effectiveness depends heavily on contextual factors, including climate, culture, economic conditions, and user expectations. Real-world applications emphasize:

- climate-responsive design strategies;
- alignment with local cultural and behavioral patterns;
- consideration of economic feasibility and maintenance capacity.

This leads to the principle of contextual intelligence, where smart systems are tailored to specific environments rather than applied as generic solutions.

9.3.10 Bridging the gap between design and operation

A persistent challenge in smart structures is the gap between design intent and operational reality. Even well-designed systems may underperform due to insufficient coordination, lack of training, or inadequate commissioning. Key issues include:

- limited collaboration between design and facility management teams;
- insufficient user training and system understanding;

- lack of postoccupancy evaluation.

Addressing this gap requires:

- comprehensive documentation and knowledge transfer;
- early involvement of operators and facility managers;
- rigorous commissioning and continuous evaluation.

This reinforces the principle of continuity between design and operation, ensuring that performance objectives are sustained over time.

The lessons derived from real-world applications collectively point toward a more grounded and systemic understanding of smart structures. Key insights include:

- integration is more critical than technological quantity;
- simplicity enhances reliability, usability, and longevity;
- human interaction is central to system effectiveness;
- performance must be continuously monitored and refined;
- collaboration is essential for system coherence;
- lifecycle thinking ensures long-term value;
- data must be actionable and accessible;
- flexibility and resilience are as important as efficiency;
- contextual responsiveness determines success.

These insights redefine smart structures not as technologically dense artifacts but as coordinated, adaptive, and human-centered systems.

Real-world applications provide an essential lens through which the ambitions of smart architecture can be critically evaluated and refined. They demonstrate that the success of smart structures depends less on the presence of advanced technologies and more on the quality of their integration, management, and alignment with human and environmental needs. The lessons outlined in this section highlight the necessity of moving beyond a purely technology-driven perspective toward a systems-oriented and human-centered approach. When guided by these principles, smart structures can fulfill their potential as responsive, efficient, and resilient environments. Ultimately, the future of smart architecture lies not in increasing technological complexity but in achieving greater coherence, adaptability, and clarity—where innovation is driven by performance, and technology operates seamlessly to enhance the lived experience of the built environment.

References

Ageli, O., Chassiakos, A., 2026. Balancing indoor environmental quality and energy efficiency in buildings using digital twins technology: scientometric review and multi-layer conceptual framework. Energy and Buildings 357, 117147. Available at https://doi.org/10.1016/j.enbuild.2026.117147.

Al Dakheel, J., Del Pero, C., Aste, N., Leonforte, F., 2020. Smart buildings features and key performance indicators: a review. Sustainable Cities and Society 61, 102328. Available at https://doi.org/10.1016/j.scs.2020.102328.

Bibri, S.E., Huang, J., 2025a. AI and AI-powered digital twins for smart, green, and zero-energy buildings: a systematic review of leading-edge solutions for advancing environmental sustainability goals. Environmental Science and Ecotechnology 28. Available at https://doi.org/10.1016/j.ese.2025.100628.

Bibri, S.E., Huang, J., 2025. Artificial intelligence of things for sustainable smart city brain and digital twin systems: pioneering environmental synergies between real-time management and predictive planning. Environmental Science and Ecotechnology 26, 100591. Available at https://doi.org/10.1016/j.ese.2025.100591.

Campodonico Avendano, I.A., Heimar Andersen, K., Erba, S., Moazami, A., Aghaei, M., Najafi, B., 2024. A novel framework for assessing the smartness and the smart readiness level in highly electrified non-residential buildings: a Norwegian case study. Energy and Buildings 314, 114234. Available at https://doi.org/10.1016/j.enbuild.2024.114234.

Escamilla, A., Melenchón, J., Monzo, C., Morán, J.A., Carrascal, J.P., 2026. Exploring a user-centered approach for movement-based features in interaction design. International Journal of Human-Computer Studies 211, 103802. Available at https://doi.org/10.1016/j.ijhcs.2026.103802.

Fang, X., Chang, R., Zuo, J., Zhang, W.E., Zou, Y., Li, K., 2025. How do environmental and operational factors impact particulate matter dynamics in building construction? - Insights from real-time sensing. Journal of Environmental Management 380, 125098. Available at https://doi.org/10.1016/j.jenvman.2025.125098.

Jones, R.V., Fuertes, A., Bilverstone, P.A., 2025. A repeated cross-sectional study of a biophilic building's effects on workers' perception of indoor environmental quality, health, well-being and productivity. Journal of Building Engineering 111, 113476. Available at https://doi.org/10.1016/j.jobe.2025.113476.

Lozinsky, C.H., Casquero-Modrego, N., Walker, I.S., 2025. The health and indoor environmental quality impacts of residential building envelope retrofits: a literature review. Building and Environment 270. Available at https://doi.org/10.1016/j.buildenv.2025.112568.

Ma, W., Liu, T., 2026. Adaptive smart energy management in buildings: a deep reinforcement learning and IoT framework for energy efficiency and thermal comfort. Energy and Buildings 351, 116659. Available at https://doi.org/10.1016/j.enbuild.2025.116659.

Mahlan, S., Francis, A., Thumuganti, V., Thomas, A., Sadick, A.-M., Tokede, O., 2024. An integrated life cycle assessment and energy simulation framework for residential building walling systems. Building and Environment 257, 111542. Available at https://doi.org/10.1016/j.buildenv.2024.111542.

Martínez, I., Cano-Suñén, E., Casas, R., Fernández, Á., 2025. Smart built environment (SBE): a challenge for internet of things (IoT) ecosystems to understand dynamic habitats and their users as complex systems. Journal of Building Engineering 116, 114620. Available at https://doi.org/10.1016/j.jobe.2025.114620.

Munaro, M.R., Tavares, S.F., Bragança, L., 2020. Towards circular and more sustainable buildings: a systematic literature review on the circular economy in the built environment. Journal of Cleaner Production 260, 121134. Available at https://doi.org/10.1016/j.jclepro.2020.121134.

Schmidt, M., Moreno, M.V., Schülke, A., Macek, K., Mařík, K., Pastor, A.G., 2017. Optimizing legacy building operation: the evolution into data-driven predictive cyber-physical systems. Energy and Buildings 148, 257–279. Available at https://doi.org/10.1016/j.enbuild.2017.05.002.

Schwegmann, L., Zabel, V., Koch, S., 2024. Adaptive structures in civil engineering – state of the art. Engineering Computations 41 (6), 1424–1440. Available at https://doi.org/10.1108/ec-12-2023-0952.

Stefanopoulou, A., Michailidis, I., Karatzinis, G., Lepidas, G., Boutalis, Y.S., Kosmatopoulos, E.B., 2025. Ensuring real-time data integrity in smart building applications: a systematic end-to-end comprehensive pipeline evaluated in numerous real-life cases. Energy and Buildings 336. Available at https://doi.org/10.1016/j.enbuild.2025.115586.

Sun, Q., Reisach, D., Langenberg, S., Dillenburger, B., 2026. Computational design methods for geometry-driven upcycling of found objects in construction. Automation in Construction 183. Available at https://doi.org/10.1016/j.autcon.2026.106803.

Vandamme, E., Passarelli, R.N., Rinke, M., 2026. Adaptability in timber construction: evaluating frameworks for flexible multistorey building systems and material-specific requirements. Building and Environment 114617. Available at https://doi.org/10.1016/j.buildenv.2026.114617.

Verma, T., Gopalakrishnan, P., 2025. Quantifying the effect of irradiance-based blind control occupant behaviour models (BC-OBMs) on daylighting performance and energy consumption in office buildings. Journal of Building Engineering 103, 112067. Available at https://doi.org/10.1016/j.jobe.2025.112067.

Verma, R., Kumar, S., Rakshit, D., Udayraj, 2026. Integrated evaluation of passive building strategies through experimental analysis, energy simulation, and life cycle assessment in the composite climate of India. Case Studies in Thermal Engineering 81, 107956. Available at https://doi.org/10.1016/j.csite.2026.107956.

Future trends in architectural design and structural engineering

10

10.1 Emerging technologies in design and engineering

The future trajectory of architectural design and structural engineering is being fundamentally reshaped by a convergence of emerging technologies that collectively redefine how the built environment is conceived, produced, and operated. Rather than functioning as isolated innovations, these technologies form an interconnected ecosystem in which computational intelligence, advanced materials, digital fabrication, and data-driven systems operate in synergy. Within this evolving paradigm, architecture and engineering are no longer sequential or discipline-bound processes; instead, they become simultaneous, data-informed, and adaptive practices. Design is increasingly understood as a dynamic system of relationships, where geometry, performance, material behavior, and user interaction are continuously negotiated through digital and physical feedback loops. This transformation marks a decisive shift from static design methodologies toward an integrated, responsive, and performance-oriented approach that aligns with the broader evolution of smart structures. As illustrated in Fig. 10.1, smart structures operate within an integrated technological ecosystem in which sensing, computation, and actuation are continuously interconnected. Rather than functioning as isolated systems, these technologies form a cohesive network that supports real-time decision-making, adaptive performance, and user-responsive environments.

Artificial intelligence (AI) plays a central role in this transformation by introducing new modes of design generation, evaluation, and prediction (Sreenivasan and Suresh, 2024). AI-driven systems enable architects and engineers to process vast datasets, identify hidden patterns, and generate optimized design alternatives based on predefined performance criteria (Nabizadeh Rafsanjani and Nabizadeh, 2023). In contrast to traditional iterative workflows, which rely heavily on human intuition and sequential refinement, generative design systems can simultaneously evaluate multiple constraints—such as structural efficiency, environmental performance, and spatial quality—producing solutions that balance competing objectives with a high degree of precision. This capability allows for the creation of highly efficient structural forms, reduction of material consumption, and enhancement of environmental responsiveness through data-informed geometries. Moreover, AI extends beyond design generation into predictive analytics, enabling designers to anticipate building performance under varying conditions over time. This shifts architectural practice toward a proactive model in which anticipated performance outcomes actively inform early design decisions, embedding intelligence at the conceptual stage rather than applying it retrospectively. Fig. 10.2 demonstrates how generative design shifts architectural practice from a linear, intuition-based process to an iterative, data-driven methodology. By embedding performance criteria directly into the design

Smart Civil and Structural Engineering. DOI: https://doi.org/10.1016/B978-0-443-30223-7.00015-2

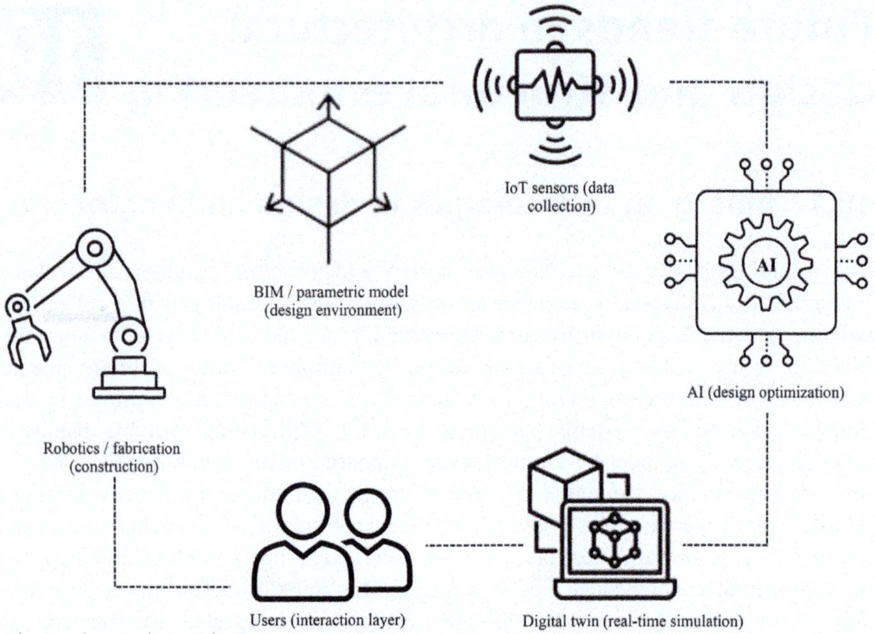

Figure 10.1 Integrated smart building technology ecosystem.

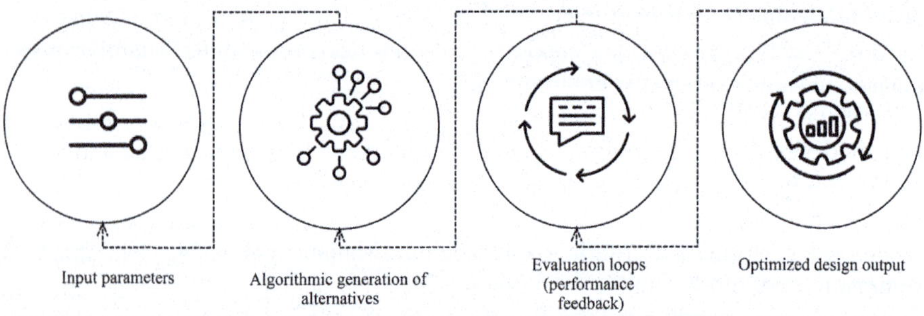

Figure 10.2 AI-driven generative design workflow (*AI*, artificial intelligence).

workflow, architects and engineers can explore a wide range of optimized solutions that respond simultaneously to multiple constraints.

Closely related to this development is the emergence of digital twin technology, which establishes a continuous connection between physical buildings and their virtual counterparts (Yang and Ng, 2024). A digital twin is not simply a static model but a dynamic, data-driven representation that evolves in real time through input from sensors and monitoring systems embedded within the building. This technology enables continuous performance tracking, predictive maintenance, and scenario-based simulation, allowing building operators and designers to anticipate system behavior, identify inefficiencies, and implement corrective strategies before failures occur. In the context of smart structures, digital twins transform buildings into cyber-physical systems, where the boundary between design and operation is effectively dissolved. The implications of this shift are profound: buildings are no longer completed at the end of construction but continue to evolve through data-driven optimization throughout their lifecycle. This ongoing feedback loop enhances long-term performance, supports informed decision-making, and contributes to the broader development of intelligent urban infrastructures.

Parallel to advances in digital intelligence, robotics, and automation, construction methodologies and material realization are being redefined (Leder and Menges, 2023; Zeng et al., 2024). Robotic fabrication systems enable unprecedented levels of precision in the production and assembly of building components, allowing complex geometries generated through computational design to be accurately translated into physical form. On-site robotic systems are increasingly employed for tasks such as material placement, structural assembly, and quality inspection, reducing human error and improving construction efficiency (Xie and Alwisy, 2026). These technologies not only enhance productivity but also expand the formal and structural possibilities of architecture by enabling the realization of geometries that would be impractical or impossible using conventional methods. Furthermore, automation contributes to improved safety conditions by minimizing human exposure to hazardous environments, while also reducing material waste through precise fabrication processes. The integration of robotics with parametric design workflows establishes a direct link between digital models and physical construction, reinforcing the continuity between design intent and built outcome (Geno et al., 2022).

Additive manufacturing, commonly referred to as 3D printing, represents another transformative development in the production of the built environment (Abdullah Agha, 2024). By enabling layer-by-layer fabrication of components and structures, this technology allows for highly efficient material usage, as material is deposited only where structurally necessary. This results in lightweight yet high-performance structural systems that challenge conventional approaches to construction. Additive manufacturing also supports a high degree of customization, allowing each component to be uniquely tailored without increasing production complexity or cost. In architectural applications, this opens new possibilities for form generation, structural optimization, and material expression. Moreover, the potential for on-site or localized production introduces a shift toward decentralized construction models, reducing transportation requirements and enabling rapid deployment in remote or resource-constrained contexts. As the technology matures, it is likely to play a significant role in redefining construction logistics and enabling more sustainable production practices (Kolářová et al., 2026).

Material innovation further expands the scope of architectural and structural intelligence by introducing responsiveness at the level of matter itself. Smart materials, such as shape-memory alloys, electroactive polymers, and bio-responsive composites, possess the ability to alter their physical properties in response to environmental stimuli, including temperature, light, or mechanical stress. This capacity introduces a temporal dimension to material performance, allowing structures to adapt passively or actively to changing conditions. Unlike conventional materials, which remain static after installation, these advanced materials enable buildings to self-adjust, reducing reliance on mechanical systems and enhancing overall efficiency. The concept of programmable matter extends this potential even further, envisioning materials that can dynamically reconfigure their geometry or behavior over time. Such innovations fundamentally challenge traditional distinctions between structure, envelope, and system, suggesting a future in which adaptability is embedded directly within the material fabric of architecture.

The proliferation of Internet of Things (IoT) technologies has enabled the widespread integration of sensor networks that serve as the informational backbone of smart structures. These distributed systems continuously collect data related to environmental conditions, structural performance, and user behavior, enabling real-time monitoring and control of building operations. Through cloud-based platforms and advanced analytics, this data is transformed into actionable insights that inform decision-making at both operational and strategic levels. IoT integration facilitates the coordination of multiple subsystems—such as lighting, heating, ventilation, and air conditioning, security, and structural monitoring—within a unified framework, enhancing overall efficiency and responsiveness (Liu et al., 2026b). As these networks become increasingly sophisticated, buildings evolve into interconnected nodes within larger urban systems, contributing to the development of smart cities characterized by data-driven infrastructure and integrated resource management.

Extended reality technologies, including virtual reality, augmented reality, and mixed reality, are transforming how architects and engineers engage with design information (Yeom et al., 2024). These tools enable immersive visualization of projects before construction, allowing stakeholders to experience spatial qualities, material conditions, and system interactions in a highly intuitive manner. This enhances communication, reduces ambiguity, and supports more informed decision-making during the design process. In construction and maintenance contexts, augmented reality can overlay digital models onto physical environments, facilitating precise installation and enabling real-time access to hidden system information. As a result, extended reality technologies act as a bridge between digital design and physical realization, improving coordination across all project phases and reducing the likelihood of errors or misinterpretations (Fig. 10.3).

In parallel, blockchain technology is emerging as a tool for enhancing transparency and trust within the construction and project management ecosystem (Rathnayake et al., 2025). By enabling secure and decentralized data exchange, blockchain supports reliable documentation of transactions, contracts, and material supply chains. This is particularly valuable in complex projects involving multiple stakeholders, where issues of accountability and data integrity can significantly impact project outcomes. Through immutable records and transparent workflows, blockchain has the potential to streamline collaboration, reduce disputes, and improve overall efficiency in project delivery.

Collectively, these emerging technologies signal a profound transformation in architectural and engineering practice. Design processes are shifting from manual and intuition-driven methods toward algorithmically informed systems that integrate

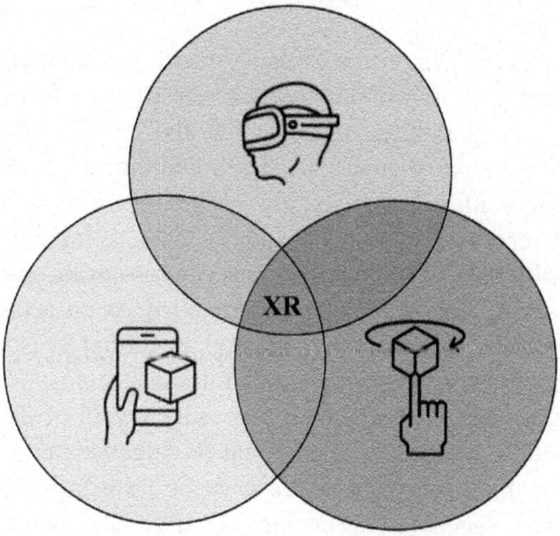

Figure 10.3 Extended reality (XR) spectrum: Intersection of virtual reality, augmented reality, and mixed reality.

performance criteria from the outset. Static representations are being replaced by dynamic models that evolve in response to real-time data, while fragmented workflows give way to integrated digital ecosystems that support collaboration across disciplines. Construction methods are transitioning from labor-intensive processes to automated and robotic systems, and materials are evolving from passive elements into active participants in building performance. These shifts collectively redefine architecture as a technologically enriched and continuously adaptive discipline.

Ultimately, the significance of these technologies lies not only in their individual capabilities but also in their capacity to operate collectively as part of a unified system. They enable the creation of buildings that are more efficient, resilient, and responsive, while also enhancing the quality of human experience. However, their successful implementation requires a rethinking of traditional professional roles, design methodologies, and collaborative frameworks. Architects and engineers must engage with new forms of knowledge, embrace interdisciplinary collaboration, and develop the capacity to integrate technological innovation into coherent design strategies.

As the built environment continues to evolve in response to global challenges such as climate change, resource scarcity, and urbanization, these emerging technologies provide a critical foundation for future development. They offer the tools necessary to create environments that are not only technically advanced but also adaptable, sustainable, and aligned with human needs. In this context, the future of design and engineering lies in the ability to harness technological innovation in a way that is both intelligent and purposeful, ensuring that the built environment remains responsive to the complexities of an ever-changing world.

10.2 Sustainable and green innovations

Sustainability has undergone a profound transformation in contemporary architectural and engineering practice, evolving from a set of performance targets into a comprehensive design paradigm that shapes the entire lifecycle of the built environment. In the context of smart structures and emerging technologies, sustainability is no longer confined to reducing energy consumption or environmental impact. Instead, it represents a holistic and systemic approach that redefines how buildings interact with natural ecosystems, material resources, and human well-being. This expanded understanding positions sustainability as an active, measurable, and adaptive framework, in which buildings are conceived not as isolated objects but as integral components of broader ecological and urban systems. Within this paradigm, environmental performance is embedded across all layers of design, from material selection and structural systems to operational intelligence and user interaction.

A critical shift in this evolving landscape is the transition from efficiency-based sustainability toward regenerative design. Traditional approaches have largely focused on minimizing negative environmental impacts, such as reducing emissions, conserving energy, or limiting resource consumption. While these strategies remain important, they are increasingly complemented—and in some cases replaced—by regenerative principles that seek to produce positive environmental outcomes. Regenerative buildings are designed to restore ecosystems, enhance biodiversity, and contribute to the health of natural systems. This may involve the integration of living landscapes within architectural frameworks, the use of building systems that purify air and water, or the generation of surplus renewable energy that can be returned to the grid. In this context, architecture becomes an active ecological agent, capable of contributing to environmental restoration rather than merely mitigating harm. Structural and spatial systems are thus reimagined as platforms for ecological interaction, reinforcing a net-positive approach to sustainability.

Material innovation plays a central role in enabling this transformation, particularly through the adoption of circular economy principles. The traditional linear model of extraction, use, and disposal is increasingly being replaced by strategies that prioritize reuse, recyclability, and long-term resource efficiency. Buildings are now designed with disassembly in mind, allowing components to be recovered, repurposed, and reintegrated into future construction cycles. This approach extends the lifespan of materials and reduces waste, transforming buildings into repositories of valuable resources rather than endpoints of consumption. The use of recycled materials, bio-based components, and low-embodied-carbon products further supports this transition, while emerging tools such as material passports enable the tracking of material origins, properties, and reuse potential over time. Structural systems are also evolving in response to these principles, with modular construction and prefabrication enabling greater flexibility, adaptability, and resource efficiency. Through these strategies, material use becomes an active component of sustainability, contributing to both environmental performance and long-term economic value.

Energy performance remains a fundamental aspect of sustainable design, yet its conceptual framework has expanded significantly. Rather than focusing solely on reducing energy demand, contemporary approaches emphasize achieving a balance between consumption and production. Net-zero energy buildings are designed to generate as much energy as they consume over the course of a year, while energy-positive buildings exceed this threshold, contributing surplus energy to external systems (Li et al., 2025). This shift is enabled by advancements in renewable energy technologies, including high-efficiency photovoltaic systems, integrated solar thermal solutions, and energy storage systems that allow for the management of fluctuating supply and demand. At the same time, architectural design strategies—such as optimized orientation, passive solar design, natural ventilation, and daylighting—play a critical role in reducing baseline energy requirements. Structural systems support these objectives by facilitating the integration of energy-generating technologies without compromising performance or esthetics. In this integrated model, buildings function not merely as consumers of energy but as active participants in decentralized energy networks.

As environmental conditions become increasingly unpredictable due to climate change, the importance of adaptability and resilience in sustainable design has become more pronounced. Climate-responsive architecture incorporates real-time environmental data and predictive modeling to enable buildings to adjust dynamically to changing conditions. Adaptive façades, responsive shading systems, and dynamic ventilation strategies allow buildings to regulate heat gain, airflow, and daylight in response to external stimuli, maintaining comfort while minimizing energy use. Structural systems are similarly designed to accommodate extreme weather events, incorporating flexibility, redundancy, and robustness to ensure long-term stability. This approach reflects a shift from static resistance to dynamic responsiveness, where buildings are capable of evolving in response to both short-term fluctuations and long-term climatic trends. The integration of adaptability into design not only enhances resilience but also extends the functional lifespan of buildings, ensuring their relevance in an uncertain future.

Water management has emerged as another critical dimension of sustainable innovation, particularly in regions facing increasing water scarcity and climate-related challenges. Water-sensitive design strategies aim to reduce consumption, enhance reuse, and integrate water systems into the architectural and urban fabric. Rainwater harvesting systems collect and store precipitation for nonpotable uses, while greywater recycling systems enable the reuse of water from sinks, showers, and other sources. Green infrastructure elements—such as bioswales, green roofs, and permeable surfaces—further support water management by facilitating natural filtration, reducing runoff, and mitigating flood risks. At the building scale, these systems operate as closed-loop networks that minimize waste and optimize resource use (Batista et al., 2023). At the urban scale, they contribute to broader water management strategies, enhancing resilience and reducing pressure on centralized infrastructure. Through these integrated approaches, water is redefined not as a passive utility but as an active component of environmental design.

Equally important is the growing recognition of human well-being as a core component of sustainability. Biophilic design strategies seek to reconnect occupants with natural systems through the integration of light, air, vegetation, and natural materials (Zhong et al., 2024). These elements enhance not only the esthetic quality of spaces but also their physiological and psychological benefits, contributing to improved health, productivity, and overall well-being. Natural ventilation and daylighting reduce reliance on artificial systems while creating more comfortable and dynamic indoor environments. The incorporation of vegetation—both within interiors and on building exteriors—improves air quality, regulates temperature, and supports biodiversity. This human-centered approach to sustainability emphasizes the interdependence between environmental performance and occupant experience, positioning well-being as an integral component of design rather than an ancillary consideration.

Structural engineering contributes significantly to sustainable innovation through the development of low-carbon and resource-efficient systems. Advances in materials such as mass timber and engineered wood products offer viable alternatives to traditional high-carbon materials, providing both structural performance and environmental benefits (Kathiravel et al., 2026). Similarly, the development of low-carbon concrete and alternative binders reduces the environmental impact of construction while maintaining durability and strength. Computational design tools enable engineers to optimize structural geometry, ensuring that material is used only where necessary to resist loads. This results in lighter, more efficient structures that minimize resource consumption without compromising safety. In this context, structural efficiency becomes a direct contributor to sustainability, demonstrating how engineering innovation can align with environmental objectives.

The integration of renewable energy systems into architectural and structural elements further reinforces the role of buildings as active environmental systems (Sheng et al., 2026). Technologies such as building-integrated photovoltaics, solar shading devices with energy-generating capabilities, and small-scale wind systems are seamlessly incorporated into façades and roofs. This integration not only enhances energy performance but also contributes to the architectural expression of the building, demonstrating that sustainability and esthetics can coexist. By embedding energy generation within the building fabric, these systems support decentralized energy production and reduce reliance on external energy sources.

Advancements in digital technologies have introduced new opportunities for continuous performance monitoring and optimization. Data-driven systems enable real-time tracking of energy use, water consumption, indoor environmental quality, and system performance. This information is analyzed to identify inefficiencies, inform operational adjustments, and maintain alignment with performance targets. Such approaches reflect the principle of adaptive sustainability, where environmental performance is not fixed at the point of design but evolves continuously through feedback and learning. Buildings become dynamic systems capable of self-improvement, responding to both internal and external conditions to optimize their performance over time. As shown in Fig. 10.4, regenerative design extends beyond efficiency by positioning buildings as active participants within ecological systems. The integration of energy, water, and biological processes enables the building to contribute positively to its environment rather than merely reducing its impact.

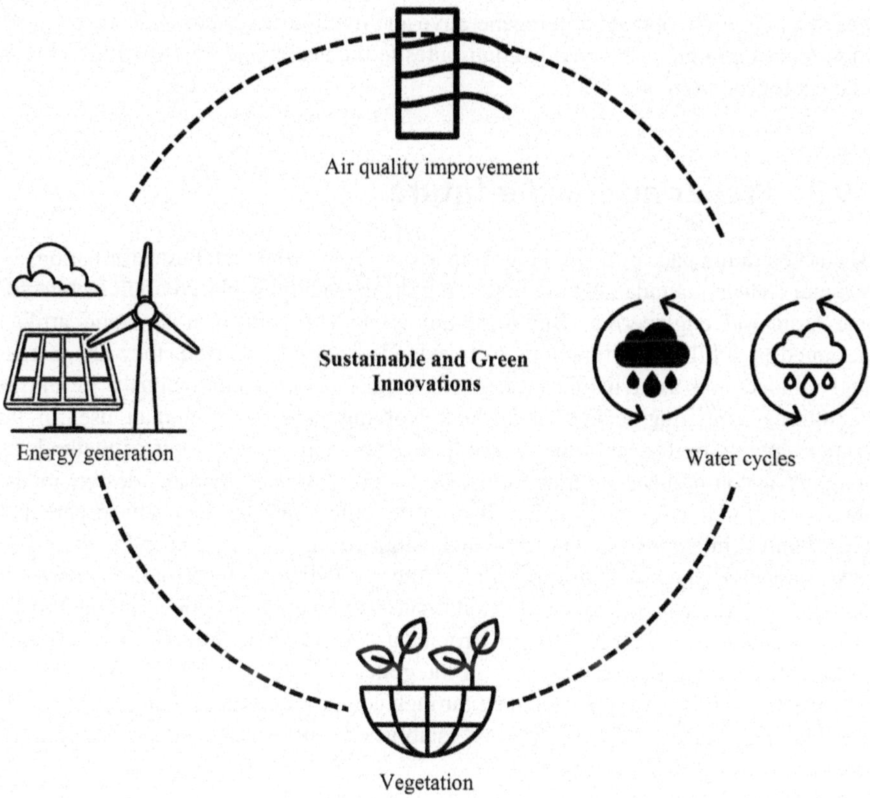

Figure 10.4 Regenerative building system diagram.

Collectively, these innovations signal a paradigm shift in sustainable design. The focus moves from reduction to regeneration, from linear consumption to circular resource flows, and from static efficiency to dynamic adaptability. Buildings are no longer conceived as passive entities but as active participants in ecological, technological, and social systems. They generate energy, manage resources, support human well-being, and contribute to environmental restoration. This integrated approach reflects a deeper understanding of sustainability as a complex and multidimensional challenge that requires coordinated solutions across disciplines and scales.

Sustainable and green innovations redefine the role of architecture and structural engineering in addressing global environmental challenges. They demonstrate that sustainability is not a limitation but a catalyst for creativity, driving the development of new materials, systems, and design methodologies. As climate change, resource scarcity, and urbanization continue to intensify, the adoption of these innovations will be essential in shaping a built environment that is resilient, adaptive, and environmentally responsible. In this emerging context, the future of architecture lies in its ability to operate as a regenerative and intelligent system—one that not only minimizes harm but also actively contributes to the health and vitality of both human and ecological communities.

10.3 Preparing for the future

The accelerating pace of technological innovation, environmental transformation, and societal change is fundamentally redefining the expectations placed upon architecture and structural engineering. Buildings are no longer conceived as static artifacts designed to fulfill fixed functions over extended periods; instead, they are increasingly understood as dynamic systems that must respond continuously to shifting conditions, emerging technologies, and evolving patterns of human use. In this context, preparing for the future is not limited to adopting new tools or methods but requires a comprehensive transformation in professional mindset, design methodology, and collaborative practice. It involves cultivating the capacity to anticipate uncertainty, integrate complexity, and align design decisions with long-term environmental and societal priorities. Future readiness, therefore, becomes an ongoing and strategic process that extends across education, practice, and the broader cultural framework of the built environment. Fig. 10.5 highlights the shift from fragmented project delivery toward an integrated lifecycle approach, where design, construction, and operation are linked through continuous data exchange. This model ensures that performance objectives established during early design stages are maintained and refined over time.

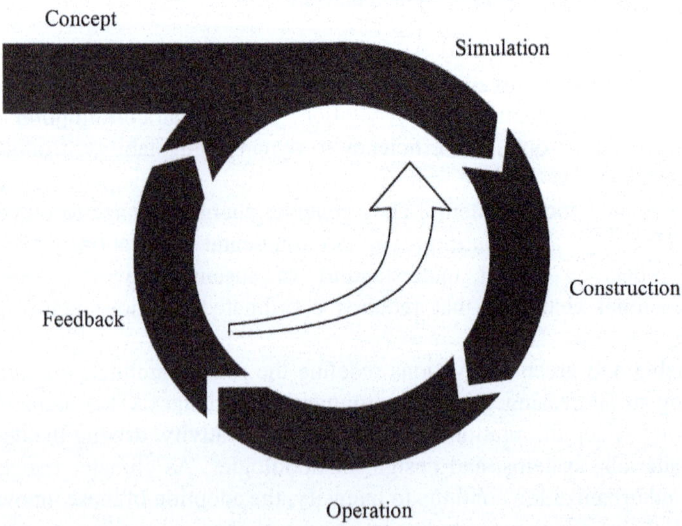

Figure 10.5 Future integrated design workflow across the building lifecycle.

A critical foundation for this transformation lies in the evolution of education and professional training. Traditional models, which often separate architectural design from engineering analysis, are increasingly inadequate in addressing the demands of integrated and intelligent systems (Liu et al., 2026a). Future-oriented education must foster interdisciplinary thinking, enabling architects to engage with structural logic and engineers to participate in spatial and experiential design. This convergence of knowledge is further supported by the growing importance of computational literacy, which includes proficiency in parametric modeling, simulation tools, data analysis, and algorithmic thinking. Equally important is the development of systems thinking, through which buildings are understood not as isolated objects but as interconnected networks of spatial, structural, environmental, and technological components. Beyond formal education, lifelong learning becomes essential, as professionals must continuously adapt to rapidly evolving technologies and methodologies. Engagement with emerging fields such as artificial intelligence, digital fabrication, advanced materials, and data science is no longer optional but necessary for maintaining relevance in an increasingly complex design landscape. As summarized in Fig. 10.6, the evolution of architecture reflects a fundamental transition from static, deterministic design toward dynamic, data-driven, and adaptive systems. This transformation is characterized by the integration of digital technologies, real-time responsiveness, and lifecycle-based performance thinking.

In parallel with educational transformation, the future of practice is characterized by the integration of workflows and the dissolution of traditional disciplinary boundaries. The complexity of smart structures and adaptive systems requires architects, engineers, data specialists, and construction professionals to collaborate within shared frameworks from the earliest stages of design. Digital platforms such as building information modeling and parametric environments enable real-time collaboration, allowing multiple disciplines to contribute simultaneously to a unified model. This shift from sequential to concurrent design processes enhances coordination, reduces conflicts, and ensures that performance considerations are embedded from the outset. Collaboration in this context is not merely a matter of coordination

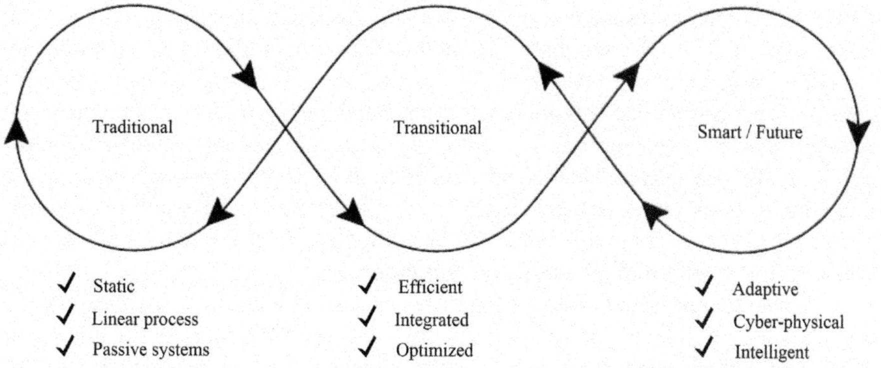

Figure 10.6 Evolution of architectural paradigms from static to intelligent systems.

but represents a form of coauthorship, where design solutions emerge from the synthesis of diverse expertise. Such integrated workflows also improve efficiency in project delivery, aligning conceptual ambitions with technical feasibility and operational requirements.

Climate change represents one of the most urgent and defining challenges shaping the future of the built environment (Hurlimann et al., 2025). Preparing for this reality requires a transition from reactive compliance toward proactive and anticipatory design strategies. Architects and engineers must design for a wide range of environmental scenarios, including extreme heat, flooding, storms, and long-term climatic shifts. This necessitates the incorporation of adaptive systems that can respond dynamically to changing conditions, as well as the use of predictive models that extend beyond historical data to consider future projections. Resilience becomes a central design objective, not as a static condition of resistance but as a dynamic capacity for adaptation and recovery. In this context, sustainability and resilience are no longer separate concerns but are integrated into a unified framework that addresses both environmental performance and long-term viability.

The increasing centrality of data in architectural and engineering practice further defines the future landscape. Buildings are becoming data-rich environments, equipped with sensors and monitoring systems that generate continuous streams of information related to performance, usage, and environmental conditions. Preparing for this shift requires the development of competencies in data analytics, interpretation, and application. Designers must be able to integrate real-time data into both design processes and operational strategies, using predictive modeling to anticipate performance outcomes and optimize system behavior. At the same time, the growing reliance on data introduces important considerations related to governance, privacy, and security. Ethical data management becomes a critical responsibility, ensuring that technological capabilities are aligned with user rights and societal values. In this sense, data is not merely a technical resource but a cultural and ethical dimension of contemporary design practice (Lee, 2025).

Adaptability emerges as a fundamental principle in preparing for an uncertain future. Buildings must be designed to accommodate change, not only in terms of daily use but also in response to long-term shifts in function, technology, and social needs. Flexible spatial configurations, modular construction systems, and adaptable structural frameworks enable buildings to evolve over time without requiring extensive reconstruction. This approach extends the lifespan of the built environment, reduces resource consumption, and enhances economic and environmental sustainability. It also reflects a broader conceptual shift, in which buildings are understood as evolving systems rather than fixed solutions. Designing for adaptability requires a balance between stability and flexibility, ensuring that structures remain robust while allowing for transformation and reconfiguration.

As technological capabilities expand, the ethical dimension of architectural and engineering practice becomes increasingly significant. Preparing for the future involves a critical engagement with questions related to environmental responsibility, social equity, accessibility, and user autonomy. Design decisions must consider not only technical and esthetic outcomes but also their broader societal implications. This includes addressing disparities in access to resources, ensuring inclusivity in design, and minimizing environmental harm. The integration of digital technologies further amplifies these concerns, particularly in relation to data privacy and the potential for technological control over user environments. Responsible innovation, therefore, becomes a guiding principle, ensuring that technological advancement serves the collective good rather than reinforcing inequalities or unintended consequences.

Innovation itself is deeply linked to a culture of experimentation and iterative development. Preparing for the future requires the creation of environments that support research, prototyping, and testing of new ideas. Experimental projects and pilot implementations provide valuable opportunities to explore emerging technologies, evaluate their performance, and refine their application before broader adoption. Collaboration with academic institutions, research organizations, and technology developers further enriches this process, bridging the gap between theoretical exploration and practical implementation. This culture of experimentation encourages calculated risk-taking and fosters the development of innovative solutions that can address complex and evolving challenges.

A holistic approach to the building lifecycle is another critical aspect of future readiness. Traditionally, design, construction, and operation have been treated as separate phases, leading to inefficiencies and performance gaps. Future-oriented practice emphasizes continuity across these stages, supported by digital models that extend from design into construction and operational management. The involvement of facility managers and operational stakeholders during the design phase ensures that performance objectives are aligned with long-term use and maintenance strategies. This integrated lifecycle approach enhances efficiency, reduces costs, and ensures that buildings continue to perform as intended over time. It also supports the development of adaptive maintenance strategies, where interventions are informed by real-time data and predictive analysis (Radman et al., 2025).

The future of architecture and engineering is also shaped by broader urban and societal dynamics. Rapid urbanization, demographic changes, and evolving patterns of work and living require new approaches to design that address density, diversity, and connectivity. Buildings must be integrated into larger urban systems, contributing to the development of smart cities characterized by interconnected infrastructure, efficient resource management, and enhanced quality of life. This requires a shift in perspective, where individual projects are understood as components of a larger network, and design decisions are evaluated in relation to their impact on the urban environment. Issues such as mobility, energy distribution, and social interaction become integral to architectural and engineering considerations, reinforcing the need for interdisciplinary and multiscalar thinking.

Collectively, these dimensions point toward a redefinition of professional practice in architecture and engineering. Preparing for the future involves embracing adaptability over rigidity, collaboration over isolation, and continuous learning over static expertise. It requires a commitment to data-informed decision-making, ethical responsibility, and systemic integration across all phases of the building lifecycle (Gilbert et al., 2025). These principles enable professionals to navigate complexity and uncertainty, ensuring that the built environment remains responsive, resilient, and relevant.

The future of architectural design and structural engineering will be shaped not only by technological advancements but also by the ability of practitioners to engage with change in a thoughtful and strategic manner. Preparing for this future requires moving beyond reactive problem-solving toward proactive foresight, where design anticipates emerging conditions and opportunities. By integrating interdisciplinary knowledge, embracing innovation, and maintaining a strong ethical foundation, architects and engineers can position themselves at the forefront of this transformation. In doing so, they contribute to the creation of built environments that are not only technologically advanced but also sustainable, resilient, and deeply attuned to human and ecological needs. In this evolving paradigm, buildings are no longer static objects but dynamic systems embedded within complex networks, capable of learning, adapting, and contributing to a continuously changing world.

References

Abdullah Agha, D., 2024. 3D paper printing for the built environment: optimization of the material behavior & production process to reach quality integration and dimensional accuracy. Mechanik, Werkstoffe und Konstruktion im Bauwesen 73. Available at https://doi.org/10.1007/978-3-658-45076-2.

Batista, L.T., Franco, J.R.Q., Fakury, R.H., Franco Porto, M., Alves, L.V.R., Kohlmann, G.S., 2023. BIM-IoT-FM integration: strategy for implementation of sustainable water management in buildings. Smart and Sustainable Built Environment 13 (5), 1096–1116. Available at https://doi.org/10.1108/sasbe-11-2022-0250.

Geno, J., Goosse, J., Van Nimwegen, S., Latteur, P., 2022. Parametric design and robotic fabrication of whole timber reciprocal structures. Automation in Construction 138, 104198. Available at https://doi.org/10.1016/j.autcon.2022.104198.

Gilbert, M., Barati, K., Shen, X., 2025. Lifecycle cost analysis of residential buildings considering thermal performance and financial uncertainties. Energy and Buildings 348, 116442. Available at https://doi.org/10.1016/j.enbuild.2025.116442.

Hurlimann, A., Moosavi, S., Warren-Myers, G., March, A., Bush, J., Cutroni, L., 2025. Facilitating climate change action across built environment life stages – perspectives from built environment professionals. Cities (London, England) 165, 106113. Available at https://doi.org/10.1016/j.cities.2025.106113.

Kathiravel, R., Hou, D., Feng, H., 2026. Optimizing structural and environmental performance in mass timber hybrid high-rise buildings through parametric design. Journal of Building Engineering 118, 115005. Available at https://doi.org/10.1016/j.jobe.2025.115005.

Kolářová, M., Kloužková, A., Lauermannová, A.M., Záleská, M., Pivák, A., Pavlíková, M., et al., 2026. Incorporation of waste glass dust with pigments containing Cd, S, and Se into magnesium oxychloride cement for sustainable production using 3D printing. Journal of Building Engineering 122. Available at https://doi.org/10.1016/j.jobe.2026.115793.

Leder, S., Menges, A., 2023. Architectural design in collective robotic construction. Automation in Construction 156. Available at https://doi.org/10.1016/j.autcon.2023.105082.

Lee, S.-H., 2025. From computation to curation: expanding the boundaries of design practice. Design Studies 101, 101357. Available at https://doi.org/10.1016/j.destud.2025.101357.

Li, S.Y., Widén, J., Shadram, F., 2025. Towards realizing life cycle net-zero energy/emissions (LCNZE) buildings: a systematic literature review. Journal of Building Engineering 114. Available at https://doi.org/10.1016/j.jobe.2025.114387.

Liu, Q., Li, X., Tan, K.H., Ah, K.H., 2026. A review of AIoT-enabled cyber-physical systems in building energy management: towards intelligent operation. Applied Energy 409, 127482. Available at https://doi.org/10.1016/j.apenergy.2026.127482.

Liu, X., Liu, R., Zhang, X., Abate, A.G., Nielsen, P.S., Keles, D., 2026b. IoT-enabled interoperability for energy-mobility integration: a systematic review of protocols, platforms, and standards. Applied Energy 411. Available at https://doi.org/10.1016/j.apenergy.2026.127611.

Nabizadeh Rafsanjani, H., Nabizadeh, A.H., 2023. Towards human-centered artificial intelligence (AI) in architecture, engineering, and construction (AEC) industry. Computers in Human Behavior Reports 11, 100319. Available at https://doi.org/10.1016/j.chbr.2023.100319.

Radman, K., Jelodar, M.B., Lovreglio, R., Ghazizadeh, E., Wilkinson, S., 2025. Real-time tracking and analysis in construction projects: a realCONs framework. Advanced Engineering Informatics 67, 103511. Available at https://doi.org/10.1016/j.aei.2025.103511.

Rathnayake, B., Gunathilake, L., Edirisinghe, R., Perera, S., 2025. EcoConstruct: a blockchain-based system for carbon trading in construction projects. Construction Innovation 25 (7), 213–234. Available at https://doi.org/10.1108/ci-08-2024-0224.

Sheng, Y., Shen, L., Wang, W., Sun, Y., 2026. Optimal planning of urban-scale rooftop PV systems for maximizing renewable energy integration by reducing grid congestion-induced curtailments: a case study. Renewable Energy 264, 125559. Available at https://doi.org/10.1016/j.renene.2026.125559.

Sreenivasan, A., Suresh, M., 2024. Design thinking and artificial intelligence: a systematic literature review exploring synergies. International Journal of Innovation Studies 8 (3), 297–312. Available at https://doi.org/10.1016/j.ijis.2024.05.001.

Xie, C., Alwisy, A., 2026. Advancing robotic automation in wood-framed construction using vision-driven adaptive control. Automation in Construction 185, 106858. Available at https://doi.org/10.1016/j.autcon.2026.106858.

Yang, J., Ng, S.T., 2024. Prospects for digital twin technology in the building modular construction and operation phases: a game theory-based analysis. Journal of Cleaner Production 470, 143344. Available at https://doi.org/10.1016/j.jclepro.2024.143344.

Yeom, S., Kim, J., Kang, H., Jung, S., Hong, T., 2024. Digital twin (DT) and extended reality (XR) for building energy management. Energy and Buildings 323. Available at https://doi.org/10.1016/j.enbuild.2024.114746.

Zeng, L., Guo, S., Wu, J., Markert, B., 2024. Autonomous mobile construction robots in built environment: a comprehensive review. Developments in the Built Environment 19, 100484. Available at https://doi.org/10.1016/j.dibe.2024.100484.

Zhong, W., Schröder, T., Bekkering, J., 2024. Implementing biophilic design in architecture through three-dimensional green spaces: guidelines for building technologies, plant selection, and maintenance. Journal of Building Engineering 92. Available at https://doi.org/10.1016/j.jobe.2024.109648.

The role of sustainability in smart architecture

11

11.1 Sustainability in architectural design

Sustainability in architectural design has undergone a significant conceptual and operational transformation, evolving from a peripheral concern into a central framework that fundamentally shapes the built environment. In contemporary practice, sustainability is no longer interpreted as a set of technical add-ons aimed at reducing environmental impact; instead, it operates as an integrative design paradigm that informs decision-making across all scales—from material selection and building systems to spatial organization and urban integration. Within the context of smart architecture, this paradigm is further expanded through the incorporation of digital intelligence, enabling buildings to respond dynamically to environmental conditions, user behavior, and long-term performance requirements. As a result, sustainability emerges not as a constraint on design but as a generative force that drives innovation, form-making, and experiential quality (Wang et al., 2024).

A critical shift in sustainable architectural thinking lies in its early integration within the design process. Rather than being addressed during later stages as a form of technical compliance, sustainability is embedded at the conceptual phase, where key decisions regarding orientation, massing, and spatial configuration are established. Climate-responsive design strategies play a fundamental role in this process (Wu et al., 2026). By analyzing solar exposure, prevailing wind patterns, seasonal variations, and microclimatic conditions, architects can shape buildings that work in harmony with their environment. This approach enables the effective use of passive environmental forces—such as natural ventilation, solar shading, and thermal inertia—thereby reducing reliance on energy intensive mechanical systems. In this sense, architectural form becomes an environmental interface, mediating between external climatic conditions and internal comfort requirements.

Passive design strategies continue to form the foundation of sustainable architecture, providing low-energy solutions that are both robust and contextually responsive. Techniques such as daylight optimization, cross-ventilation, and the strategic use of thermal mass enhance environmental performance while maintaining simplicity and reliability. The integration of orientation, solar control, and natural ventilation strategies forms the foundation of climate-responsive design, where architectural form operates as an environmental mediator rather than a static enclosure (Fig. 11.1). However, contemporary sustainable design increasingly operates through hybrid systems that integrate passive strategies with advanced active technologies. High-efficiency heating, ventilation, and air conditioning systems, intelligent lighting controls, and responsive façade elements complement passive approaches, enabling buildings to adapt to fluctuating environmental conditions. This integration reflects a transition toward performance-driven design,

Smart Civil and Structural Engineering. DOI: https://doi.org/10.1016/B978-0-443-30223-7.00004-8

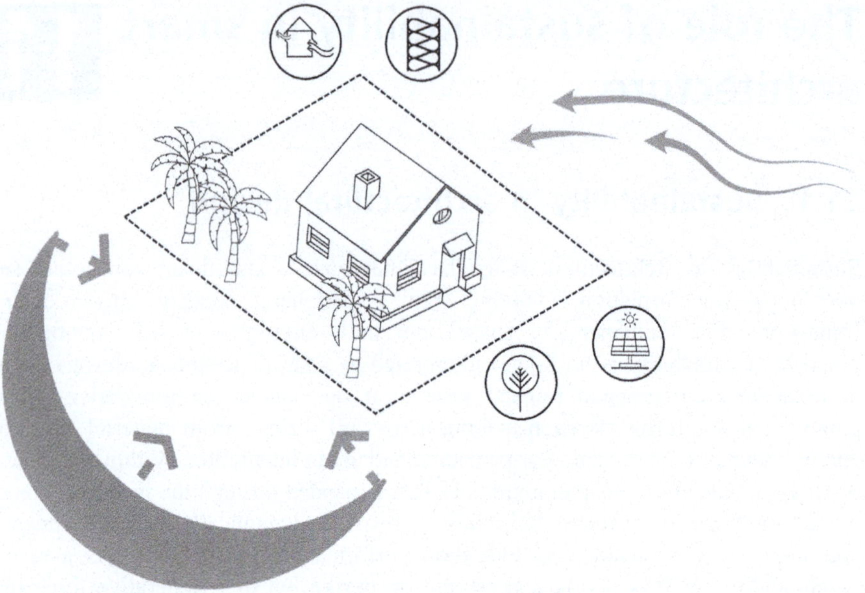

Figure 11.1 Climate-responsive design strategies.

where buildings are conceived as holistic environmental systems rather than isolated assemblies of components. The emphasis shifts from individual efficiency measures to the orchestration of multiple systems working in synergy.

Material selection represents another critical dimension of sustainability, as the environmental impact of buildings is closely linked to the extraction, production, and lifecycle performance of construction materials (Janpetch et al., 2026). Contemporary architectural practice increasingly prioritizes low-embodied-carbon materials, renewable resources, and recyclable components, aligning with broader goals of reducing carbon emissions and resource depletion. At the same time, the concept of material circularity is gaining prominence, encouraging the design of buildings that can be disassembled, reconfigured, and reused (Ranasinghe et al., 2024). Modular construction systems, reversible connections, and material passports support this approach, transforming buildings into repositories of valuable resources rather than sources of waste (Haigh and Li, 2026). This perspective not only enhances environmental performance but also influences architectural expression, as materials become visible indicators of sustainability and innovation.

Beyond environmental efficiency, sustainability in architectural design encompasses the enhancement of human well-being. The built environment plays a crucial role in shaping physical health, psychological comfort, and social interaction. Access to natural daylight, high indoor air quality, thermal comfort, acoustic performance, and visual connectivity to the external environment are essential components of a healthy indoor environment. Biophilic design strategies, which integrate natural elements such as vegetation, water, and natural materials, further strengthen the relationship between occupants and their surroundings (Di Giuseppe et al., 2024). These approaches contribute to improved cognitive function, reduced stress levels, and increased productivity, reinforcing the idea that sustainability is inherently linked to human-centered design. In this context, environmental performance and user experience are not separate objectives but mutually reinforcing dimensions of architectural quality.

Water and resource management introduce an additional layer of complexity and opportunity within sustainable design (Booth and Charlesworth, 2014). As global water scarcity becomes an increasingly critical issue, buildings are being designed to incorporate integrated water systems that reduce consumption and enhance reuse. Rainwater harvesting, greywater recycling, and efficient distribution systems enable buildings to operate with greater autonomy and resilience. At the same time, green infrastructure strategies—such as permeable surfaces, green roofs, and bioswales—support stormwater management and contribute to urban ecological balance. These systems reflect a shift toward closed-loop resource management, where water and materials are continuously cycled within the built environment, reducing dependency on external infrastructure and minimizing environmental impact.

Lifecycle thinking is central to the long-term effectiveness of sustainable architecture. Traditional approaches often focused on initial construction performance, neglecting the operational and end-of-life phases of buildings (Berto et al., 2026). In contrast, contemporary sustainable design adopts a lifecycle-oriented perspective that considers durability, maintenance, adaptability, and eventual deconstruction. Designing for longevity and flexibility allows buildings to accommodate changing functional requirements, technological advancements, and environmental conditions over time. This reduces the need for demolition and reconstruction, conserving resources and extending the useful life of built assets. Lifecycle-oriented strategies also emphasize the importance of maintenance accessibility and system upgradability, ensuring that buildings remain efficient and relevant throughout their existence.

The integration of smart technologies significantly enhances the capacity of architecture to achieve and maintain sustainable performance. Through the deployment of sensor networks, data analytics, and automated control systems, buildings can monitor environmental conditions, energy use, and occupancy patterns in real-time. This information enables dynamic adjustments to lighting, ventilation, heating, and cooling systems, optimizing performance continuously. Unlike traditional buildings, where performance is largely predetermined, smart buildings operate as adaptive systems that learn from data and refine their behavior over time. This introduces the concept of adaptive sustainability, where environmental performance is not fixed at the design stage but evolves through ongoing interaction between users, systems, and environmental conditions.

Sustainability in architectural design must also be understood within a broader urban and ecological context (Wu, 2014). Buildings are not isolated entities; they are components of larger systems that include transportation networks, energy infrastructures, water cycles, and ecological landscapes. Designing with an awareness of these interdependencies allows architecture to contribute to sustainable urban development and environmental resilience. For example, buildings can support decentralized energy production, reduce urban heat island effects, and enhance biodiversity through integrated landscape strategies. This systemic perspective positions architecture as a mediator between human activity and natural processes, reinforcing its role in shaping sustainable and resilient cities (Bleicher and Kumar, 2025).

Sustainability in architectural design represents a comprehensive and evolving paradigm that integrates environmental responsibility, technological innovation, and human well-being. It challenges architects to move beyond conventional design approaches and engage with complex, dynamic systems that operate across multiple scales and timeframes. By embedding sustainability into the core of design thinking, architecture can contribute to the creation of built environments that are not only efficient and resilient but also adaptive, regenerative, and deeply responsive to both human and ecological needs. In this expanded framework, sustainability is not merely a goal to be achieved but an ongoing process that defines the future trajectory of architectural practice.

11.2 Green building and sustainable practices

Green building and sustainable practices represent the applied implementation layer of environmental responsibility within the built environment, focusing on how ecological objectives are operationalized through coordinated systems, measurable benchmarks, and lifecycle management strategies. While sustainability in design establishes environmental intentions, green building practices ensure that these ambitions are translated into verifiable performance through construction methods, operational governance, technological systems, and long-term management protocols. This perspective shifts sustainability from conceptual aspiration to measurable execution, emphasizing that environmental responsibility must be continuously validated through operational outcomes rather than assumed through design intent alone (Yang et al., 2025).

A central advancement in this field is the emergence of performance governance frameworks, which define how environmental targets are established, tracked, and enforced. These frameworks rely on clearly defined metrics, such as operational carbon intensity, system efficiency ratios, and resource consumption benchmarks, which are embedded into project delivery contracts and operational protocols. Rather than being treated as optional goals, sustainability targets become binding performance criteria that guide decision-making across all stages, from procurement to facility management. This institutionalization of performance ensures accountability among stakeholders and aligns environmental objectives with project delivery mechanisms.

The commissioning process has also evolved into a critical phase for ensuring that sustainable systems function as intended. Beyond traditional system testing, advanced commissioning practices involve iterative calibration, performance validation under real operating conditions, and integration of control systems. Postoccupancy commissioning extends this process further, allowing systems to be fine-tuned based on actual usage patterns and environmental conditions. This approach addresses one of the most persistent challenges in building performance, the gap between design expectations and operational outcomes, by embedding verification into the early stages of occupancy.

A significant development within green building practice is the adoption of digital operational infrastructures, which enable continuous interaction between physical systems and data environments. These infrastructures integrate sensors, control systems, and analytics platforms into a unified network that supports real-time decision-making. Rather than operating independently, building subsystems, such as environmental controls, energy distribution, and occupancy management, are coordinated through centralized platforms that optimize performance dynamically. This integration transforms buildings into responsive operational environments, where adjustments are made continuously in response to changing internal and external conditions.

Within this framework, predictive analytics plays an increasingly important role. By analyzing historical and real-time data, building systems can anticipate future conditions and adjust proactively. For example, occupancy trends can inform spatial conditioning strategies, while weather forecasts can influence system preconditioning or load balancing. This predictive capability reduces inefficiencies, prevents system overloads, and enhances overall operational stability. It represents a shift from reactive management to anticipatory performance control, where systems are designed to respond before inefficiencies occur.

Another critical dimension is the integration of maintenance intelligence into building operations. Traditional maintenance strategies often rely on fixed schedules, which can lead to unnecessary interventions or delayed responses to system failures. In contrast, contemporary approaches utilize condition-based and predictive maintenance models, where system performance is continuously monitored and analyzed. Maintenance actions are triggered based on actual performance data, such as deviations in system efficiency or early indicators of component wear. This approach not only reduces operational costs but also extends the lifespan of building systems and minimizes downtime.

The role of user-centered operational design has become increasingly prominent in determining the effectiveness of sustainable practices (Abbakyari et al., 2023). Buildings are now designed to facilitate seamless interaction between occupants and environmental systems, ensuring that user behavior aligns with performance objectives. Interfaces are developed to be intuitive and informative, providing occupants with clear feedback on environmental conditions and system responses. In some cases, adaptive learning algorithms analyze user preferences and adjust system behavior accordingly, creating personalized environmental conditions without compromising overall efficiency (Almatared et al., 2026). This integration of human behavior into operational logic underscores the importance of aligning technological systems with real-world usage patterns (Schaumann, 2024).

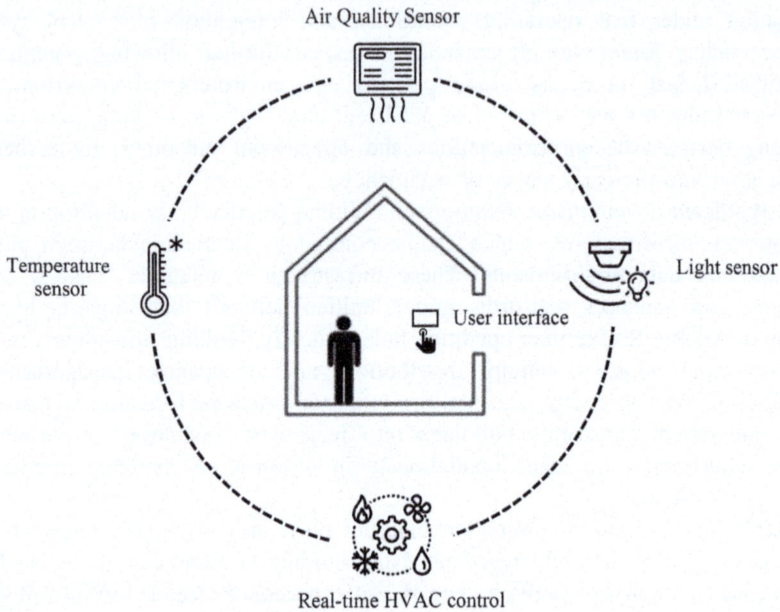

Figure 11.2 Indoor environmental quality and user interaction.

As shown in Fig. 11.2, indoor environmental quality in smart buildings is achieved through a continuous interaction between sensing technologies, user interfaces, and responsive building systems (Minassian et al., 2025). Environmental data collected from distributed sensors is processed in real-time, enabling automated adjustments in ventilation, thermal conditions, and lighting levels. At the same time, occupants can influence these parameters through intuitive control interfaces, ensuring that environmental performance aligns with individual comfort preferences. This feedback-driven model transforms indoor environments into adaptive systems, where human experience and technological intelligence are closely interconnected (Varol et al., 2026).

Supply chain management has also undergone a transformation, incorporating sustainability verification mechanisms that extend beyond the building site. Digital documentation systems enable stakeholders to track material sourcing, transportation, and processing, ensuring compliance with environmental and ethical standards. This level of transparency supports informed decision-making and reduces the risk of unintended environmental impacts. Furthermore, the integration of supply chain data into digital building models allows for more accurate lifecycle assessments, linking material choices directly to long-term performance outcomes (Gharaibeh et al., 2023).

The concept of resource optimization networks further expands the scope of green building practices. Buildings are increasingly designed to function as nodes within larger infrastructure systems, interacting with external energy grids, water networks, and waste management systems. Smart grid integration, for example, allows buildings to adjust energy consumption in response to grid demand, contributing to overall system stability. Similarly, buildings can participate in district-level resource sharing, where excess energy or recovered resources are redistributed within a network. This interconnected approach enhances efficiency at both the building and urban scales.

Economic performance is deeply intertwined with these operational strategies. The integration of lifecycle financial modeling enables stakeholders to evaluate the long-term economic implications of sustainable practices. By considering factors such as energy savings, maintenance costs, system durability, and potential upgrades, lifecycle models provide a comprehensive understanding of value over time. This approach shifts decision-making away from short-term cost minimization toward long-term performance optimization, reinforcing the economic viability of sustainable strategies.

Risk mitigation and operational resilience are also central to advanced green building practices. Buildings must be capable of maintaining functionality under conditions of uncertainty, including fluctuating resource availability, technological disruptions, or changing usage patterns. Strategies such as decentralized system configurations, adaptive control hierarchies, and redundancy planning ensure that buildings can continue to operate effectively even when individual components are compromised. This capacity for resilience enhances both environmental and operational stability, supporting long-term performance continuity.

A further layer of sophistication is introduced through continuous commissioning and performance auditing, where building systems are periodically evaluated and recalibrated to maintain optimal operation. This process is supported by data analytics platforms that identify trends, anomalies, and opportunities for improvement. By embedding evaluation into routine operation, buildings can adapt to evolving conditions and maintain alignment with performance targets over time.

Collaboration remains a critical enabler of these practices, but its nature has evolved toward integrated digital coordination. Stakeholders operate within shared digital environments that provide real-time access to design data, operational metrics, and performance indicators. This transparency facilitates informed decision-making and ensures that all participants are aligned with project objectives. The integration of operational stakeholders, such as facility managers, into early project phases further strengthens this coordination, ensuring that long-term performance considerations are embedded within design and construction decisions.

At a systemic level, green building practices contribute to the redefinition of buildings as active participants within larger ecological and infrastructural systems. Rather than functioning as isolated entities, buildings engage in continuous exchange with their surroundings, contributing to resource efficiency, environmental stability, and urban resilience. This perspective reflects a broader shift toward systems thinking, where the performance of individual buildings is understood in relation to wider networks.

At the institutional level, sustainability assessment systems such as Leadership in Energy and Environmental Design, Building Research Establishment Environmental

Assessment Method, WELL, Living Building Challenge, and net-zero certification frameworks remain essential mechanisms for establishing standardized environmental and operational benchmarks. However, their contemporary role has evolved substantially beyond conventional design-stage certification. Rather than functioning solely as symbolic indicators of environmental intent or one-time project achievements, these frameworks increasingly operate as dynamic accountability structures that require continuous performance validation. Advanced building management systems, integrated sensor networks, and real-time analytics platforms now enable operational data to be directly linked with certification criteria, allowing compliance to be monitored, audited, and recalibrated throughout the building's lifecycle.

This transformation fundamentally alters the purpose of certification. Environmental standards are no longer static checkpoints achieved upon project completion; instead, they become embedded within ongoing operational governance. Buildings are expected to demonstrate sustained performance across energy consumption, carbon emissions, indoor environmental quality, water management, and occupant wellness over extended periods. Through this shift, certification systems increasingly support adaptive management models where deviations from sustainability targets can be identified early, corrective measures can be implemented rapidly, and long-term environmental integrity can be preserved. This operational continuity enhances transparency for stakeholders while ensuring that sustainability remains a measurable reality rather than an aspirational designation.

Moreover, digital compliance infrastructures strengthen this process by enabling automated reporting, predictive benchmarking, and scenario-based performance forecasting. Facilities can continuously compare real-world operations against certification objectives, identify inefficiencies, and optimize systems proactively. This capability is particularly important in large-scale or high-performance developments where operational complexity can otherwise undermine sustainability goals. Certification thus becomes an integrated component of strategic building governance, aligning environmental stewardship with facility management, financial planning, and long-term asset resilience.

In this expanded framework, green building practices transcend traditional environmental mitigation strategies and emerge as sophisticated systems of operational intelligence. They integrate governance structures, digital ecosystems, resource optimization methodologies, and predictive management into a unified performance model that governs how buildings function over time. Sustainability is therefore reframed not as a final design outcome but as a continuously managed operational condition requiring verification, adaptation, and accountability.

Green building and sustainable practices establish a comprehensive architectural paradigm centered on measurable ecological performance, technological integration, and lifecycle resilience. Their significance extends beyond reducing environmental burdens to redefining how buildings are conceived, delivered, and managed as accountable systems. By embedding sustainability into governance protocols, maintenance intelligence, and adaptive operational frameworks, these practices transform the built environment into a responsive and verifiable ecosystem—one capable of evolving alongside environmental challenges, technological advancements, and human expectations. Through this progression, architecture becomes not only environmentally responsible but operationally intelligent, resilient, and continuously aligned with broader sustainability imperatives.

11.3 Smart structures as sustainable solutions

Smart structures redefine sustainability by shifting the focus from predefined environmental targets to continuous operational intelligence, where buildings actively regulate, evaluate, and recalibrate their performance over time. Rather than depending on static assumptions embedded during design or construction, these systems operate through an ongoing interaction between data, decision-making, and physical response. This transformation introduces a fundamentally different model of sustainability—one that is dynamic, iterative, and context-sensitive, allowing buildings to maintain relevance and efficiency under changing conditions.

A critical advancement lies in the development of multilayered sensing and interpretation frameworks. Smart structures do not merely collect data; they organize it into meaningful patterns that inform operational logic. Inputs related to spatial occupancy, temporal usage rhythms, internal system loads, and external environmental fluctuations are continuously synthesized. This layered understanding allows buildings to differentiate between transient conditions and persistent trends, enabling more precise and stable responses. As a result, operational strategies evolve from reactive adjustments to informed, context-aware regulation, significantly reducing inefficiencies associated with oversimplified control models.

Equally important is the emergence of distributed decision-making architectures within building systems. Instead of relying on a single centralized control point, smart structures often operate through interconnected subsystems that communicate and coordinate actions. This decentralized intelligence enhances resilience, as localized responses can occur independently while still contributing to overall system balance. For example, different zones within a building may adjust conditions based on localized needs while remaining aligned with global performance objectives. This structure allows for scalable adaptability, where the building can respond simultaneously at multiple levels without operational conflict.

The concept of temporal layering of performance further distinguishes smart structures from conventional systems. Buildings are no longer optimized for a single moment or condition; instead, they are calibrated across multiple time horizons. Short-term responsiveness ensures immediate comfort and efficiency, while medium-term adjustments account for recurring patterns such as daily or seasonal cycles. Long-term adaptation incorporates gradual changes in usage, occupancy behavior, or external conditions. This multitemporal framework allows buildings to operate with a high degree of stability while remaining flexible, effectively reducing performance degradation over time.

Another significant dimension is the integration of predictive and anticipatory capabilities. By analyzing historical data and identifying recurring patterns, smart structures can forecast future conditions and adjust system behavior in advance. This anticipatory mode reduces the need for reactive corrections, which are often less efficient and more resource-intensive. For instance, systems can precondition spaces based on expected occupancy or adjust operational strategies in anticipation of external fluctuations. This forward-looking approach enhances both efficiency and reliability, establishing a proactive model of sustainability where performance is guided by prediction rather than correction.

Smart structures also introduce a new level of operational granularity, where control is applied with high spatial and functional precision. Instead of treating buildings as uniform environments, systems differentiate between zones, activities, and user groups. This fine-grained control minimizes unnecessary resource allocation and allows for highly targeted responses. The ability to manage performance at this level of detail is particularly valuable in complex or large-scale buildings, where variability in use is significant. By aligning system behavior with specific conditions, smart structures achieve optimized resource distribution without compromising overall coherence.

The role of feedback loops is central to maintaining continuous performance refinement. Data collected during operation is not only used for immediate adjustments but also fed back into system logic to improve future responses. This iterative process enables buildings to "learn" from their own performance, gradually enhancing efficiency and stability. Over time, the building develops a form of operational memory, allowing it to respond more effectively to recurring conditions. This learning capacity distinguishes smart structures as evolving systems, capable of improving their sustainability performance beyond initial expectations. As illustrated in Fig. 11.3, smart structures operate through a continuous feedback loop in which sensing, analysis, and response are tightly integrated. This closed-cycle mechanism allows building systems to adjust dynamically based on real-time conditions while continuously evaluating the effectiveness of those adjustments. Through this iterative process, performance is not fixed but progressively refined, enabling a high level of precision and adaptability in building operation.

In addition to internal optimization, smart structures facilitate adaptive alignment between infrastructure and usage patterns. Rather than requiring users to conform to rigid system configurations, the building adjusts to accommodate varying behaviors and preferences. This alignment reduces friction between occupants and systems, leading to more effective and consistent performance outcomes. Importantly, it also

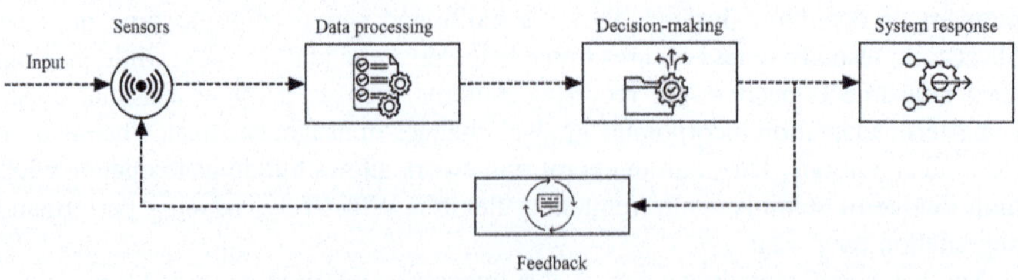

Figure 11.3 Smart structure feedback loop system.

supports a more inclusive and flexible built environment, where diverse patterns of use can be accommodated without compromising efficiency.

Another critical aspect is the integration of redundancy and adaptive resilience within system design. Smart structures are capable of maintaining operational continuity even when individual components experience disruption. Through real-time monitoring and system coordination, functions can be redistributed or reconfigured to maintain performance. This capacity for internal adjustment enhances reliability and reduces vulnerability to failure, contributing to long-term sustainability by ensuring consistent operation under uncertain conditions.

Smart structures also extend sustainability through their ability to support continuous recalibration of performance benchmarks. Instead of relying on fixed targets, performance criteria can evolve based on changing conditions, technological advancements, or updated operational priorities. This flexibility ensures that buildings remain aligned with current standards and expectations without requiring major interventions. It also allows for incremental improvement, where performance is gradually enhanced through ongoing optimization rather than periodic upgrades.

At the urban and infrastructural level, smart structures contribute to sustainability by enabling synchronized interaction across networks (Gu et al., 2025). Through digital connectivity, buildings can coordinate their operations with external systems, participating in broader strategies for resource distribution and demand management. This networked intelligence allows for more balanced and efficient use of resources at larger scales, supporting systemic resilience and reducing localized inefficiencies. In this context, individual buildings act as responsive nodes within a larger, interconnected environment. As shown in Fig. 11.4, smart buildings extend their functionality beyond individual boundaries by interacting with larger infrastructural

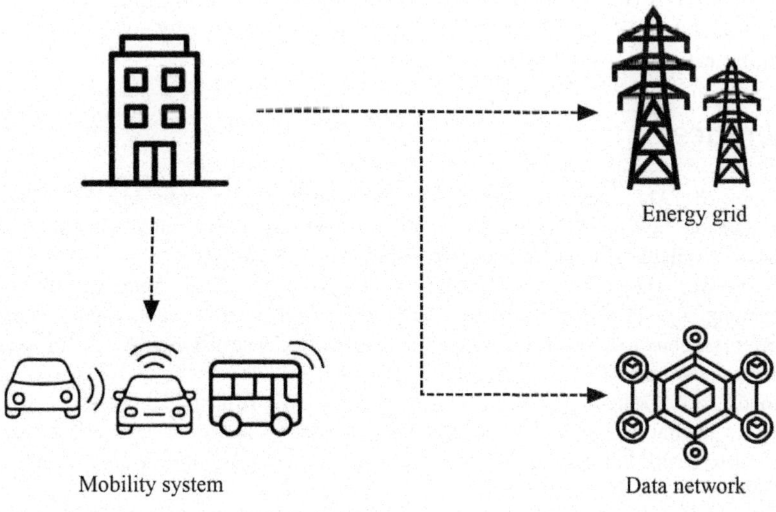

Figure 11.4 Smart building as an urban node.

networks. Through continuous exchange of energy, data, and mobility-related information, buildings contribute to coordinated urban performance, supporting more efficient resource distribution and system-wide adaptability. This networked integration reinforces the role of smart structures as active components within complex urban ecosystems (Lestari et al., 2026).

Despite these advantages, the effectiveness of smart structures depends on the careful management of system complexity and long-term operability. Excessive reliance on intricate or opaque control mechanisms can hinder usability and reduce reliability. Therefore, successful implementations prioritize clarity in system architecture, ensuring that complexity is managed internally while remaining accessible to operators and users. This balance between sophistication and usability is essential for maintaining long-term performance.

Furthermore, the integration of intelligent systems requires attention to the durability and adaptability of digital infrastructure (Argyroudis et al., 2022). As technologies evolve, systems must be capable of integration, upgrading, or replacement without disrupting overall functionality. Designing for technological continuity ensures that smart structures remain effective over extended periods, avoiding obsolescence and maintaining sustainability performance.

Smart structures establish a model in which sustainability is embedded within continuous operational behavior, rather than being defined solely by initial design decisions or static performance metrics (Sánchez-García et al., 2025). Their ability to sense, interpret, predict, and adapt allows buildings to function as responsive systems that evolve alongside their environments and users. Smart structures represent a mature and forward-looking approach to sustainability, characterized by adaptability, precision, and long-term responsiveness. By integrating real-time intelligence with coordinated system behavior and predictive capabilities, they enable buildings to maintain high levels of performance under changing conditions. This approach transforms the built environment into an active and evolving system, capable of sustaining efficiency, resilience, and relevance over time.

References

Abbakyari, M., Abuzeinab, A., Adefila, A., Whitehead, T., Oyinlola, M., 2023. Designing sustainable housing using a user-centred approach: paipe case study. Buildings 13 (10), 2496. Available at https://doi.org/10.3390/buildings13102496.

Almatared, M., Abuhussain, M., Andleeb, Z., Bashir, F.M., 2026. Adaptive reinforcement learning algorithm for real-time energy optimization in building digital twins with heterogeneous IoT sensor networks. Automation in Construction 182, 106714. Available at https://doi.org/10.1016/j.autcon.2025.106714.

Argyroudis, S.A., Mitoulis, S.A., Chatzi, E., Baker, J.W., Brilakis, I., Gkoumas, K., et al., 2022. Digital technologies can enhance climate resilience of critical infrastructure. Climate Risk Management 35, 100387. Available at https://doi.org/10.1016/j.crm.2021.100387.

Berto, R., Bedon, C., Mio, A., Mazelli, A., Rosato, P., 2026. Integrating life cycle thinking in structural design for historic building retrofit: a multi-criteria decision analysis approach.

Journal of Cultural Heritage 79, 199–209. Available from. Available at https://doi.org/10.1016/j.culher.2026.03.008, https://www.sciencedirect.com/science/journal/12962074.

Bleicher, J., Kumar, P., 2025. A novel framework integrating resilient and sustainable urbanism for designing climate resilient sustainable urban neighbourhoods. Sustainable Cities and Society 130, 106577. Available at https://doi.org/10.1016/j.scs.2025.106577.

Booth, C.A., Charlesworth, S.M., 2014. Water resources in the built environment: management issues and solutions. Water Resources in the Built Environment: Management Issues and Solutions 1–421. Available from. Available at https://doi.org/10.1002/9781118809167, http://www.wiley.com/remtitle.cgi?isbn=0470670916.

Di Giuseppe, E., Marcelli, L., Latini, A., D'Orazio, M., 2024. Evaluating biophilic design strategies in immersive virtual indoor environments: a systematic review on the implications for buildings occupants. Journal of Building Engineering 98 Available from. Available at https://doi.org/10.1016/j.jobe.2024.111127, https://www.sciencedirect.com/science/journal/23527102.

Gharaibeh, L., Matarneh, S., Eriksson, K., Lantz, B., 2023. Digital transformation of the wood construction supply chain through building information modelling: current state of practice. Construction Innovation 24 (7), 273–291. Available from. Available at https://doi.org/10.1108/CI-05-2023-0124, http://www.emeraldinsight.com/info/journals/ci/ci.jsp.

Gu, Q., Sing, M.C.P., Jefferies, M., Kanjanabootra, S., 2025. Bridging the gap between smart cities and sustainability: current practices and future trends. Cities (London, England) 159 Available from. Available at https://doi.org/10.1016/j.cities.2025.105799, https://www.sciencedirect.com/science/journal/02642751.

Haigh, R., Li, L., 2026. Decade review of modular construction systems in the building and construction industry: research trends toward the development and utilisation of timber materials within the context of a circular economy. Wood Material Science and Engineering Available from. Available at https://doi.org/10.1080/17480272.2025.2611132, http://www.tandf.co.uk/journals/titles/17480272.asp.

Janpetch, N., Trakolkul, C., Jantarachot, K., Thepjunthra, K., Thongchart, S., Plitsiri, I., et al., 2026. Life cycle carbon emissions and thermal performance of unfired bricks for sustainable construction materials. Case Studies in Construction Materials 24, e05873. Available at https://doi.org/10.1016/j.cscm.2026.e05873.

Lestari, F., Amelia, D., Karenina, A., Dorodjatoen, A.M.H., 2026. Urban transformation towards a smart city: the ecosystem of tangerang LIVE. Urban Governance Available from. Available at https://doi.org/10.1016/j.ugj.2026.02.001, https://www.sciencedirect.com/science/journal/26643286.

Minassian, R., Mihăiţă, A.-S., Shirazi, A., 2025. Optimizing indoor environmental prediction in smart buildings: a comparative analysis of deep learning models. Energy and Buildings 327, 115086. Available at https://doi.org/10.1016/j.enbuild.2024.115086.

Ranasinghe, N., Domingo, N., Kahandawa, R., 2024. Enhancing building material circularity: a systematic review on prerequisites, obstacles and the critical role of data traceability. Journal of Building Engineering 98, 111136. Available at https://doi.org/10.1016/j.jobe.2024.111136.

Sánchez-García, E., Martínez-Falcó, J., Marco-Lajara, B., Zakimova, A., 2025. Achieving environmentally responsible production through green managerial awareness, human resource management and employee behavior. Journal of Environmental Management 373, 123739. Available at https://doi.org/10.1016/j.jenvman.2024.123739.

Schaumann, D., 2024. Human behavior adaptability in responsive buildings: an exploratory study in workplace settings. Buildings 14 (6) Available from. Available at https://doi.org/10.3390/buildings14061830, www.mdpi.com/journal/buildings.

Varol, A., Hossein Motlagh, N., Leino, M., Tarkoma, S., Virkki, J., 2026. Creation of AI-driven smart spaces for enhanced indoor environments – a survey. Internet of Things 36, 101876. Available at https://doi.org/10.1016/j.iot.2026.101876.

Wang, X., Teigland, R., Hollberg, A., 2024. Identifying influential architectural design variables for early-stage building sustainability optimization. Building and Environment 252, 111295. Available at https://doi.org/10.1016/j.buildenv.2024.111295.

Wu, J., 2014. Urban ecology and sustainability: the state-of-the-science and future directions. Landscape and Urban Planning 125, 209–221. Available from. Available at https://doi.org/10.1016/j.landurbplan.2014.01.018, www.elsevier.com/inca/publications/store/5/0/3/3/4/7.

Wu, Z., Li, J., Zhang, Y., Zhang, Z., Xu, G., Zhao, H., 2026. Learning sustainable wisdom from the past: a generative framework for extracting climate-responsive design strategies from vernacular architecture. Energy and Buildings 350 Available from. Available at https://doi.org/10.1016/j.enbuild.2025.116669, https://www.sciencedirect.com/science/journal/03787788.

Yang, S., Yi, W., Zong, H., 2025. A systematic review of retrofitting buildings for sustainability from a lifecycle perspective. Sustainable Cities and Society 131, 106702. Available at https://doi.org/10.1016/j.scs.2025.106702.

Digital design integration: BIM, parametric design, and AI in smart architecture

12

12.1 Computational design and parametric modeling

The emergence of computational design and parametric modeling represents one of the most transformative developments in contemporary architecture and structural engineering (Caetano et al., 2020). Over the last several decades, the design process has evolved from a largely manual and representational practice into a data-driven, performance-oriented methodology in which geometry, structure, material behavior, environmental performance, and user requirements can be integrated within a unified digital environment. This transformation has fundamentally altered the relationship between architectural design and engineering analysis, enabling a level of collaboration, optimization, and innovation that was previously unattainable.

Within the broader context of smart architecture, computational design serves as a critical foundation for the development of intelligent, adaptive, and responsive structures. Rather than treating design as a linear sequence of decisions, computational methods establish dynamic relationships between parameters, allowing design solutions to evolve through continuous feedback and evaluation. Consequently, buildings can be conceived not merely as static objects but as complex systems capable of responding to environmental conditions, performance requirements, and changing user needs.

12.1.1 Evolution of digital design in architecture and engineering

The history of digital design in architecture and engineering reflects a gradual transition from manual drafting techniques to sophisticated computational environments. For centuries, architects and engineers relied on hand drawings, physical models, and empirical calculations to communicate design intent and assess structural feasibility. While these methods produced many remarkable buildings, they imposed limitations on complexity, precision, and interdisciplinary coordination.

The introduction of computer-aided design (CAD) systems during the latter part of the twentieth century marked the first major shift toward digital workflows. CAD technologies replaced manual drafting with digital representations, improving accuracy, efficiency, and documentation processes. Designers could generate, modify, and reproduce drawings more rapidly than ever before, significantly reducing the time required for design development and construction documentation.

Despite these advantages, early CAD systems largely replicated traditional drafting practices in digital form. The software functioned primarily as a drawing tool rather than a design intelligence platform. Geometries remained static, and modifications

Smart Civil and Structural Engineering. DOI: https://doi.org/10.1016/B978-0-443-30223-7.00001-2

often required extensive manual revisions. As architectural ambitions grew and projects became increasingly complex, the limitations of conventional CAD workflows became more apparent.

The next major evolution occurred with the emergence of computational and algorithmic design methodologies. Rather than directly drawing geometric elements, designers began defining relationships, rules, and parameters that could generate geometry automatically. This shift transformed the role of the architect and engineer from geometric authors to system designers capable of controlling design outcomes through computational logic.

Parametric modeling platforms such as Rhino and Grasshopper enabled designers to establish associative relationships between geometric elements, performance criteria, and structural constraints. Modifying a single parameter could trigger updates throughout the entire model, allowing rapid exploration of alternative solutions. This capability fundamentally changed the design process, encouraging experimentation, optimization, and iterative development (Goldbach and Lázaro, 2024).

Simultaneously, advances in simulation technologies introduced performance-based design methodologies. Environmental analysis, structural simulations, daylight studies, airflow modeling, and energy assessments could now be integrated directly into the design process (Shi and Yang, 2013). Instead of evaluating performance after a design was completed, architects and engineers could use simulation results to guide design decisions from the earliest conceptual stages. As a result, digital design evolved from a representational tool into a predictive and analytical environment capable of informing decision-making throughout the project lifecycle.

12.1.2 *Principles of computational design*

Computational design is founded upon the concept that architecture can be generated through relationships, rules, and algorithms rather than through isolated geometric manipulations. At its core, computational design seeks to establish logical systems capable of producing, modifying, and optimizing design solutions. One of the most important principles is rule-based design. In this approach, designers define a set of rules governing geometric relationships, spatial organization, environmental responses, or structural behavior. Rather than manually constructing every component, the system generates solutions according to predefined parameters (Grima et al., 2026).

Rule-based methodologies are particularly valuable for complex projects where numerous variables must be coordinated simultaneously. Structural requirements, environmental performance, circulation patterns, material constraints, and fabrication considerations can all be embedded within the computational framework, allowing the design to respond dynamically to changing conditions. Another fundamental concept is associative geometry. Traditional geometric models often consist of independent elements with limited relationships. In contrast, parametric models establish dependencies between components (Eltaweel and SU, 2017). Changes made to one element automatically propagate throughout the system, ensuring consistency and coordination.

Associative geometry supports a highly iterative design process. Designers can quickly evaluate multiple alternatives by adjusting parameters rather than rebuilding geometry from scratch. This flexibility is especially valuable in smart architectural projects where performance criteria frequently evolve during design development.

Computational design also relies heavily on design logic and algorithms. Algorithms provide the procedural framework through which geometry is generated and modified. These logical sequences may be simple or highly sophisticated, incorporating mathematical relationships, optimization routines, environmental simulations, and artificial intelligence (AI) techniques. The integration of algorithms enables designers to explore forms and configurations that would be difficult or impossible to develop using traditional methods. Consequently, computational design expands the creative possibilities available to architects and engineers while simultaneously enhancing analytical rigor. As illustrated in Fig. 12.1, the process incorporates a decision loop that evaluates whether the proposed solution satisfies the desired performance objectives. If the criteria are not met, the input parameters are adjusted, and the design is regenerated until an optimized outcome is achieved.

12.1.3 Rhino and Grasshopper in architectural practice

Among the various computational design platforms currently available, Rhino and Grasshopper have emerged as some of the most widely adopted tools in architecture and engineering. Their popularity stems from their flexibility, accessibility, and capacity to integrate geometric modeling with performance analysis and optimization (Chew et al., 2024).

Rhino provides a robust three-dimensional modeling environment capable of handling complex geometries with high precision. Unlike many conventional CAD platforms,

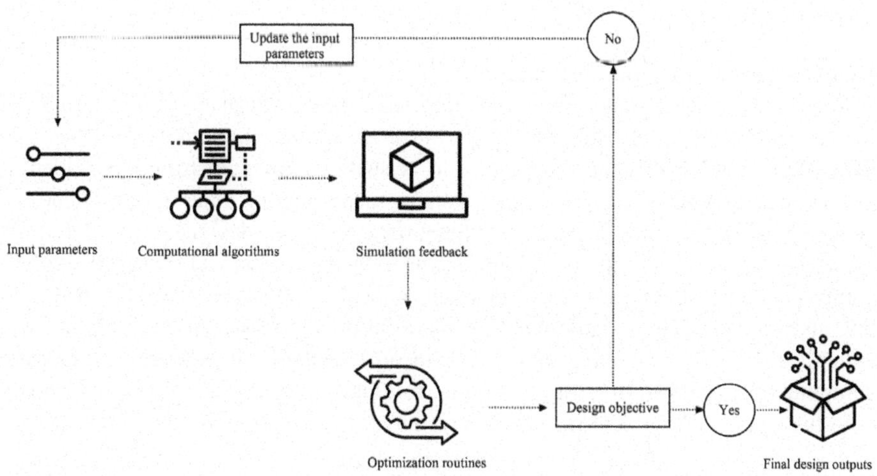

Figure 12.1 Parametric design workflow illustrating the iterative relationship between design objectives, input parameters, computational algorithms, simulation feedback, and optimization routines.

Rhino excels at representing free-form surfaces, double-curved geometries, and intricate spatial configurations. Grasshopper extends Rhino's capabilities through a visual programming interface. Instead of writing traditional code, users construct algorithms by connecting graphical components representing mathematical operations, geometric functions, and data relationships. This visual approach makes computational design accessible to architects and engineers without extensive programming backgrounds.

The workflow typically begins with the definition of key design parameters. These may include dimensions, environmental factors, structural constraints, or user requirements. Grasshopper then establishes relationships between these parameters, generating geometry based on predefined logic. One of the most powerful aspects of Grasshopper is its capacity for real-time modification. As parameters are adjusted, the model updates instantly, allowing designers to observe the consequences of design decisions immediately. This rapid feedback loop supports experimentation, optimization, and interdisciplinary collaboration. Moreover, Grasshopper can connect to a wide range of analysis tools, including structural simulation software, environmental modeling platforms, daylight analysis programs, and fabrication systems. These integrations transform the design environment into a comprehensive platform for performance-driven decision-making.

12.1.4 Parametric design and structural optimization

One of the most significant contributions of parametric design is its ability to support structural optimization (Yu et al., 2022). Traditional design processes often separate architectural form generation from structural evaluation. Parametric methodologies, however, allow these activities to occur simultaneously within a shared digital framework. Form-finding techniques represent a particularly important application. Rather than imposing arbitrary geometries, designers can generate forms that naturally respond to structural forces. Digital simulations allow geometries to evolve according to load paths, material properties, and boundary conditions, producing highly efficient structural configurations.

Load path optimization further enhances structural performance by ensuring that forces travel through the structure in the most direct and efficient manner possible. Parametric models enable engineers to visualize force distributions, identify inefficiencies, and refine structural systems accordingly. Material efficiency is another critical benefit. Through computational analysis, designers can determine precisely where material is required and where it can be reduced without compromising performance. This approach not only improves structural efficiency but also contributes to sustainability by minimizing resource consumption and embodied carbon. As environmental concerns become increasingly important, computational optimization plays a crucial role in developing structures that achieve both high-performance and responsible resource use.

12.1.5 Parametric design for smart structures

The integration of parametric design with smart technologies has given rise to a new generation of adaptive and responsive architectural systems. These systems use computational intelligence to modify their behavior in response to environmental conditions, occupancy patterns, and performance requirements. Responsive façades represent one of the most prominent applications. Parametric models can control shading devices, façade panels, and glazing systems that adjust according to solar exposure, daylight levels, and thermal conditions. These adaptive systems improve occupant comfort while reducing energy consumption.

Adaptive building envelopes extend this concept further by incorporating dynamic components capable of changing configuration, transparency, or thermal performance (Vedrtnam et al., 2025). Such systems allow buildings to respond continuously to environmental fluctuations throughout the day and across seasons.

Kinetic systems provide perhaps the most visible expression of smart architecture. Through the integration of sensors, actuators, and computational controls, kinetic structures can physically move and transform in response to changing conditions. Parametric design serves as the underlying framework governing these movements, ensuring that geometric transformations remain coordinated, efficient, and structurally feasible.

12.1.6 Case studies

12.1.6.1 The Helix Bridge, Singapore

The Helix Bridge in Singapore represents one of the most remarkable examples of computational design and parametric modeling applied to contemporary infrastructure. Officially opened in 2010, the bridge was designed through a collaboration between Cox Architecture, Architects 61, and Arup Engineers. Spanning approximately 280 meters across Marina Bay, the pedestrian bridge serves not only as a transportation link between key urban destinations but also as an architectural landmark that demonstrates the potential of computational methodologies in creating structurally efficient and visually distinctive forms.

The conceptual inspiration for the bridge originates from the molecular structure of deoxyribonucleic acid (DNA). The designers sought to transform the double-helix configuration of DNA into a large-scale architectural structure capable of supporting pedestrian movement while creating a memorable urban experience. Translating such a complex biological geometry into a buildable structure presented considerable challenges that would have been difficult to resolve using traditional drafting and engineering methods. Computational design tools enabled architects and engineers to generate, analyze, and refine the geometry while maintaining precise control over structural performance and fabrication requirements.

Parametric modeling formed the foundation of the design process. Rather than manually defining each structural component, designers established a series of geometric relationships and algorithms governing the curvature, spacing, and orientation of the steel elements. This approach allowed the entire bridge geometry to be controlled through adjustable parameters. Modifications to one aspect of the design automatically propagated throughout the digital model, ensuring consistency and reducing the likelihood of coordination errors. Such flexibility proved particularly valuable during the iterative design process, where architectural intentions and structural requirements needed to be continuously balanced.

The structural system consists of two intertwined steel helices connected by a network of transverse and diagonal members. Together, these elements create a highly efficient load-bearing system capable of supporting pedestrian traffic while resisting environmental forces such as wind and thermal expansion. Computational structural analysis enabled engineers to simulate multiple loading scenarios and optimize the arrangement of members to achieve maximum efficiency with minimum material consumption. Through digital simulation, the design team evaluated stress distribution, deformation behavior, and dynamic performance, ensuring that the bridge satisfied safety and serviceability requirements.

An important aspect of the project was the integration of architectural expression and structural functionality. The double-helix geometry is not merely decorative; it contributes directly to the bridge's structural behavior. Computational tools facilitated the development of a form in which aesthetic considerations and engineering performance became inseparable. This integration reflects one of the key principles of contemporary computational design, namely the ability to generate forms whose geometry emerges from multiple performance criteria rather than purely visual preferences.

The bridge also incorporates environmental and experiential considerations. Viewing platforms positioned along the structure provide panoramic perspectives of Marina Bay and the surrounding skyline. At night, an advanced lighting system illuminates the steel framework, emphasizing the intricate geometry and enhancing the bridge's visual identity within the urban landscape. Computational visualization tools played an important role in assessing lighting effects and user experience during the design phase.

From a construction perspective, the Helix Bridge demonstrates the advantages of digital fabrication and advanced manufacturing techniques. The structure comprises thousands of steel components, many of which possess unique geometries. Parametric models were linked directly to fabrication processes, enabling precise production through computer-controlled manufacturing systems. This digital workflow minimized errors, improved construction efficiency, and facilitated the assembly of the complex structural network.

The Helix Bridge illustrates how computational design can transform a conceptual idea inspired by nature into a functional, efficient, and iconic piece of infrastructure. It demonstrates the capacity of parametric methodologies to manage geometric complexity while supporting interdisciplinary collaboration among architects, engineers, fabricators, and contractors.

Fig. 12.2 presents the Helix Bridge through three-dimensional exterior views and a sectional representation. These illustrations highlight the distinctive double-helix configuration, the structural arrangement of steel members, and the spatial relationship between the pedestrian pathway and the surrounding structural framework. The sectional view further demonstrates how the intertwined structural elements create both stability and architectural identity.

12.1.6.2 The Metropol Parasol, Seville, Spain

The Metropol Parasol in Seville, Spain, stands as one of the most influential examples of computational design applied to large-scale timber architecture. Designed by

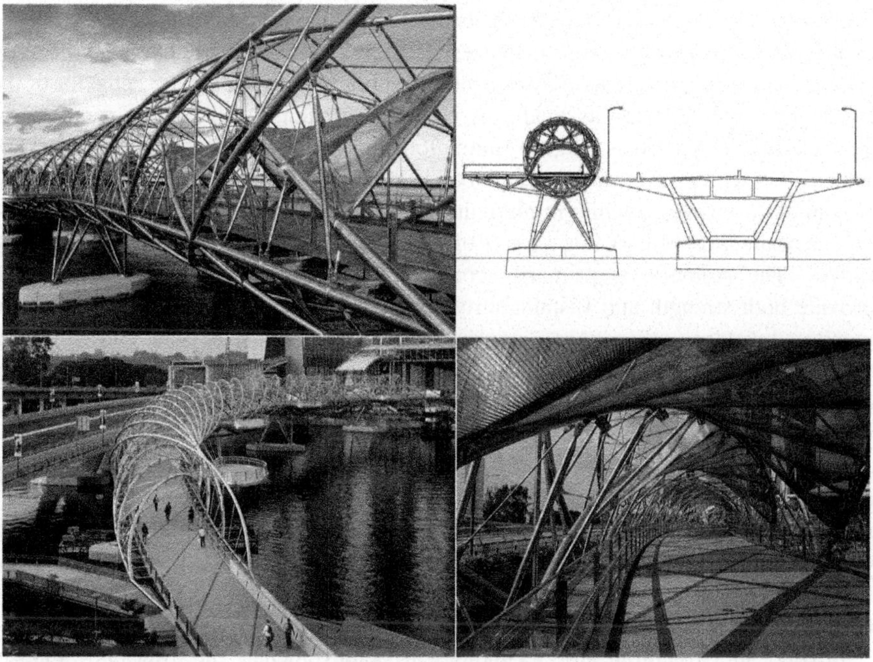

Figure 12.2 The Helix Bridge, Singapore: exterior views and section illustrating the DNA-inspired double-helix structural system developed through parametric modeling and computational design. *DNA*, deoxyribonucleic acid.

Jürgen Mayer H. and completed in 2011, the project occupies a prominent urban site in the historic center of Seville. The intervention combines multiple functions, including public plazas, elevated walkways, archeological exhibition spaces, restaurants, and observation platforms, creating a vibrant civic destination that reconnects different layers of the city.

The architectural concept is characterized by a series of interconnected parasol-like forms that create a vast canopy over the public square. The resulting structure possesses a highly complex geometry that would have been extremely difficult to design, analyze, and construct using conventional methods. Computational design tools enabled the development of an organic and continuous form while maintaining structural feasibility and construction efficiency.

At the heart of the project lies an extensive use of parametric modeling. The canopy consists of a lattice-like network of timber members arranged according to carefully defined geometric rules. Designers employed algorithmic techniques to generate and control the relationships among structural elements, allowing the overall geometry to evolve in response to architectural, structural, and functional requirements. Through parametric methods, the design team was able to explore numerous alternatives while maintaining coherence across the entire structure.

Structural optimization played a critical role in the realization of the project. The expansive canopy spans large distances while maintaining a relatively lightweight appearance. Computational analysis enabled engineers to evaluate the behavior of the timber lattice under various loading conditions, including self-weight, wind loads, and crowd-induced forces. Digital simulations informed decisions regarding member dimensions, connection details, and load distribution strategies, ensuring adequate structural performance while minimizing material consumption.

One of the most remarkable aspects of the Metropol Parasol is its use of engineered timber. The structure is primarily composed of laminated timber elements that provide both strength and sustainability benefits. Timber offers significant environmental advantages compared to conventional construction materials, particularly in terms of embodied carbon and renewable resource potential. Computational optimization contributed further to sustainability objectives by reducing unnecessary material use and improving structural efficiency.

The project also highlights the growing relationship between computational design and digital fabrication. The lattice system comprises thousands of components, many of which are geometrically unique. Advanced computational models generated detailed fabrication data for each element, enabling accurate production through computer-controlled manufacturing technologies. Computer numerical control (CNC) machining ensured precise cutting and shaping of timber components, facilitating efficient assembly on site.

Beyond its structural and technological achievements, the Metropol Parasol demonstrates the social and urban potential of computational design. The canopy provides extensive shading in Seville's hot climate, enhancing outdoor comfort and encouraging public activity throughout the year. Elevated walkways offer new perspectives of the historic city, while the spaces beneath the structure accommodate markets, cultural events, and everyday social interactions. Computational methodologies thus supported not only the creation of an innovative form but also the enhancement of urban functionality and public experience.

The project has become a widely recognized example of how digital technologies can expand the possibilities of timber construction. By integrating architectural creativity, structural engineering, environmental performance, and advanced fabrication, the Metropol Parasol demonstrates the transformative impact of computational design on contemporary architecture.

Fig. 12.3 illustrates the Metropol Parasol through plan drawings, sectional representations, bird's-eye perspectives, and interior views. Together, these images reveal the spatial organization of the project, the relationship between the canopy and the public spaces beneath it, and the complexity of the timber lattice structure. The bird's-eye view emphasizes the organic geometry of the parasols, while the interior views demonstrate the quality of light, shade, and spatial enclosure generated by the structural system.

12.1.6.3 Beijing National Aquatics Center (Water Cube), Beijing, China

The Beijing National Aquatics Center, widely known as the Water Cube, represents one of the most celebrated achievements in computational architecture and structural

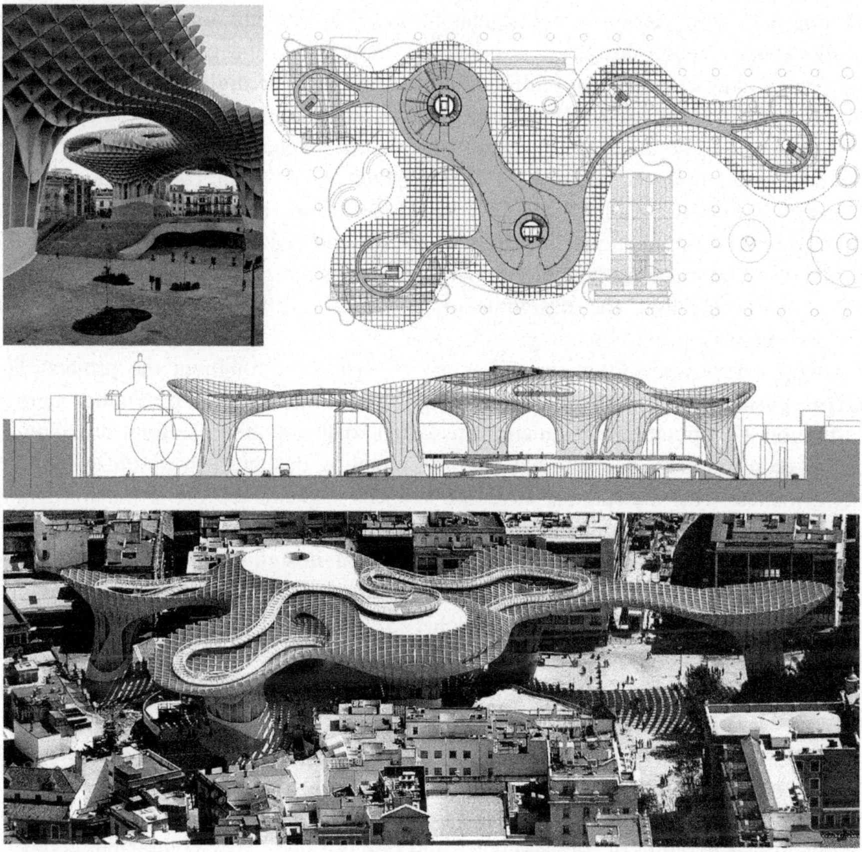

Figure 12.3 Metropol Parasol, Seville, Spain: plan, section, aerial view, and interior view showing the computationally generated timber lattice structure and multifunctional urban space.

engineering. Completed in 2008 for the Beijing Olympic Games, the project was designed through a collaboration involving PTW Architects (formerly Peddle Thorp & Walker), Arup, China State Construction Engineering Corporation (CSCEC), and China Construction Design International (CCDI). The building quickly became an international symbol of innovation, demonstrating how computational design can integrate mathematics, structural performance, environmental efficiency, and architectural expression.

The design concept emerged from the study of natural foam formations and cellular structures. Rather than relying on conventional geometric patterns, the architects and engineers employed the mathematical principles of the Weaire–Phelan structure, a highly efficient arrangement of irregular polyhedral cells. This geometry provided the basis for both the building's architectural identity and its structural organization.

Computational modeling was indispensable in translating these mathematical principles into a practical building system. The complex cellular geometry consists of thousands of interconnected elements that vary in size and orientation. Parametric design tools enabled designers to generate, manipulate, and optimize the structure while maintaining consistency throughout the project. The digital model served as a platform for coordinating architectural, structural, and environmental requirements.

The Water Cube utilizes a three-dimensional steel space-frame structure enclosed by ethylene tetrafluoroethylene (ETFE) cushions. The steel framework provides structural stability, while the ETFE envelope offers exceptional environmental performance. Computational structural analysis enabled engineers to evaluate the behavior of the space-frame under gravity, wind, seismic, and thermal loads. Through iterative optimization, the design team achieved a structure that combines high strength with material efficiency.

Environmental performance was a major objective throughout the project. The ETFE cushions permit significant daylight penetration while providing thermal insulation. Computational simulations assessed solar radiation, daylight distribution, energy consumption, and thermal comfort, enabling designers to optimize building performance. The transparent envelope reduces dependence on artificial lighting during daytime hours, contributing to lower operational energy demands.

The building demonstrates how computational design can support sustainable architecture. By integrating structural optimization and environmental analysis within a unified digital workflow, the project achieved substantial reductions in material use and energy consumption. The cellular geometry itself contributes to efficiency by distributing loads effectively throughout the structure while minimizing the amount of material required.

Another important aspect of the Water Cube is its iconic architectural expression. The façade creates the visual impression of a cluster of soap bubbles, transforming a scientific principle into a memorable architectural form. Computational tools enabled the precise generation of this geometry and ensured that every component could be fabricated and assembled accurately. The project therefore illustrates the ability of computational methodologies to bridge the gap between abstract mathematical concepts and tangible architectural experiences.

The Water Cube has had a lasting influence on the development of computational architecture. It demonstrated that advanced digital tools could be used not only to create visually striking buildings but also to improve structural efficiency, environmental performance, and construction feasibility. As a result, the project is frequently cited as a milestone in the evolution of performance-driven design.

Fig. 12.4 presents various exterior and interior views of the Water Cube. The exterior images emphasize the distinctive cellular façade and its resemblance to natural foam structures, while the interior views reveal the spatial qualities generated by the translucent ETFE envelope. Together, these illustrations demonstrate the successful integration of computational geometry, structural innovation, and environmental design that defines the project.

Collectively, the Helix Bridge, the Metropol Parasol, and the Beijing National Aquatics Center illustrate the transformative role of computational design and parametric modeling in contemporary architecture and engineering. Although differing in function, scale, and materiality, all three projects demonstrate how

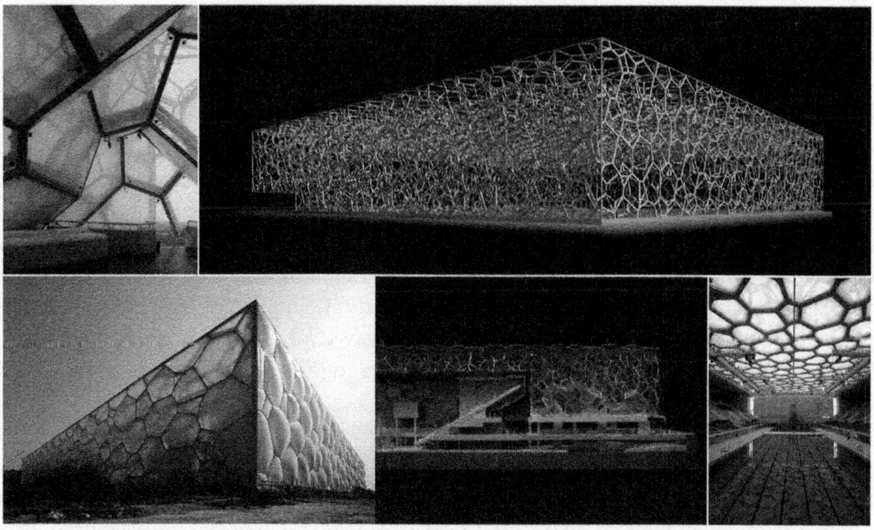

Figure 12.4 Beijing National Aquatics Center (Water Cube), China: exterior and interior views highlighting the computationally generated cellular façade, ETFE envelope, and performance-driven structural design. *ETFE*, ethylene tetrafluoroethylene.

digital methodologies enable the creation of structures that integrate architectural expression, structural performance, environmental responsiveness, and construction efficiency. These projects serve as influential precedents for the continued development of smart structures and adaptive architectural systems in the twenty-first century.

12.2 Building information modeling and integrated information management

The rapid digitalization of the architecture, engineering, and construction (AEC) industry has transformed the way buildings and infrastructure are conceived, designed, constructed, operated, and maintained. Among the most influential developments in this transformation is building information modeling (BIM), a methodology that extends far beyond three-dimensional digital representation and serves as a comprehensive platform for information generation, exchange, management, and collaboration throughout the building lifecycle. BIM has emerged as a fundamental component of smart architecture and engineering, enabling stakeholders to integrate design intelligence, performance data, construction information, operational requirements, and maintenance strategies into a unified digital environment. As buildings become increasingly complex and technologically sophisticated, BIM provides the framework necessary to coordinate multidisciplinary knowledge, support informed decision-making, and facilitate the creation of intelligent and sustainable built environments.

The origins of BIM can be traced to the evolution of CAD technologies during the latter decades of the 20th century. Traditional CAD systems revolutionized architectural and engineering documentation by replacing manual drafting with digital drawing processes. However, CAD primarily focused on geometric representation and lacked the capacity to manage relationships among building components, performance characteristics, and lifecycle information. The emergence of object-oriented modeling techniques during the 1990s introduced a new paradigm in which building elements could be represented as intelligent objects containing geometric, physical, functional, and performance attributes. This shift marked the beginning of BIM as a methodology that integrates graphical and nongraphical information into a coordinated digital model.

Today, BIM represents far more than a software application. It is a collaborative process that enables multiple stakeholders to contribute to and access a centralized information repository throughout the lifecycle of a project. In contrast to conventional drawings, which often result in fragmented and isolated information, BIM creates a shared digital environment in which architects, engineers, contractors, facility managers, and owners can work simultaneously while maintaining data consistency and accuracy. This capability has significantly improved project coordination, reduced errors and conflicts, enhanced communication, and increased efficiency across the AEC industry.

The development of BIM has also led to the concept of multidimensional modeling, commonly described through BIM dimensions ranging from 3D to 7D. Three-dimensional BIM represents the geometric model of a building, providing a visual and spatial representation of architectural, structural, and building systems. The fourth dimension (4D) incorporates time-related information, enabling project scheduling and construction sequencing simulations. By integrating construction schedules with the digital model, project teams can visualize construction processes, identify logistical challenges, and optimize project delivery strategies.

The fifth dimension (5D) extends BIM capabilities by incorporating cost information into the model. Through the integration of quantities, material specifications, and construction activities, stakeholders can perform cost estimation, budget analysis, and financial forecasting more accurately than traditional methods. Cost implications of design modifications can be evaluated in real-time, supporting more informed decision-making throughout the project lifecycle.

The sixth dimension (6D) focuses on sustainability and environmental performance. Energy consumption, daylight availability, carbon emissions, material efficiency, thermal performance, and environmental impacts can be analyzed directly within the BIM environment. Designers can evaluate alternative design strategies and optimize building performance before construction begins, contributing to more sustainable and resource-efficient projects.

The seventh dimension (7D) addresses facility management and operational performance. Information related to maintenance schedules, equipment specifications, asset management, operational procedures, and lifecycle costs can be integrated into the BIM model, providing facility managers with a comprehensive digital resource for long-term building operation. As a result, BIM evolves from a design and construction tool into a lifecycle information management platform that supports decision-making throughout the entire existence of a building or infrastructure asset.

The true value of BIM becomes evident through its role as a collaborative platform. Historically, the architecture, engineering, and construction industries have operated through fragmented workflows in which architects, engineers, contractors, and consultants worked independently and exchanged information through drawings, reports, and meetings. This fragmentation often resulted in coordination errors, design conflicts, information loss, and project delays. BIM fundamentally changes this process by establishing a shared digital environment where information is continuously updated and accessible to all project participants.

For architects, BIM provides a powerful platform for design development, visualization, and coordination. Architectural concepts can be developed in a parametric environment where modifications automatically update related building components and documentation. Design alternatives can be evaluated rapidly, enabling architects to explore multiple solutions while maintaining consistency across drawings, schedules, and specifications. Visualization capabilities further improve communication with clients and stakeholders by providing realistic representations of design proposals. BIM serves as a centralized platform for managing and exchanging information throughout the entire building lifecycle, facilitating coordination among design, engineering, construction, and operational stakeholders (Fig. 12.5).

Structural engineers benefit from BIM through enhanced coordination between architectural and structural systems. Structural models can be directly integrated with architectural models, allowing engineers to evaluate load paths, structural performance, and construction feasibility while minimizing conflicts between disciplines. The exchange of digital information improves accuracy and reduces the need for repetitive modeling efforts, thereby increasing productivity and design quality.

Mechanical, electrical, and plumbing engineers similarly benefit from coordinated BIM environments. Complex building services can be modeled and analyzed within the digital framework, enabling clash detection, system optimization, and performance evaluation before construction begins. Conflicts that would traditionally emerge during construction can be identified and resolved during design stages, reducing costs and delays.

Contractors and construction managers utilize BIM for construction planning, logistics management, scheduling, and resource allocation. Construction sequences can be simulated through 4D BIM models, allowing teams to visualize site operations and identify potential challenges before they occur. Material quantities can be extracted directly from the model, improving procurement accuracy and reducing waste. Safety planning, site logistics, and equipment deployment can also be optimized through BIM-based simulations.

The increasing integration of BIM with smart structures represents one of the most significant developments in contemporary architectural and engineering practice. Smart structures are characterized by their ability to sense environmental conditions, process information, and respond adaptively to changing circumstances. These

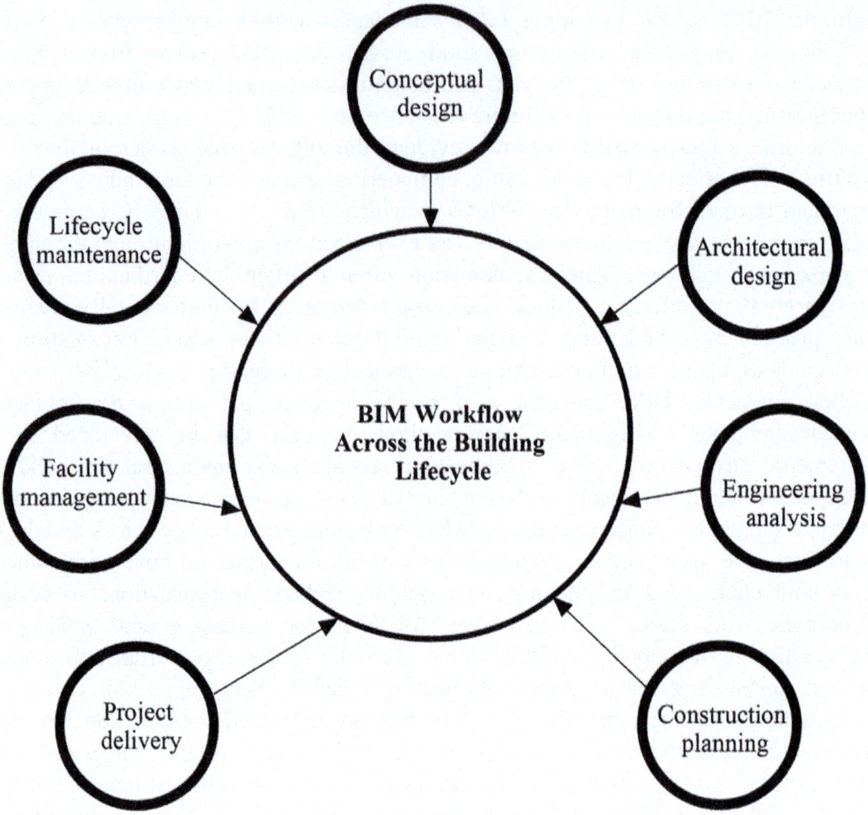

Figure 12.5 BIM workflow across the building lifecycle. *BIM*, building information modeling.

capabilities require extensive data management and information exchange, making BIM an ideal platform for supporting intelligent building systems (Hu et al., 2024).

Sensors play a central role in smart structures by collecting information related to structural performance, environmental conditions, occupancy patterns, energy consumption, and equipment operation. Traditionally, sensor data existed independently from design and construction information. However, BIM enables sensor information to be linked directly to digital building models, creating a comprehensive framework for monitoring and managing building performance.

For example, structural health monitoring systems can collect real-time data regarding deformation, vibration, stress, and temperature within structural components. When integrated with BIM, this information can be visualized within the digital model, allowing engineers and facility managers to identify potential issues and evaluate structural performance more effectively. Similar approaches can be applied to environmental monitoring systems, energy management platforms, and adaptive façade technologies.

Operational data integration further enhances the capabilities of BIM-enabled smart structures. Building systems continuously generate large volumes of information related to HVAC operation, lighting performance, indoor environmental quality, occupancy behavior, and resource consumption. By connecting these data streams to BIM environments, stakeholders can evaluate building performance in real-time and make informed decisions regarding operation and maintenance strategies.

Lifecycle management represents another critical advantage of BIM integration within smart structures. Buildings are increasingly viewed as long-term assets that require continuous monitoring, maintenance, and adaptation throughout their operational lives. BIM provides a digital repository containing detailed information about building components, systems, and performance histories, supporting predictive maintenance strategies and reducing operational costs. Rather than relying on reactive maintenance approaches, facility managers can use BIM-integrated data to anticipate failures, optimize maintenance schedules, and extend asset lifespans.

The relationship between BIM and sustainability has become increasingly important as the construction industry seeks to address environmental challenges and reduce carbon emissions (Waqar et al., 2023). Buildings account for a significant portion of global energy consumption and greenhouse gas emissions, making sustainable design and operation essential objectives for contemporary architecture and engineering.

One of the most significant contributions of BIM to sustainability lies in its support for energy simulation and performance analysis. Building geometry, material properties, occupancy schedules, and environmental conditions can be integrated into simulation models that evaluate energy consumption, thermal comfort, daylight performance, and renewable energy potential. These analyzes can be conducted during early design stages, allowing project teams to optimize building performance before construction begins.

Carbon assessment has also become a key application of BIM technology. The environmental impacts associated with material extraction, manufacturing, transportation, construction, operation, and end-of-life disposal can be evaluated through lifecycle assessment methodologies integrated within BIM environments. Designers can compare alternative materials and construction strategies based on embodied carbon and environmental performance metrics, supporting more sustainable decision-making processes (Su et al., 2020).

Facility management represents another area where BIM contributes to sustainability objectives. Operational energy consumption often exceeds construction-related impacts over a building's lifecycle. BIM-enabled facility management systems provide detailed information regarding building assets, maintenance schedules, equipment performance, and energy usage patterns. This information supports efficient building operation, reduces resource consumption, and extends asset longevity.

The evolution of BIM has contributed directly to the emergence of digital twins, which represent one of the most transformative concepts in contemporary building lifecycle management. A digital twin can be defined as a dynamic digital representation of a physical asset that continuously receives and processes real-time information from sensors and operational systems. While BIM provides a static or periodically updated information model, digital twins establish a live connection between physical and virtual environments.

Although BIM and digital twins share common foundations, important distinctions exist between them. BIM primarily serves as an information-rich model supporting design, construction, and asset management processes. Digital twins, by contrast, incorporate continuous data streams that reflect the current state of the physical asset. This capability enables real-time monitoring, predictive analytics, operational optimization, and autonomous decision support.

Real-time monitoring is one of the defining characteristics of digital twin technology. Sensor networks continuously collect information related to structural behavior, environmental conditions, equipment performance, energy consumption, and occupant activities. These data streams are transmitted to the digital twin environment, where advanced analytics and AI algorithms process the information and generate actionable insights.

The integration of BIM and digital twin technologies creates a powerful framework for lifecycle management. BIM provides the foundational information model, while digital twins extend this model through continuous operational feedback. Together, these technologies support proactive asset management, predictive maintenance, energy optimization, resilience planning, and long-term sustainability objectives. Understanding the distinctions between these two technologies is essential for appreciating their respective roles within smart architecture and infrastructure systems. A comparative overview of their characteristics, functions, and applications is presented in Table 12.1.

Several large-scale projects demonstrate the transformative potential of BIM and integrated information management. One of the most notable examples is the Crossrail project in London, one of Europe's largest infrastructure developments. The project utilized BIM extensively to coordinate information among numerous stakeholders involved in the design and construction of railway tunnels, stations, and associated infrastructure. The complexity of the project required sophisticated information management systems capable of integrating architectural, structural, mechanical, electrical, and operational data. BIM facilitated coordination across multiple disciplines, improved communication among project teams, and supported efficient construction planning. The digital information environment developed for Crossrail also established a foundation for future operational management and maintenance activities.

Singapore provides another significant example through its Smart Nation initiatives and digitally enabled urban development strategies. The city-state has embraced BIM as a key component of its digital transformation agenda, requiring BIM implementation for many large-scale projects and integrating building information with broader smart city infrastructure. BIM models are increasingly linked with sensor networks, urban data platforms, and digital twin technologies to support building management, energy optimization, transportation planning, and environmental monitoring. These initiatives demonstrate how BIM can function not only at the building scale but also as part of an integrated urban information ecosystem.

The expansion and modernization of Heathrow Airport terminals further illustrate the value of BIM in complex infrastructure environments. Airport facilities involve

Table 12.1 Comparison between BIM and digital twin technologies.

Aspect	BIM	Digital twin
Primary purpose	Design, construction, and information management	Real-time operation and performance optimization
Data nature	Static or periodically updated	Dynamic and continuously updated
Lifecycle focus	Design for facility management	Entire lifecycle with live operational feedback
Information source	Design models and project documentation	BIM data combined with real-time sensor inputs
Monitoring capability	Limited monitoring	Continuous real-time monitoring
Predictive analytics	Limited	Extensive predictive capabilities
Sensor integration	Optional	Essential component
Decision support	Design and construction decisions	Operational and strategic decisions
Maintenance strategy	Planned maintenance	Predictive maintenance
Relationship to physical asset	Digital representation	Living digital counterpart of the physical asset

BIM, building information modeling.

intricate interactions among architectural spaces, structural systems, mechanical services, security infrastructure, transportation networks, and operational requirements. BIM-enabled project teams to coordinate these diverse systems within a unified digital environment, reducing conflicts and improving project delivery efficiency. The resulting digital models continue to support facility management, maintenance planning, and operational decision-making long after construction completion.

As the AEC industry advances toward increasingly intelligent and interconnected built environments, BIM continues to evolve from a design and construction methodology into a comprehensive information management framework. The integration of BIM with sensors, smart systems, AI, digital twins, and advanced analytics is transforming how buildings and infrastructure are conceived, delivered, and operated. Rather than serving merely as a digital representation of physical assets, BIM has become a central platform for knowledge integration, collaboration, sustainability assessment, operational optimization, and lifecycle management.

The convergence of BIM and integrated information management represents one of the most important technological shifts in contemporary architecture and engineering. By enabling stakeholders to create, share, analyze, and utilize information throughout the lifecycle of buildings and infrastructure, BIM supports the development of smarter, more sustainable, and more resilient built environments. As digital technologies continue to mature, BIM will remain a foundational element in the realization of intelligent architecture and smart structural systems, bridging the gap between physical assets and their digital counterparts while facilitating informed decision-making across all stages of the building lifecycle.

12.3 Artificial intelligence and performance-driven design

The rapid advancement of digital technologies has significantly transformed the AEC industry, introducing innovative methods for conceiving, evaluating, and managing the built environment. Among these technologies, AI has emerged as one of the most influential developments, fundamentally changing how buildings are designed, constructed, and operated (Brozovsky et al., 2024). As contemporary architectural projects become increasingly complex and performance requirements continue to expand, conventional design methodologies often struggle to process the vast amount of information necessary to achieve optimal outcomes. AI addresses these challenges by providing computational systems capable of learning from data, recognizing patterns, predicting future conditions, and supporting decision-making processes throughout the building lifecycle.

Within smart architecture, AI represents more than a technological advancement; it signifies a paradigm shift toward evidence-based, data-driven, and performance-oriented design approaches (Manmatharasan et al., 2025). Rather than relying solely on intuition and experience, architects and engineers can leverage computational intelligence to evaluate multiple alternatives simultaneously, optimize design parameters, and predict building performance before construction begins. The integration of AI within architectural workflows relies on continuous feedback between data analysis, simulation, and optimization processes, enabling informed and performance-oriented decision-making (Fig. 12.6).

Historically, digital technologies in architecture were primarily associated with drafting, visualization, and documentation. Early CAD systems improved productivity by replacing manual drafting techniques, while subsequent developments in BIM enabled more sophisticated digital representations of buildings. However, AI extends beyond representation and documentation by introducing capabilities related to reasoning, learning, adaptation, and prediction. These capabilities enable computational systems to analyze extensive datasets derived from environmental measurements, simulation outputs, operational records, sensor networks, and user interactions, thereby supporting more intelligent design decisions (Albukhari, 2025).

AI can be broadly defined as the ability of computational systems to perform tasks that traditionally require human intelligence, including learning, reasoning, problem-solving, pattern recognition, and decision-making. In architectural applications, AI systems process information from multiple sources and transform raw data into actionable knowledge. This transformation enables designers to identify relationships between variables, anticipate future outcomes, and develop solutions that respond effectively to environmental, functional, and operational requirements.

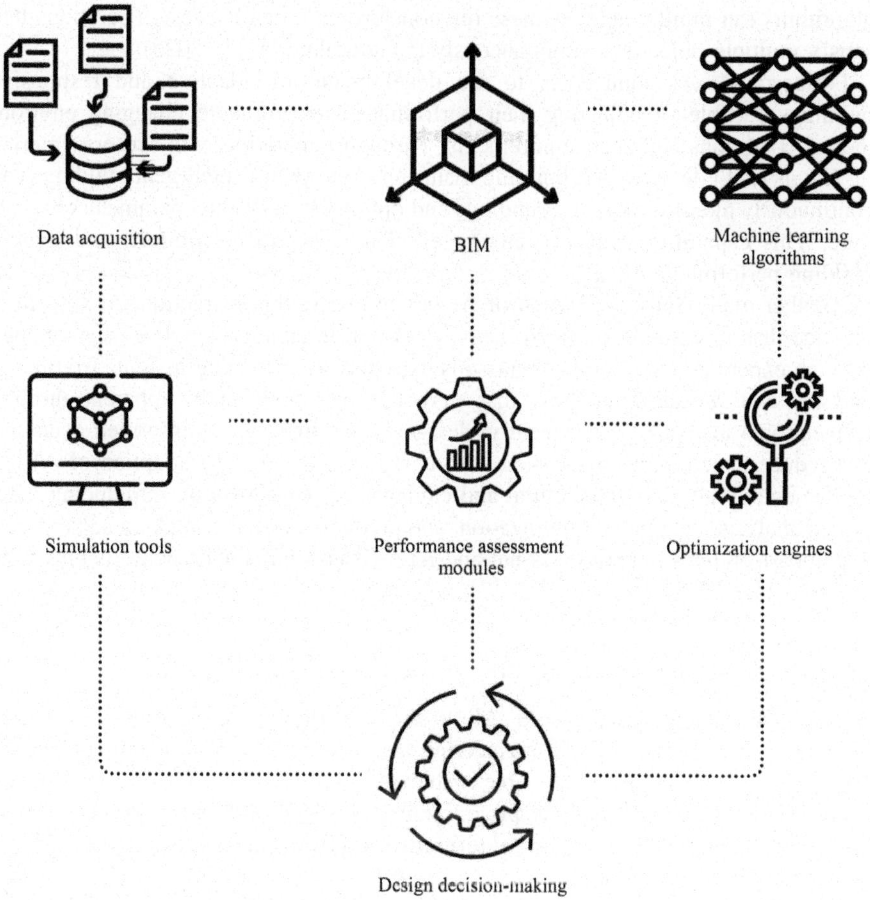

Figure 12.6 AI-assisted design workflow integrating data analytics and performance evaluation. *AI*, artificial intelligence.

12.3.1 Artificial intelligence in architecture

The application of AI in architecture encompasses a wide range of activities, from conceptual design and performance simulation to construction management and facility operation. Contemporary buildings generate enormous quantities of information, both during design and throughout their operational life. AI provides the analytical framework necessary to manage and utilize these data effectively.

One of the primary advantages of AI in architecture is its capacity to process information at a scale far beyond human capabilities. Modern building projects involve numerous interconnected variables, including geometry, structure, materials, climate, energy systems, occupant behavior, costs, and regulatory requirements. Evaluating the interactions among these variables manually is often impractical. AI algorithms can rapidly analyze these relationships and identify design solutions that satisfy multiple objectives simultaneously (Elmousalami et al., 2025).

Furthermore, AI contributes to the development of adaptive and responsive buildings capable of adjusting their performance according to changing environmental conditions and user requirements. Through integration with sensors, Internet of Things (IoT) devices, and building management systems, intelligent buildings can continuously monitor their surroundings and optimize operational parameters in real-time. This capability enhances energy efficiency, occupant comfort, and overall building performance.

Another important contribution of AI lies in knowledge extraction. Architectural practice has traditionally relied heavily on professional expertise accumulated through experience. AI complements this expertise by extracting insights from large datasets and revealing patterns that may not be immediately apparent through conventional analysis. Consequently, designers can make more informed decisions and reduce uncertainties associated with complex projects. The adoption of AI has fundamentally altered architectural and engineering workflows by introducing data-driven analysis, automated optimization, and predictive decision-making capabilities. A comparison between conventional and AI-assisted design approaches is presented in Table 12.2.

12.3.2 Machine learning and data-driven intelligence

Among the various branches of AI, machine learning has become particularly influential in architectural and engineering applications. Machine learning refers to computational techniques that enable systems to improve their performance through experience rather than explicit programming. Instead of relying exclusively on predefined rules, machine learning algorithms learn from data and continuously refine their predictive capabilities.

In architecture, machine learning is increasingly employed to analyze historical building data and predict future performance outcomes. Large datasets containing information about energy consumption, indoor environmental quality, maintenance records, occupancy patterns, and environmental conditions serve as training material for predictive models. Once trained, these models can estimate how new buildings are likely to perform under similar conditions.

Energy prediction represents one of the most common applications of machine learning in the built environment. Traditional energy simulations require extensive modeling and computational resources. Machine learning models can complement these simulations by rapidly estimating energy consumption based on design characteristics, climate conditions, and operational parameters. Such capabilities are particularly valuable during the early stages of design when detailed information may not yet be available (Ali et al., 2023).

Occupant behavior modeling constitutes another significant application (Mylonas et al., 2024). Human activities strongly influence building performance, yet occupant behavior remains difficult to predict using conventional methods. Machine learning algorithms can analyze patterns of occupancy, equipment usage, lighting preferences,

Table 12.2 Comparison between traditional design approaches and AI-assisted design methodologies.

Aspect	Traditional design approach	AI-assisted design methodology
Design process	Sequential and largely manual process based on designer experience and iterative revisions.	Data-driven and computationally enhanced process that supports rapid generation and evaluation of multiple design alternatives.
Decision-Making	Relies primarily on professional judgment, precedent studies, and limited performance analysis.	Integrates machine learning, predictive analytics, and simulation-based assessments to support evidence-based decisions.
Design exploration	Limited number of alternatives due to time and resource constraints.	Thousands of alternatives can be generated and assessed automatically through generative design algorithms.
Optimization capability	Optimization is often performed manually and focuses on a limited number of variables.	Multiobjective optimization simultaneously evaluates numerous performance criteria and identifies balanced solutions.
Data utilization	Uses relatively small datasets and designer interpretation.	Processes large datasets from BIM models, sensors, simulations, operational records, and environmental databases.
Performance evaluation	Performance assessments are typically conducted at later design stages.	Continuous performance evaluation occurs throughout the design process using integrated simulation and AI tools.
Energy analysis	Energy performance is often evaluated after key design decisions have been made.	AI enables early stage energy prediction, optimization, and continuous performance refinement.
Daylight analysis	Manual adjustments and iterative simulation studies are required to improve daylight performance.	Automated optimization identifies façade, glazing, and shading configurations that maximize daylight while minimizing glare.
Structural design	Structural systems are developed through engineering calculations and conventional optimization techniques.	AI-assisted structural optimization explores alternative geometries, material distributions, and load paths to improve efficiency.
Simulation integration	Simulation tools are often used independently and at discrete project stages.	Simulation tools are integrated within a continuous feedback loop that supports real-time design refinement.

(*Continued*)

Table 12.2 (Continued)

Aspect	Traditional design approach	AI-assisted design methodology
Predictive capabilities	Limited ability to forecast future performance and operational behavior.	Predictive models estimate energy use, occupant behavior, maintenance needs, and system performance.
Adaptability	Design modifications require substantial manual effort and redesign.	AI systems rapidly adapt solutions to changing project requirements and environmental conditions.
Construction planning	Scheduling and resource allocation depend largely on manual coordination.	AI optimizes schedules, logistics, risk assessment, and resource management through predictive analytics.
Building operation	Building management systems typically operate using predefined schedules and fixed control rules.	Intelligent systems continuously learn from occupancy patterns and environmental data to optimize operations.
Maintenance strategy	Maintenance is generally reactive or scheduled at fixed intervals.	Predictive maintenance identifies potential failures before they occur, reducing costs and downtime.
Occupant-centered design	Occupant preferences are often estimated through standards and assumptions.	Machine learning analyzes user behavior and comfort patterns to support personalized environments.
Collaboration	Coordination among disciplines may involve fragmented workflows and information exchange.	AI and BIM integration facilitate interdisciplinary collaboration through shared digital platforms.
Lifecycle perspective	Emphasis is frequently placed on design and construction phases.	Considers the entire building lifecycle, including operation, maintenance, adaptation, and end-of-life performance.
Innovation potential	Innovation depends largely on the designer's expertise and available project time.	Computational exploration uncovers novel forms, systems, and performance solutions beyond conventional approaches.
Overall outcome	Functional designs achieved through iterative refinement and professional experience.	High-performance, adaptive, and optimized buildings informed by data analytics, simulation, and intelligent decision support.

AI, artificial intelligence; *BIM*, building information modeling.

and thermal comfort responses to develop more accurate behavioral models. These insights improve the reliability of performance predictions and support the development of user-centered design strategies.

Machine learning also contributes to predictive maintenance. Building systems inevitably experience wear, degradation, and performance deterioration over time. By analyzing sensor data and operational records, machine learning algorithms can identify early indicators of potential failures and recommend maintenance actions before problems become critical. This proactive approach reduces operational costs, minimizes downtime, and extends the lifespan of building components.

12.3.3 Predictive systems and intelligent forecasting

Predictive systems represent one of the most valuable applications of AI in smart architecture. These systems utilize historical information, real-time sensor data, and advanced analytical models to forecast future conditions and behaviors. Within building operations, predictive algorithms can estimate future energy demands based on weather forecasts, occupancy schedules, equipment performance, and environmental conditions (Alanne and Sierla, 2022). Facility managers can use these forecasts to optimize energy procurement, schedule maintenance activities, and improve operational efficiency. Similarly, predictive models can anticipate peak loads and support demand-response strategies that reduce energy costs and improve grid stability.

Structural health monitoring provides another important example. Modern structures increasingly incorporate sensors that continuously measure strain, vibration, displacement, and environmental exposure. AI-based predictive systems analyze these measurements to detect anomalies, assess structural condition, and forecast maintenance requirements. By identifying deterioration trends at an early stage, engineers can implement preventative measures that enhance safety and reduce lifecycle costs.

The importance of predictive systems extends beyond efficiency and maintenance. Climate change, urbanization, and resource scarcity are introducing unprecedented uncertainties into the built environment. Predictive analytics enable architects and engineers to anticipate future challenges and develop resilient design solutions capable of adapting to evolving conditions.

12.3.4 Generative design and computational creativity

Perhaps the most transformative application of AI in architecture is generative design (Li et al., 2025). Unlike conventional design methodologies, where architects manually develop and evaluate alternatives, generative design employs computational algorithms to automatically generate large numbers of potential solutions based on predefined objectives and constraints. In a generative design workflow, designers establish performance goals, spatial requirements, environmental constraints, and evaluation criteria. The computational system then explores thousands of possible configurations, assessing each alternative according to the specified objectives. Generative design systems explore numerous design alternatives through iterative

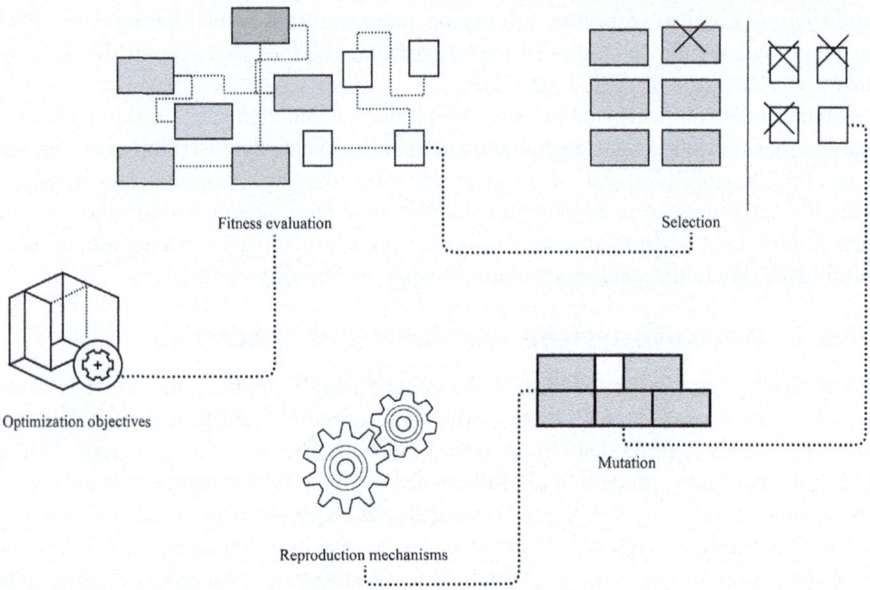

Figure 12.7 Generative design evolution through multiobjective optimization and evolutionary algorithms.

computational processes, progressively improving performance according to predefined objectives and constraints (Fig. 12.7).

This approach fundamentally alters the design process. Architects transition from being sole creators of individual solutions to becoming evaluators and curators of computationally generated possibilities. The result is a broader exploration of design space and increased opportunities for innovation. Generative design is particularly effective when addressing complex problems involving multiple interacting variables. Building geometry, structural systems, daylight availability, energy performance, material consumption, and user experience can all be incorporated into a unified optimization framework. The ability to simultaneously evaluate these factors enables designers to identify solutions that achieve balanced performance across multiple objectives (Chew et al., 2024).

12.3.5 Multiobjective optimization

Contemporary architectural projects rarely involve a single design objective. Buildings must satisfy diverse and often conflicting requirements, including energy efficiency, structural safety, thermal comfort, visual quality, environmental sustainability, and economic feasibility. Multiobjective optimization provides a computational framework for managing these competing criteria. Rather than optimizing one variable at the expense of others, AI algorithms seek solutions that achieve acceptable performance across all objectives simultaneously.

For example, the design of a building façade may involve balancing daylight penetration, solar heat gain, thermal insulation, visual transparency, construction cost, and aesthetic considerations. Conventional design approaches often rely on sequential evaluations and subjective judgments. In contrast, AI-based optimization algorithms can analyze thousands of alternatives and identify configurations that provide optimal trade-offs among competing objectives. The resulting solutions are often represented through Pareto front analyzes, where designers can visualize the relationships between different performance metrics and select alternatives that best align with project priorities.

12.3.6 Evolutionary algorithms

Evolutionary algorithms constitute one of the most widely used optimization techniques in generative design (Gradišar et al., 2022). Inspired by biological evolution, these algorithms mimic processes such as natural selection, mutation, crossover, and adaptation. The optimization process begins with a population of design alternatives. Each alternative is evaluated according to predefined performance criteria. Higher-performing solutions are selected and combined to generate new generations of designs. Through successive iterations, the population evolves toward increasingly effective solutions.

Evolutionary algorithms have demonstrated particular effectiveness in structural engineering applications. Structural forms, member dimensions, support conditions, and load paths can be optimized to reduce material consumption while maintaining safety and performance requirements. This approach often produces highly efficient geometries that resemble forms found in nature, where material distribution has evolved to maximize efficiency.

Architectural applications include roof structures, façade systems, space frames, bridges, and complex free-form geometries. The resulting designs frequently exhibit innovative characteristics that would be difficult to achieve using traditional methodologies.

12.3.7 Design exploration and innovation

Beyond optimization, generative design serves as a powerful tool for design exploration. Architectural creativity often emerges through the investigation of multiple possibilities and the discovery of unexpected relationships among design variables. Traditional workflows typically limit exploration due to time and resource constraints. Designers may evaluate only a small number of alternatives before selecting a preferred solution. Generative design dramatically expands this capacity by enabling the rapid generation and assessment of thousands of design variations.

This expanded design space encourages experimentation and innovation. Computational systems can reveal solutions that challenge conventional assumptions and introduce novel approaches to spatial organization, structural expression, environmental responsiveness, and material efficiency. As a result, AI functions not only as an analytical tool but also as a catalyst for creativity, supporting the development of innovative architectural solutions that combine performance and design excellence.

12.3.8 Simulation tools for performance-driven design

The emergence of performance-driven architecture has transformed the way buildings are conceived, evaluated, and refined throughout the design process. Traditionally, many design decisions were based on precedent studies, professional intuition, and simplified calculations. Although these approaches remain valuable, contemporary architectural projects increasingly require precise evaluation of environmental, structural, economic, and operational performance. The growing complexity of building systems, combined with the demand for sustainability, energy efficiency, occupant well-being, and climate resilience, has elevated simulation technologies to a central role in architectural practice.

Simulation tools provide virtual environments in which designers can test the behavior of buildings before physical construction begins. These digital analyzes allow architects and engineers to predict how a proposed design will perform under varying conditions, identify potential shortcomings, and evaluate alternative solutions (Wen and Zhou, 2025). AI significantly enhances these capabilities by accelerating analytical processes, improving predictive accuracy, and enabling automated optimization. As a result, simulation is no longer viewed merely as a verification tool used at the end-of-design development; instead, it has become an integral component of an iterative and continuous design methodology.

In AI-assisted workflows, simulations generate large quantities of performance data that are subsequently analyzed by machine learning algorithms and optimization engines. The information obtained from these analyzes informs design modifications, creating a feedback loop in which performance outcomes continuously guide design evolution. This dynamic relationship between simulation, analysis, and optimization forms the foundation of contemporary performance-driven design methodologies (Fig. 12.8).

12.3.9 Energy simulation

Energy simulation remains one of the most important analytical techniques in sustainable architecture. Buildings account for a substantial proportion of global energy consumption and greenhouse gas emissions, making energy performance a critical consideration during design and operation (Pan et al., 2023). Energy modeling software enables architects and engineers to estimate heating and cooling demands, lighting energy consumption, equipment loads, renewable energy generation, and overall building efficiency.

Modern energy simulation platforms can represent complex interactions between building envelopes, mechanical systems, climatic conditions, occupancy schedules, and operational strategies. These models provide valuable insights into how design decisions influence annual energy consumption and carbon emissions. Through comparative analysis, designers can evaluate alternative solutions and select strategies that improve environmental performance while maintaining functional requirements.

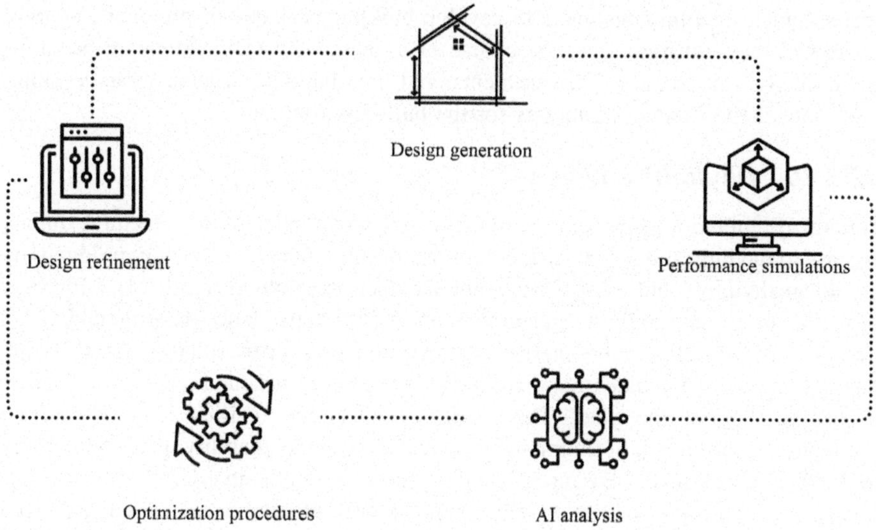

Figure 12.8 Simulation-based feedback loop for performance-driven design.

AI significantly expands the capabilities of conventional energy modeling. Machine learning algorithms can analyze operational data collected from existing buildings and use these datasets to calibrate simulation models. Calibration improves the reliability of predictions by ensuring that simulated behavior more closely reflects actual building performance. This capability is particularly important because discrepancies often exist between predicted and measured energy consumption due to variations in occupant behavior, equipment operation, and environmental conditions.

AI also enables rapid assessment of multiple design scenarios. Instead of manually evaluating individual alternatives, intelligent systems can automatically explore numerous combinations of design variables, including insulation levels, glazing ratios, shading configurations, HVAC systems, and renewable energy technologies. By examining the relationships between these variables and energy outcomes, AI algorithms identify optimal solutions that balance performance, cost, and environmental impact.

Furthermore, predictive models developed through machine learning can estimate energy performance during the earliest stages of design, when detailed information may not yet be available. This capability supports informed decision-making at a point where modifications are relatively inexpensive and have the greatest influence on long-term building efficiency. Consequently, architects can incorporate sustainability objectives from the outset rather than attempting to improve performance after major design decisions have already been finalized.

The integration of AI and energy simulation also supports the development of net-zero and positive-energy buildings. Intelligent systems can optimize the interaction between energy demand reduction measures and renewable energy production technologies, enabling designers to develop buildings capable of producing as much energy as they consume over the course of a year. Such capabilities are becoming increasingly important as governments and international organizations establish ambitious carbon reduction targets for the built environment.

12.3.10 Daylight analysis

Natural illumination represents a fundamental component of architectural quality. Access to daylight influences visual comfort, psychological well-being, productivity, circadian rhythms, and overall occupant satisfaction (Kota et al., 2014). Moreover, effective use of natural light can significantly reduce dependence on artificial lighting systems, thereby lowering energy consumption and operational costs. Daylight simulation tools evaluate the distribution of natural illumination within interior spaces throughout the year. These analyzes examine metrics such as daylight autonomy, useful daylight illuminance, spatial daylight availability, glare probability, and solar penetration. By understanding how sunlight interacts with building geometry, architects can create environments that maximize visual quality while minimizing adverse effects.

The application of AI enhances daylight analysis by introducing automated optimization capabilities. Intelligent algorithms can evaluate thousands of potential design configurations involving building orientation, window dimensions, façade composition, room proportions, and shading systems. Each configuration is assessed according to multiple performance criteria, enabling designers to identify solutions that achieve balanced outcomes.

For example, increasing the window size may improve daylight availability but simultaneously increase glare and solar heat gain. AI systems can evaluate these competing effects and determine appropriate combinations of glazing characteristics and shading devices that maintain visual comfort while reducing cooling loads. Such analyzes are particularly valuable in large commercial buildings, educational facilities, healthcare environments, and residential developments where daylight quality directly affects occupant experience.

Another advantage of AI-assisted daylight analysis is its ability to support adaptive façade design. Dynamic façade systems incorporating movable louvers, electrochromic glazing, or responsive shading devices can alter their configuration in response to changing environmental conditions. AI can predict solar conditions and optimize façade operation to maintain desirable indoor lighting levels throughout the day. The integration of daylight simulations with energy analyzes further strengthens performance-driven design. Since lighting and thermal performance are closely interconnected, intelligent systems can simultaneously evaluate visual and energy-related criteria. This holistic approach enables the development of solutions that improve both occupant comfort and overall building efficiency.

12.3.11 Computational fluid dynamics

Air movement plays a crucial role in determining indoor environmental quality, thermal comfort, ventilation effectiveness, and energy consumption. Understanding airflow behavior is therefore essential for the design of healthy and sustainable buildings. Computational fluid dynamics (CFD) provides a sophisticated numerical method for analyzing the movement of air and other fluids within and around buildings.

CFD simulations allow designers to visualize airflow patterns, temperature distributions, pressure differentials, pollutant dispersion, and ventilation effectiveness. These analyzes support a wide range of applications, including natural ventilation design, smoke management systems, thermal comfort assessments, urban wind studies, and microclimate investigations. In naturally ventilated buildings, CFD enables architects to evaluate how openings, atria, courtyards, and ventilation shafts influence air movement. By understanding these airflow dynamics, designers can improve indoor air quality while reducing reliance on mechanical cooling systems. Such strategies are particularly valuable in regions where climatic conditions support passive ventilation approaches (Hajdukiewicz et al., 2024).

Urban applications of CFD have also become increasingly important as cities continue to densify. High-rise developments significantly influence local wind conditions, solar access, and outdoor thermal comfort. CFD analyzes help planners assess the impact of proposed developments on surrounding public spaces and identify strategies that enhance pedestrian comfort and environmental quality. Despite its analytical power, CFD traditionally requires substantial computational resources and extended processing times. AI addresses this limitation by accelerating simulation workflows. Machine learning models can be trained using large CFD datasets, enabling them to predict airflow behavior with considerably lower computational requirements. These surrogate models provide rapid performance estimates that facilitate exploration of a broader range of design alternatives.

AI-driven CFD workflows also support automated optimization processes. Instead of manually adjusting design parameters and rerunning simulations, intelligent systems can systematically modify geometric configurations and evaluate resulting airflow performance. Through repeated iterations, optimal ventilation strategies can be identified more efficiently than through conventional approaches. The combination of CFD and AI is particularly valuable in addressing challenges associated with climate change and urban heat islands. By optimizing natural ventilation and microclimatic conditions, designers can reduce cooling demands and improve resilience in increasingly warm urban environments.

12.3.12 Structural simulation

Structural performance constitutes a fundamental requirement of every building project. Structures must safely withstand gravity loads, wind forces, seismic activity, thermal stresses, and other environmental influences throughout their operational lifespan (Salgado and Guner, 2021). Structural simulation tools provide engineers with the ability to predict how buildings will respond to these loading conditions before construction begins. Finite element analysis (FEA) is among the most widely used structural simulation techniques. By dividing a structure into smaller computational elements, FEA enables detailed assessment of stresses, strains, deformations, and failure mechanisms. Engineers use these analyzes to verify safety, improve efficiency, and refine structural systems.

AI has expanded the scope and effectiveness of structural simulation. AI-assisted optimization algorithms can automatically explore alternative structural configurations and identify solutions that achieve superior performance. Parameters such as member dimensions, material distribution, support locations, and geometric forms can be adjusted systematically to improve structural behavior while minimizing material consumption. This capability is particularly relevant in sustainable construction, where reducing embodied carbon has become a major objective. Since structural materials often account for a substantial proportion of a building's environmental impact, optimizing structural efficiency can significantly reduce resource consumption and associated emissions.

Machine learning techniques are also being applied to structural health monitoring. Sensors embedded within buildings and infrastructure continuously collect data related to vibration, displacement, strain, and environmental exposure. AI algorithms analyze these measurements to detect anomalies, identify deterioration patterns, and predict future maintenance requirements. Such predictive capabilities improve safety while reducing inspection and repair costs.

The integration of AI with structural simulation has contributed to the emergence of innovative architectural forms characterized by high efficiency and material economy. Complex geometries that would be difficult to design using conventional approaches can now be optimized through computational processes inspired by natural systems and evolutionary principles.

12.3.13 Artificial intelligence-assisted decision making

AI increasingly functions as a decision support framework throughout the entire building lifecycle. From conceptual planning to operational management, AI systems provide analytical insights that help stakeholders evaluate alternatives and select appropriate strategies. During early design phases, decisions regarding site planning, building orientation, massing, envelope characteristics, and structural systems have long-term consequences for energy performance, occupant comfort, environmental impact, and lifecycle costs. AI enables these decisions to be informed by quantitative analysis rather than intuition alone (Gao et al., 2025).

By integrating information related to climate conditions, solar exposure, wind patterns, land use constraints, transportation networks, and regulatory requirements, intelligent systems can generate recommendations that align with project objectives. This capability allows architects to identify opportunities and constraints at the beginning of the design process, reducing the likelihood of costly revisions later.

Construction management represents another area where AI delivers substantial benefits. Large projects involve complex interactions among schedules, resources, logistics, contractors, and suppliers. AI can analyze project information to identify potential delays, optimize sequencing, improve resource allocation, and enhance coordination among stakeholders. Risk management is similarly enhanced through predictive analytics. By examining historical project data and real-time site information, AI systems can identify patterns associated with accidents, delays, and cost overruns. These insights enable project teams to implement preventative measures and improve overall project performance.

12.3.14 Intelligent building operations

The influence of AI extends well beyond design and construction into the operational phase of buildings. Modern facilities are increasingly equipped with sensors, IoT networks, smart meters, and building management systems that continuously generate information regarding environmental conditions, occupancy patterns, equipment performance, and energy use. AI processes these extensive datasets to optimize operational efficiency in real-time. Intelligent control systems dynamically adjust heating, ventilation, air conditioning, lighting, and security systems according to actual building conditions rather than predetermined schedules. This adaptive approach improves resource utilization while maintaining occupant comfort.

Machine learning algorithms continuously refine operational strategies by learning from historical patterns and user behavior. As a result, building systems become progressively more efficient over time. Predictive maintenance functions further enhance operational performance by identifying potential equipment failures before they occur, thereby reducing downtime and maintenance expenditures.

12.3.15 Future directions of artificial intelligence in architecture

The future of architecture will be shaped by the convergence of AI with BIM, digital twins, robotics, sensor technologies, cloud computing, and advanced simulation platforms (Diaz Schery et al., 2026). Together, these innovations are creating integrated digital ecosystems capable of supporting continuous optimization throughout the lifecycle of buildings and infrastructure. Digital twins represent one of the most significant developments in this context. A digital twin is a dynamic virtual representation of a physical asset that continuously receives data from the real environment. By integrating simulation models, sensor information, and operational analytics, digital twins enable ongoing performance assessment and informed decision-making.

Human-AI collaboration is expected to define the next generation of architectural practice. Rather than replacing professional expertise, AI serves as an augmentation technology that expands analytical capacity and automates repetitive tasks. Architects and engineers remain responsible for creativity, cultural interpretation, ethical considerations, and human-centered design values.

As these technologies mature, buildings will increasingly function as adaptive systems capable of responding intelligently to environmental conditions and user needs. The built environment will evolve from static physical structures into responsive ecosystems characterized by continuous learning, optimization, and adaptation. Through the integration of simulation technologies, predictive analytics, intelligent control systems, and data-driven decision-making, AI is establishing a new paradigm for architecture—one focused on performance, sustainability, resilience, and long-term value creation.

References

Alanne, K., Sierla, S., 2022. An overview of machine learning applications for smart buildings. Sustainable Cities and Society 76. Available at https://doi.org/10.1016/j.scs.2021.103445, http://www.elsevier.com/wps/find/journaldescription.cws_home/724360/description#description.

Albukhari, I.N., 2025. The role of artificial intelligence (AI) in architectural design: a systematic review of emerging technologies and applications. Journal of Umm Al-Qura University for Engineering and Architecture 16 (4), 1457–1476. Available at https://doi.org/10.1007/s43995-025-00186-1, https://link.springer.com/journal/43995.

Ali, A., Jayaraman, R., Mayyas, A., Alaifan, B., Azar, E., 2023. Machine learning as a surrogate to building performance simulation: predicting energy consumption under different operational settings. Energy and Buildings 286, 112940. Available at https://doi.org/10.1016/j.enbuild.2023.112940.

Brozovsky, J., Labonnote, N., Vigren, O., 2024. Digital technologies in architecture, engineering, and construction. Automation in Construction 158, 105212. Available at https://doi.org/10.1016/j.autcon.2023.105212.

Caetano, I., Santos, L., Leitão, A., 2020. Computational design in architecture: defining parametric, generative, and algorithmic design. Frontiers of Architectural Research 9 (2), 287–300. Available at https://doi.org/10.1016/j.foar.2019.12.008, http://www.keaipublishing.com/en/journals/frontiers-of-architectural-research/.

Chew, Z.X., Wong, J.Y., Tang, Y.H., Yip, C.C., Maul, T., 2024. Generative design in the built environment. Automation in Construction 166, 105638. Available at https://doi.org/10.1016/j.autcon.2024.105638.

Diaz Schery, C.A., Caiado, R.G.G., Ivson, P., Santos, R.S., Thadeu Corseuil, E., 2026. An integrative framework for BIM, digital twin and generative AI convergence in construction: a multi-dimensional analysis of synergies and enabling factors. Smart and Sustainable Built Environment 1–39. Available at https://doi.org/10.1108/SASBE-08-2025-0502, www.emeraldinsight.com/products/journals/journals.htm?id=sasbe.

Elmousalami, H., Maxy, M., Hui, F.K.P., Aye, L., 2025. AI in automated sustainable construction engineering management. Automation in Construction 175. Available at https://doi.org/10.1016/j.autcon.2025.106202, https://www.sciencedirect.com/science/journal/09265805.

Eltaweel, A., SU, Y., 2017. Parametric design and daylighting: a literature review. Renewable and Sustainable Energy Reviews 73, 1086–1103. Available at https://doi.org/10.1016/j.rser.2017.02.011, https://www.journals.elsevier.com/renewable-and-sustainable-energy-reviews.

Gao, M.Y., Li, C., Petzold, F., Tiong, R.L.K., Yang, Y., 2025. Lifecycle framework for AI-driven parametric generative design in industrialized construction. Automation in Construction 174. Available at https://doi.org/10.1016/j.autcon.2025.106146, https://www.sciencedirect.com/science/journal/09265805.

Goldbach, A.K., Lázaro, C., 2024. CAD-integrated parametric design and analysis of lightweight shell structures. Structures 64. Available at https://doi.org/10.1016/j.istruc.2024.106566, https://www.sciencedirect.com/science/journal/23520124.

Gradišar, L., Klinc, R., Turk, Ž., Dolenc, M., 2022. Generative design methodology and framework exploiting designer-algorithm synergies. Buildings 12 (12), 2194. Available at https://doi.org/10.3390/buildings12122194.

Grima, G., Pelanconi, M., Santecchia, E., Ortona, A., 2026. Beyond lattices: a geometry-first computational design framework for additive manufacturing of porous architectures. Additive Manufacturing Letters 18. Available at https://doi.org/10.1016/j.addlet.2026.100396, https://www.sciencedirect.com/science/journal/27723690.

Hajdukiewicz, M., González Gallero, F.J., Mannion, P., Loomans, M.G.L.C., Keane, M.M., 2024. A narrative review to credible computational fluid dynamics models of naturally ventilated built environments. Renewable and Sustainable Energy Reviews 198. Available at https://doi.org/10.1016/j.rser.2024.114404, https://www.sciencedirect.com/science/journal/13640321.

Hu, X., Olgun, G., Assaad, R.H., 2024. An intelligent BIM-enabled digital twin framework for real-time structural health monitoring using wireless IoT sensing, digital signal processing, and structural analysis. Expert Systems with Applications 252, 124204. Available at https://doi.org/10.1016/j.eswa.2024.124204.

Kota, S., Haberl, J.S., Clayton, M.J., Yan, W., 2014. Building information modeling (BIM)-based daylighting simulation and analysis. Energy and Buildings 81, 391–403. Available at https://doi.org/10.1016/j.enbuild.2014.06.043.

Li, C., Zhang, T., Du, X., Zhang, Y., Xie, H., 2025. Generative AI models for different steps in architectural design: a literature review. Frontiers of Architectural Research 14 (3), 759–783. Available at https://doi.org/10.1016/j.foar.2024.10.001.

Manmatharasan, P., Bitsuamlak, G., Grolinger, K., 2025. AI-driven design optimization for sustainable buildings: a systematic review. Energy and Buildings 332, 115440. Available at https://doi.org/10.1016/j.enbuild.2025.115440.

Mylonas, A., Tsangrassoulis, A., Pascual, J., 2024. Modelling occupant behaviour in residential buildings: a systematic literature review. Building and Environment 265, 111959. Available at https://doi.org/10.1016/j.buildenv.2024.111959.

Pan, Y., Zhu, M., Lv, Y., Yang, Y., Liang, Y., Yin, R., et al., 2023. Building energy simulation and its application for building performance optimization: a review of methods, tools, and case studies. Advances in Applied Energy 10, 100135. Available at https://doi.org/10.1016/j.adapen.2023.100135.

Salgado, R.A., Guner, S., 2021. A structural performance-based environmental impact assessment framework for natural hazard loads. Journal of Building Engineering 43. Available at https://doi.org/10.1016/j.jobe.2021.102908, http://www.journals.elsevier.com/journal-of-building-engineering/.

Shi, X., Yang, W., 2013. Performance-driven architectural design and optimization technique from a perspective of architects. Automation in Construction 32, 125–135. Available at https://doi.org/10.1016/j.autcon.2013.01.015.

Su, S., Wang, Q., Han, L., Hong, J., Liu, Z., 2020. BIM-DLCA: an integrated dynamic environmental impact assessment model for buildings. Building and Environment 183, 107218. Available at https://doi.org/10.1016/j.buildenv.2020.107218.

Vedrtnam, A., Kalauni, K., Soares, N., Salazar, B., Ostertag, C.P., Taylor, H.K., 2025. Phase change materials for climate-adaptive buildings: a review framed by tropical and Mediterranean climates. Journal of Energy Storage 125. Available at https://doi.org/10.1016/j.est.2025.116938, https://www.sciencedirect.com/science/journal/2352152X.

Waqar, A., Othman, I., Saad, N., Azab, M., Khan, A.M., 2023. BIM in green building: enhancing sustainability in the small construction project. Cleaner Environmental Systems 11, 100149. Available at https://doi.org/10.1016/j.cesys.2023.100149.

Wen, J., Zhou, M., 2025. Virtual simulation system-driven optimization approach for architectural environmental art design. International Journal of e-Collaboration 21 (1), 1–21. Available at https://doi.org/10.4018/ijec.378251.

Yu, Z., Dai, H., Shi, Z., 2022. Structural form-finding of bending components in buildings by using parametric tools and principal stress lines. Frontiers of Architectural Research 11 (3), 561–573. Available at https://doi.org/10.1016/j.foar.2021.11.004, http://www.keaipublishing.com/en/journals/frontiers-of-architectural-research/.

Challenges, ethics, and commercialization of smart structures

13

13.1 Technical and regulatory challenges

The growing integration of digital technologies into buildings and infrastructure has fundamentally transformed the conception, design, construction, and operation of the built environment. Smart structures, characterized by the incorporation of sensing technologies, intelligent materials, communication systems, data analytics platforms, and automated control mechanisms, offer significant opportunities to enhance building performance, operational efficiency, resilience, sustainability, and user experience. Through real-time monitoring and adaptive responses, these systems enable structures to interact dynamically with their environments, creating built environments capable of learning from operational data and responding to changing conditions.

Despite their considerable potential, the widespread adoption of smart structures remains constrained by numerous technical, organizational, economic, and regulatory challenges. Unlike conventional buildings that primarily depend on passive systems, smart structures operate through interconnected networks of physical and digital components. Their successful implementation requires coordination among diverse stakeholders, integration of multiple technologies, long-term operational planning, and compliance with evolving regulatory requirements. Consequently, the transition from traditional construction approaches to intelligent infrastructure involves much more than the installation of sensors or automation systems (Frighi, 2023). It represents a comprehensive transformation of design methodologies, construction processes, asset management strategies, and governance frameworks. As smart technologies continue to advance, understanding these challenges becomes essential for architects, engineers, policymakers, facility managers, and infrastructure owners seeking to realize the full benefits of intelligent buildings and infrastructure systems.

13.1.1 Technical complexity and system integration

One of the primary barriers to the implementation of smart structures is the inherent complexity associated with integrating diverse technologies into a unified operational environment. Conventional buildings are generally composed of architectural, structural, mechanical, electrical, and plumbing systems that function largely independently. Smart structures introduce additional layers of connectivity, intelligence, and automation that require continuous interaction among multiple subsystems.

Smart Civil and Structural Engineering. DOI: https://doi.org/10.1016/B978-0-443-30223-7.00008-5

The integration process involves sensors, actuators, controllers, communication networks, cloud platforms, digital twins, artificial intelligence applications, and building management systems (Jia et al., 2019). Each component generates, transmits, processes, or responds to information. The effectiveness of the overall system depends not only on the performance of individual technologies but also on their ability to communicate and operate seamlessly (Bellini et al., 2022). This complexity creates significant challenges during design and construction. Architects and engineers must consider information flows alongside traditional concerns such as structural safety, spatial organization, and environmental performance. Furthermore, technology providers often employ proprietary platforms and communication protocols, making integration difficult and increasing dependence on specific vendors.

The challenge becomes more pronounced in large projects where hundreds or thousands of devices must operate simultaneously. Even minor incompatibilities between software platforms or communication standards can lead to operational inefficiencies, increased maintenance requirements, and unexpected system failures. Table 13.1 summarizes the principal technical barriers that arise during the integration of smart technologies into buildings and infrastructure systems. These challenges primarily stem from the heterogeneous nature of smart structure components, which include sensors, actuators, communication networks, software platforms, and data processing systems developed by different vendors and operating under different protocols.

A key issue is interoperability, where systems fail to communicate effectively due to incompatible standards or proprietary architectures. This fragmentation results in isolated data environments, often referred to as data silos, which limit the ability of integrated systems to function as a unified whole (Himeur et al., 2023). Closely related to this is the challenge of communication reliability, where delays, interruptions, or losses in data transmission can significantly affect the responsiveness

Table 13.1 Major technical integration challenges in smart structures.

Challenge	Description	Potential consequences
Interoperability	Incompatibility among systems from different vendors	Data silos and reduced functionality
Communication reliability	Disruptions in data transmission networks	Delayed responses and system downtime
Data synchronization	Inconsistent information across platforms	Decision-making errors
Software compatibility	Conflicts between applications and updates	Operational interruptions
System complexity	A large number of interconnected technologies	Increased maintenance requirements
Scalability	Difficulty expanding systems over time	Limited future adaptability

of automated building systems, leading to inefficiencies or temporary system failures (Pantelic et al., 2022).

Another critical issue is data synchronization, which refers to inconsistencies in information exchanged between subsystems. When data is not properly synchronized across platforms, decision-making processes may be based on incomplete or outdated information, reducing the reliability of automated control strategies (Nour et al., 2019). The challenge of software compatibility further complicates integration, as frequent updates and differences in application architectures may generate conflicts between systems. These incompatibilities can disrupt operational continuity and increase maintenance demands. Finally, system complexity and scalability limitations represent long-term barriers. As the number of interconnected devices increases, system architecture becomes more difficult to manage, requiring advanced coordination strategies, higher computational capacity, and robust maintenance frameworks. Without scalable design principles, smart systems may become increasingly difficult to expand or adapt over time, limiting their long-term viability.

13.1.2 Reliability and operational robustness

Reliability is a fundamental requirement for any building or infrastructure system. In smart structures, however, reliability extends beyond physical components to include digital infrastructure, communication networks, and software applications. Unlike traditional buildings, where failures are often localized, failures within smart systems may propagate through interconnected networks. A malfunctioning sensor may generate inaccurate data, which can influence automated control decisions and affect the performance of multiple systems simultaneously. Therefore reliability must be considered at both the component and system levels.

Environmental conditions further influence system robustness. Sensors installed in bridges, façades, industrial facilities, or outdoor infrastructure are exposed to temperature fluctuations, humidity, vibration, dust, and mechanical stress. These conditions may reduce accuracy and shorten service life. Consequently, smart structures require redundancy strategies, backup systems, and continuous monitoring to ensure operational continuity (Tariq et al., 2024). Reliability challenges are particularly critical in hospitals, airports, transportation facilities, and emergency-response infrastructure, where failures may directly affect safety, operational continuity, and public welfare.

13.1.3 Data management and information overload

Smart structures continuously generate large volumes of data related to environmental conditions, structural performance, occupancy patterns, equipment operation, and resource consumption. While these data provide valuable insights, they also introduce significant management challenges (Hernández et al., 2024). Data collection alone does not create value. Organizations must establish processes for storing, validating, processing, interpreting, and utilizing information effectively. Without clear data management strategies, stakeholders may become overwhelmed by excessive quantities of information that provide limited operational benefit.

Data quality is another important concern. Inaccurate, incomplete, or inconsistent data can compromise analytical results and reduce confidence in automated decision-making systems. Sensor calibration, data verification, and quality-control procedures therefore become essential components of smart building operation (Othman et al., 2026). Moreover, the increasing use of machine learning and predictive analytics depends heavily on the availability of reliable datasets. Poor data quality can significantly reduce the effectiveness of AI-driven systems and generate misleading recommendations.

13.1.4 Cybersecurity risks

As buildings become increasingly connected through the Internet of Things (IoT), cybersecurity has emerged as one of the most important concerns associated with smart structures (Mulero-Palencia and Monzon Baeza, 2023). Modern buildings often contain thousands of connected devices that exchange information through local networks and cloud-based platforms. While connectivity enhances operational efficiency, it also increases vulnerability to cyber threats. Unauthorized access to building systems may compromise security functions, disrupt operations, manipulate environmental controls, or expose sensitive information. Potential cybersecurity risks include:

- unauthorized access to building management systems;
- manipulation of environmental controls;
- interruption of communication networks;
- data theft and privacy breaches;
- ransomware attacks targeting critical infrastructure;
- compromise of emergency-response systems.

Cybersecurity strategies must therefore be incorporated from the earliest design stages rather than treated as an afterthought. Effective protection requires secure communication protocols, network segmentation, encryption, access control mechanisms, and continuous security monitoring.

13.1.5 Retrofitting existing buildings

Although many discussions focus on newly constructed smart buildings, a significant proportion of future implementation will occur within existing building stock (Peiris et al., 2023). Retrofitting older buildings presents unique challenges that differ substantially from those encountered in new construction. Many existing structures lack adequate electrical infrastructure, communication pathways, equipment spaces, and digital connectivity. Installing sensors and automation systems often requires modifications that may disrupt building operations and increase project costs.

Historic buildings present additional constraints. Preservation regulations frequently limit physical interventions, requiring innovative solutions that minimize impacts on heritage value while introducing modern technological capabilities. Retrofitting becomes particularly important in sustainability strategies because extending the service life of existing buildings often generates lower environmental impacts than demolition and reconstruction. Consequently, developing efficient retrofit methodologies remains a priority for the future implementation of smart structures.

13.1.6 Economic challenges and lifecycle value

Economic considerations remain among the most significant factors influencing technology adoption. Smart structures generally require higher initial investments than conventional alternatives due to expenditures associated with hardware, software, networking infrastructure, commissioning, and specialized expertise. The challenge lies in balancing short-term costs against long-term benefits (Monaco et al., 2024). Many advantages of smart systems, including predictive maintenance, operational optimization, reduced downtime, and improved resource efficiency, accumulate gradually over the building lifecycle. Stakeholders focused primarily on initial capital expenditures may therefore underestimate overall value. Lifecycle assessment approaches provide a more comprehensive evaluation framework by considering:

- initial capital costs;
- operating expenses;
- maintenance costs;
- energy consumption;
- asset longevity;
- risk reduction benefits;
- productivity improvements;
- end-of-life considerations.

Research increasingly demonstrates that lifecycle benefits often outweigh higher upfront investments, particularly for large commercial buildings, healthcare facilities, airports, and critical infrastructure (Dwaikat and Ali, 2018).

13.1.7 Scalability across different building types

The scalability of smart technologies remains a significant implementation challenge. Solutions that perform effectively in large commercial developments may not be economically viable for smaller projects (Toderean et al., 2025). Residential buildings often operate under limited budgets, making advanced monitoring systems difficult to justify. Conversely, large infrastructure systems may generate enormous quantities of data that require sophisticated management strategies and specialized operational personnel.

Smart cities introduce an even greater level of complexity by connecting buildings, transportation networks, utility systems, communication infrastructure, and public services into integrated digital ecosystems. Achieving interoperability across these diverse systems remains one of the most significant challenges facing urban digital transformation.

Table 13.2 illustrates how scalability challenges in smart structures vary depending on building type and application context. Rather than representing a uniform issue,

Table 13.2 Scalability challenges across different applications.

Application type	Primary challenge	Key consideration
Residential buildings	Cost sensitivity	Simplified solutions
Commercial buildings	System integration	Return on investment (ROI) optimization
Healthcare facilities	Reliability requirements	Continuous operation
Airports	Data volume	Real-time coordination
Bridges and infrastructure	Environmental exposure	Long-term durability
Smart cities	Multistakeholder integration	Governance frameworks

scalability is highly dependent on operational scale, budget constraints, environmental exposure, and governance complexity.

In residential buildings, the primary limitation is cost sensitivity. The economic justification for advanced smart technologies is often weak in small-scale projects, leading to simplified or partial implementations that reduce system capabilities (Hwang et al., 2022). In contrast, commercial buildings tend to prioritize return on investment and system integration efficiency, requiring carefully balanced solutions that optimize both performance and operational cost. Healthcare facilities present a different challenge, where reliability and continuous operation are critical. Any system failure may directly affect patient safety, making redundancy and robustness essential design requirements. Similarly, airports and transport hubs face challenges associated with extremely high data volumes and the need for real-time coordination across multiple operational systems, including security, logistics, and passenger flow management.

For bridges and large-scale infrastructure, environmental exposure becomes the dominant factor influencing scalability. Systems must be designed for long-term durability under harsh environmental conditions, including temperature variation, moisture, vibration, and structural stress. At the highest level of complexity, smart cities introduce multistakeholder integration challenges. Here, scalability is not only a technical issue but also a governance problem, requiring coordination between public authorities, private operators, utility providers, and digital infrastructure managers. Interoperability, regulatory alignment, and data governance frameworks become critical to ensuring coherent system-wide functionality across urban networks. The table highlights that scalability in smart structures cannot be understood as a single technical parameter; rather, it must be interpreted as a context-dependent challenge shaped by economic, environmental, organizational, and regulatory conditions.

13.1.8 Maintenance and operational requirements

The long-term performance and reliability of smart structures depend significantly on effective maintenance and operational management strategies (Bokrantz et al., 2020).

Unlike conventional buildings, which primarily require periodic inspections and maintenance of physical components, smart structures necessitate the integrated management of both physical and digital infrastructures. Sensors require regular calibration, validation, and replacement to ensure data accuracy, while communication networks must be continuously monitored to maintain system connectivity and performance. Software platforms demand periodic updates to address functional improvements and cybersecurity vulnerabilities, whereas artificial intelligence algorithms require ongoing retraining and validation to maintain decision-making accuracy under evolving operating conditions. In addition, data storage and processing systems require continuous maintenance, backup procedures, and security management to ensure data integrity and resilience. These requirements shift building management practices from predominantly reactive maintenance approaches toward proactive and continuous operational oversight (Larsson et al., 2026). Consequently, facility managers increasingly assume responsibilities that extend beyond traditional building operations, functioning as technology managers who oversee digital assets alongside conventional building systems. The growing complexity of these interconnected requirements underscores the importance of incorporating lifecycle planning, maintenance strategies, and technological adaptability into the earliest stages of design and project development.

13.1.9 Workforce and skills challenges

The successful operation of intelligent buildings depends on personnel possessing expertise that extends beyond traditional architectural and engineering disciplines. Modern facility managers increasingly require knowledge of:

- building automation systems;
- data analytics;
- cloud computing;
- cybersecurity;
- artificial intelligence;
- network management;
- digital twin technologies.

However, many organizations face shortages of professionals with these interdisciplinary skills. Educational institutions and professional training programs are still adapting to rapidly evolving technological requirements. The workforce challenge therefore extends beyond technical implementation and represents a broader issue of industry transformation (Daher-Armache et al., 2025).

13.1.10 Regulatory and code compliance challenges

Regulatory frameworks often lag behind technological innovation. Most building codes were developed for static systems with predictable behavior patterns. Smart structures introduce adaptive characteristics that may not fit easily within existing regulatory frameworks (Xenakis et al., 2025). Traditional codes typically specify prescriptive requirements regarding materials, dimensions, and construction methods.

Smart systems, however, frequently rely on dynamic responses that change according to environmental conditions and operational needs. As a result, regulatory authorities may encounter difficulties evaluating:

- adaptive façades;
- intelligent materials;
- autonomous control systems;
- AI-assisted operational decisions;
- digital twin applications;
- predictive maintenance methodologies.

These uncertainties may increase approval times and create additional administrative burdens for project teams.

13.1.11 Data governance, privacy, and ethics

The widespread deployment of smart structures and intelligent building systems has introduced significant challenges related to data governance, privacy, and ethics. Through the continuous operation of sensors, monitoring devices, and digital management platforms, smart buildings generate and process large volumes of information concerning occupancy patterns, space utilization, environmental preferences, resource consumption, and user behavior. Although these data streams provide valuable insights that support performance optimization, energy efficiency, predictive maintenance, and occupant comfort, they also raise important concerns regarding privacy, informed consent, data ownership, transparency, and security (Banad et al., 2025).

Effective governance frameworks are therefore essential to ensure the responsible management of building-generated data throughout its lifecycle. Such frameworks should clearly define ownership rights, establish access control mechanisms, specify data retention and deletion policies, regulate information-sharing procedures, and implement appropriate privacy and cybersecurity protections. In addition, governance structures should identify accountability mechanisms that clarify the responsibilities of building owners, facility managers, technology providers, and other stakeholders involved in data collection and management.

Ethical considerations become increasingly important as artificial intelligence and automated decision-making systems assume a greater role in building operations. Algorithms may influence environmental controls, space allocation, energy management strategies, and other operational decisions that directly affect occupants. Consequently, issues related to algorithmic transparency, fairness, explainability, and potential bias must be carefully addressed to ensure that automated systems operate in a manner that respects user rights and promotes trust among building occupants. As smart structures continue to evolve, balancing technological innovation with ethical responsibility will remain a fundamental challenge for designers, operators, and policymakers.

13.1.12 Performance-based regulatory approaches

Many experts view performance-based regulation as a promising solution to the challenges posed by intelligent infrastructure. Rather than prescribing specific design solutions, performance-based approaches focus on measurable outcomes related to safety, energy efficiency, resilience, and occupant well-being. This flexibility enables innovation while maintaining regulatory oversight. Emerging technologies such as digital twins and continuous monitoring systems support performance-based compliance by providing real-time evidence of building performance. Future regulatory systems may increasingly rely on operational data rather than periodic inspections alone.

13.1.13 Future directions

The challenges associated with smart structures are significant but not insurmountable. Advances in communication technologies, artificial intelligence, cybersecurity frameworks, interoperability standards, and digital asset management continue to improve implementation feasibility. Future developments are expected to include:

- greater standardization across technology platforms;
- enhanced cybersecurity regulations;
- broader adoption of digital twins;
- automated compliance verification;
- increased use of performance-based codes;
- improved interoperability standards;
- expansion of workforce training programs.

As these developments mature, smart structures are likely to transition from innovative exceptions to mainstream components of the built environment.

The implementation of smart structures represents one of the most significant transformations in the history of architecture and engineering. While intelligent technologies offer substantial opportunities for improving sustainability, resilience, efficiency, and user experience, their successful deployment requires overcoming numerous technical and regulatory challenges.

Issues related to integration, reliability, cybersecurity, data management, scalability, workforce development, and regulatory adaptation must be addressed through coordinated efforts among designers, engineers, technology providers, facility managers, policymakers, and regulatory authorities. The future success of smart structures depends not only on technological innovation but also on the development of governance frameworks, professional expertise, and regulatory systems capable of supporting increasingly intelligent and adaptive built environments (Labib, 2025). Ultimately, overcoming these challenges will enable the realization of buildings and infrastructure that are not only technologically advanced but also safe, resilient, sustainable, and responsive to the evolving needs of society.

13.2 Human factors and ethical considerations

The integration of smart technologies into buildings and infrastructure has fundamentally redefined the relationship between humans and the built environment. Architecture is no longer limited to providing static physical shelter but increasingly operates as an adaptive, data-driven, and interactive system capable of sensing, interpreting, and responding to human presence, environmental conditions, and behavioral patterns. This transformation represents a structural shift in the ontological nature of architecture itself, where buildings transition from passive spatial containers into active socio-technical systems (Martínez Ruiz et al., 2025).

In this emerging paradigm, the built environment is no longer defined solely by material form, structural logic, or spatial composition. Instead, it is increasingly shaped by computational processes that continuously translate human behavior into operational decisions. The result is an architectural condition in which space becomes dynamic, contingent, and partially autonomous, mediated by layers of embedded intelligence distributed across physical and digital infrastructures.

This transformation has introduced a fundamental paradigm shift in architectural thinking. Buildings are no longer understood as fixed entities with stable performance characteristics but as evolving systems that continuously adjust their behavior in response to real-time inputs. Through embedded sensing networks, machine learning algorithms, automation platforms, and responsive environmental controls, smart buildings establish a continuous feedback relationship between occupants and spatial systems. In this configuration, architecture becomes an intermediary interface between human activity and environmental performance rather than a static background for occupation.

From a systems perspective, this shift can be understood as the emergence of cyber-physical architectural environments, where physical space and computational intelligence are inseparably integrated. In such environments, spatial performance is no longer predetermined solely at the design stage but is continuously produced through interaction between users, systems, and environmental variables. This introduces a temporal dimension to architecture in which building behavior unfolds dynamically over time rather than remaining fixed after construction.

While this evolution creates significant opportunities for comfort, energy efficiency, sustainability, and operational performance, it simultaneously introduces complex human, ethical, psychological, and social challenges. The increasing mediation of human experience through automated systems raises questions about how much control users retain over their environments and how much authority is delegated to algorithmic processes. These concerns are not merely technical but relate directly to the lived experience of space, shaping how individuals perceive comfort, autonomy, and agency within intelligent environments.

At the human level, the introduction of adaptive and responsive systems alters the traditional relationship between occupants and spatial stability. In conventional architectural environments, users develop stable expectations regarding thermal conditions, lighting levels, acoustic behavior, and spatial consistency. Smart environments, however, introduce variability as a core operational principle. While this variability is designed to optimize performance, it can also disrupt habitual patterns of spatial understanding, requiring occupants to continuously adjust their behavioral and cognitive expectations.

This dynamic interaction places increased emphasis on human-centered design logic, where technological systems must be aligned with the cognitive, behavioral, and perceptual capacities of users (Ageli and Chassiakos, 2026). The effectiveness of smart architecture is therefore not determined solely by computational performance or energy efficiency but also by how intuitively occupants can understand, anticipate, and engage with system behavior. In this sense, intelligibility becomes as important as intelligence.

A further implication of this transformation is the redefinition of architectural responsibility. In traditional design frameworks, responsibility is largely associated with material performance, structural safety, and functional adequacy. In smart environments, however, responsibility extends into the realm of algorithmic behavior, system responsiveness, and data-mediated decision-making. This expands the ethical domain of architecture beyond physical construction into the governance of technological behavior embedded within built systems.

Within this expanded framework, several foundational human factors become critical. First, the notion of user agency becomes central. Occupants must retain the ability to influence or override automated environmental decisions, ensuring that technological optimization does not eliminate personal preference or contextual variability. Second, system transparency becomes essential, as occupants need to understand how and why environmental changes occur. Third, adaptive trust formation emerges as a key psychological process, where users develop confidence in systems based on the consistency, predictability, and perceived fairness of responses.

These dimensions collectively indicate that smart architecture is not only a technological challenge but also a cognitive and experiential one. The success of intelligent environments depends on their ability to align computational logic with human interpretive frameworks. Ethically, this transformation requires a broader reconsideration of the role of architecture in shaping human behavior (Krakowski, 2025). As buildings become capable of influencing occupancy patterns, movement flows, energy consumption habits, and even behavioral routines, they begin to participate in subtle forms of behavioral conditioning. This introduces important questions about the extent to which architectural systems should guide or influence human action versus simply respond to it.

In addition, the increasing interdependence between human activity and computational systems creates a new form of relational dependency, where occupants rely on automated systems for environmental stability, while systems rely on human-generated data for operational optimization. This mutual dependency reshapes traditional hierarchies between users and built environments, establishing a coevolutionary relationship between human behavior and architectural performance.

At a broader societal level, these developments also raise questions regarding inclusivity and accessibility. Different user groups may experience smart environments in significantly different ways depending on their technological literacy, cognitive abilities, age, cultural background, and familiarity with automation systems. Ensuring equitable access to intelligible and controllable environments becomes a key design and ethical requirement.

Ultimately, the increasing integration of smart technologies into architecture demands a redefinition of what constitutes successful design. Performance can no longer be measured solely through energy efficiency, spatial optimization, or structural efficiency. Instead, it must also account for human experience, cognitive clarity, behavioral adaptability, and ethical alignment. Architecture must therefore operate simultaneously as a technical system, a behavioral framework, and a socio-ethical construct.

Understanding human factors in smart architecture is therefore essential for ensuring that technological innovation enhances rather than undermines human well-being. The challenge is not simply to make buildings smarter, but to ensure that intelligence is meaningfully aligned with human perception, autonomy, and dignity. In this sense, the future of smart architecture depends on achieving a careful equilibrium between computational optimization and human-centered spatial experience, where technology serves as an enabler of enriched human life rather than a replacement for human agency.

Human-centered smart architecture is grounded in the principle that technological systems must serve human needs rather than dominate them. In this approach, intelligence is not considered an autonomous goal or technological endpoint, but rather a mediating capacity that enhances the quality, comfort, safety, and meaningfulness of human life within built environments. The value of smart systems is therefore measured not only through computational efficiency or environmental optimization but also through their ability to support human experience in a way that remains coherent, intuitive, and contextually appropriate (Shah et al., 2025).

Within this framework, the built environment is understood as a human-first adaptive system, where technological infrastructure is embedded to extend human capabilities rather than replace them. Smart architecture is expected to augment perception, reduce cognitive effort in managing environmental conditions, and create environments that are responsive without becoming intrusive. This requires a fundamental shift from technology-driven design to experience-driven design, where human needs define the operational logic of intelligent systems. Smart systems in human-centered environments are typically designed to enhance multiple interrelated dimensions of spatial experience. These include:

- thermal comfort, through adaptive heating, cooling, and ventilation strategies that respond to occupancy patterns and microclimatic variations;
- visual comfort, through dynamic daylight control, adaptive artificial lighting, and glare reduction systems;
- acoustic quality, through active noise control, spatial sound management, and adaptive material systems;
- spatial usability, by enabling flexible layouts, responsive spatial configurations, and context-aware space allocation;
- cognitive and emotional well-being, through environments that reduce stress, support concentration, and align with human psychological preferences;
- energy efficiency without user burden, where optimization occurs in the background without requiring continuous user intervention or behavioral adaptation.

What distinguishes human-centered smart architecture from purely technological approaches is not the presence of these systems, but the way they are orchestrated in relation to human perception and interaction. The objective is not simply to automate environmental control but to ensure that automation remains supportive, predictable, and contextually sensitive. However, achieving this balance requires careful calibration between technological autonomy and human agency. As systems become more intelligent, they are increasingly capable of making autonomous decisions regarding environmental regulation, spatial adaptation, and operational optimization. While this autonomy can significantly improve performance outcomes, it also introduces the risk of diminishing human control over immediate environmental conditions.

Excessive automation may result in environments that feel overly regulated, where occupants experience limited ability to adjust or influence their surroundings. In such cases, the built environment may become perceptually opaque, meaning users are unable to understand the logic governing system behavior. This lack of transparency can weaken trust and reduce emotional attachment to space, ultimately undermining user satisfaction even if technical performance indicators improve.

On the other hand, insufficient automation limits the potential of smart systems to respond dynamically to changing conditions. Without adequate automation, buildings may fail to optimize energy consumption, adapt to occupancy fluctuations, or respond effectively to environmental variations. This creates inefficiencies and reduces the overall value proposition of intelligent building systems. The challenge, therefore lies not in maximizing automation but in calibrating its degree and scope in relation to human expectations and contextual requirements.

Human-centered design in smart architecture therefore emphasizes the concept of balanced intelligence. In this model, technological systems operate as supportive infrastructures that enhance human decision-making rather than replace it. Automation is strategically distributed across functions where it adds clear value, while critical or preference-sensitive decisions remain accessible to human control. This hybrid configuration allows systems to operate efficiently while preserving human autonomy and adaptability.

A further essential principle in human-centered smart architecture is system legibility and interpretability. For occupants to meaningfully engage with smart environments, they must be able to form understandable mental models of how building systems operate. Legibility refers to the clarity with which system behavior can be perceived, while interpretability refers to the ability to understand the underlying logic behind system actions.

In practical terms, this means that occupants should be able to recognize why environmental changes occur, such as variations in lighting intensity, temperature adjustments, or spatial reconfigurations. Equally important is the ability to understand how individual actions—such as movement, occupancy, or device interaction—contribute to system responses. Without this feedback loop, users may perceive smart environments as unpredictable or externally controlled, leading to discomfort or disengagement.

To support legibility, smart systems often require carefully designed interaction interfaces and feedback mechanisms. These may include visual dashboards, ambient indicators, mobile applications, or subtle environmental cues embedded within the architecture itself. The goal is not to overwhelm users with technical data, but to provide meaningful, context-sensitive information that supports intuitive understanding.

Another important dimension of human-centered smart architecture is the recognition that users are not homogeneous. Different occupants possess varying levels of technological literacy, cognitive capacity, cultural expectations, and behavioral adaptability. A well-designed smart environment must therefore accommodate a wide spectrum of user profiles, ensuring that both technologically experienced and nonexperienced individuals can interact comfortably with the system.

This introduces the need for adaptive interaction design, where systems adjust not only environmental conditions but also communication styles and levels of complexity based on user preferences or behavior. In such systems, experienced users may access advanced controls and detailed system data, while other users may interact through simplified, automated interfaces that require minimal input.

Ultimately, human-centered smart architecture reframes intelligence in buildings as a relational quality rather than a purely computational one. Intelligence is not defined solely by the sophistication of algorithms or the density of sensors but by the quality of interaction between systems and occupants. The effectiveness of a smart environment is therefore measured by its ability to support human comfort, autonomy, understanding, and emotional engagement in a coherent and balanced manner.

In this sense, human-centered design establishes a critical boundary condition for technological development in architecture: smart systems must remain aligned with human perception, cognitive capacity, and experiential expectations. The goal is not to create environments that operate independently of human input but to design spaces where human presence remains central, meaningful, and empowered within an increasingly intelligent built environment (Bibri et al., 2026).

One of the defining characteristics of smart architecture is its capacity to respond dynamically to changing environmental conditions, operational requirements, and patterns of human activity. Unlike conventional buildings, where environmental conditions remain relatively stable unless manually adjusted, intelligent environments continuously modify their behavior through automated control systems, predictive analytics, and real-time monitoring. As a result, occupants are no longer interacting with a fixed physical setting but with an environment that evolves over time in response to both internal and external stimuli.

This dynamic nature fundamentally alters the way individuals perceive, understand, and engage with architectural space. Occupants must adapt not only to physical environments but also to the behaviors of the technological systems embedded within them. Adaptation therefore becomes a two-way process: while smart buildings are designed to learn from user behavior and environmental conditions, users simultaneously learn how the building responds to their actions and changing circumstances.

Examples of such adaptive behavior include:

- lighting systems that modify illumination levels according to daylight availability, occupancy status, or user preferences;
- heating, ventilation, and air-conditioning systems that regulate indoor conditions based on occupancy density, weather forecasts, and energy demand;

- dynamic façade and shading systems that respond to solar exposure, glare conditions, and seasonal climatic variations;
- flexible and reconfigurable spaces that alter layouts or operational functions according to changing patterns of use.

These adaptive mechanisms contribute to improved environmental performance, reduced resource consumption, and enhanced operational efficiency. However, the introduction of continuous environmental variability also creates new behavioral and psychological demands for occupants. Users must become accustomed to spaces that may not behave in exactly the same manner from one moment to the next, even when physical conditions appear similar.

The process of adaptation is particularly important during the initial stages of occupancy. When users first encounter intelligent environments, they often bring expectations formed through experiences with conventional buildings. These expectations may not align with the operational logic of smart systems. For example, occupants accustomed to manually adjusting thermostats may initially find automated climate control systems unfamiliar or restrictive. Similarly, changes in lighting intensity throughout the day may be interpreted as system malfunctions rather than intentional responses to environmental conditions.

Successful adaptation therefore depends on the development of familiarity between occupants and building systems. Over time, users gradually learn how environmental controls function, how spaces respond to occupancy patterns, and how system behavior aligns with broader performance objectives. This learning process contributes to the formation of confidence and comfort within intelligent environments.

Adaptability is influenced by multiple dimensions of human behavior, including cognitive, emotional, social, and cultural factors. Cognitive adaptability refers to an individual's ability to understand and interpret system behavior. Emotional adaptability relates to comfort with environmental variability and changing conditions. Social adaptability concerns the ways groups of occupants negotiate shared environmental settings, while cultural adaptability reflects differing attitudes toward technology, automation, and environmental control.

Research in environmental psychology and human–building interaction consistently demonstrates that occupants develop internal representations, often referred to as mental models, of how buildings function. These mental models allow individuals to predict environmental responses and make informed decisions regarding their behavior within a space. In conventional buildings, such models are generally simple because environmental relationships are relatively direct and predictable. Opening a window increases ventilation, adjusting a thermostat changes temperature, and switching on a light provides illumination (Han and Mo, 2026).

In smart buildings, however, the relationships between actions and outcomes may be more complex. Environmental changes often result from combinations of sensor data, predictive algorithms, and optimization processes that are not immediately visible to occupants. Consequently, users may struggle to understand why specific changes occur, particularly when multiple variables influence system behavior simultaneously. For this reason, adaptability depends heavily on three fundamental conditions:

13.2.1 Predictability

Predictability refers to the extent to which occupants can anticipate system responses based on observable conditions and prior experience. When users can reasonably predict how a building will behave, they develop confidence in the system and are more likely to accept automated interventions. Predictability does not require that systems remain static. Rather, it requires that changes occur according to understandable patterns. Occupants are generally willing to accept adaptive behavior when the logic behind system actions appears coherent and rational. Conversely, unpredictable environmental changes may create uncertainty and reduce satisfaction, even when objective performance metrics improve.

13.2.2 Consistency

Consistency concerns the reliability with which systems apply their operational logic over time. Adaptive systems must avoid erratic or contradictory responses that undermine user understanding. If similar conditions repeatedly produce different outcomes without a clear explanation, occupants may perceive the system as unreliable or arbitrary. Consistency is particularly important in shared environments such as offices, educational facilities, healthcare buildings, and public institutions, where multiple users interact with common environmental systems. Stable operational patterns enable occupants to develop expectations that support comfort, productivity, and efficient use of space. At the same time, consistency must be balanced with adaptability. Systems should respond appropriately to changing conditions while maintaining a recognizable operational framework that users can understand and trust.

13.2.3 Feedback clarity

Feedback clarity refers to the effectiveness with which systems communicate their status, actions, and intentions to occupants. Adaptive environments function most successfully when users receive meaningful information regarding environmental changes and system responses.

Feedback may be delivered through various mechanisms, including:

- digital dashboards and mobile interfaces;
- ambient visual indicators integrated into architectural elements;
- user notifications explaining environmental adjustments;
- interactive control systems providing real-time performance information.

Clear feedback reduces uncertainty and supports informed interaction between occupants and building systems. It also helps users understand the relationship between their actions and resulting environmental outcomes, reinforcing a sense of participation rather than passive observation. Beyond these factors, occupant adaptability is strongly influenced by perceived control. Numerous studies have shown that individuals are more likely to accept automated systems when they retain some degree of influence over environmental conditions. The availability of override options, preference settings, or personalized control interfaces can significantly improve user acceptance, even if such controls are used infrequently.

The concept of perceived control is particularly important because satisfaction is often shaped less by actual control than by the knowledge that control is available when needed. Occupants who feel empowered to influence their surroundings typically demonstrate higher levels of trust and comfort than those who perceive themselves as subject to uncontrollable automated processes.

Adaptability is also linked to learning and behavioral evolution over time. Smart environments create opportunities for reciprocal adaptation, where both occupants and systems continuously adjust to one another. Advanced building management platforms may learn user preferences, occupancy schedules, and environmental sensitivities, while occupants gradually become familiar with system capabilities and operational patterns. This coadaptive relationship represents a significant departure from traditional building-user interactions and forms one of the defining characteristics of intelligent environments.

Another important consideration is the diversity of occupant populations. Adaptation rates vary considerably among individuals based on factors such as age, technological literacy, professional background, cognitive style, and previous experience with digital systems. Younger users and technologically experienced occupants may adapt rapidly to intelligent environments, whereas other groups may require additional support, training, or simplified interfaces. Inclusive design strategies must therefore accommodate varying levels of adaptability and ensure that smart environments remain accessible to all users.

In organizational settings, occupant adaptability can also influence the overall effectiveness of smart building initiatives. Buildings may achieve exceptional technical performance according to energy or operational metrics, yet fail to deliver intended benefits if users do not understand or accept system behavior. Postoccupancy evaluations frequently reveal that user satisfaction depends as much on behavioral integration and communication strategies as on technical system performance.

Consequently, the implementation of smart architecture should not be viewed solely as a technological deployment process. It should also be understood as a process of behavioral transition, requiring education, engagement, and ongoing interaction between occupants and intelligent systems. Designers, facility managers, and technology providers must therefore consider occupant adaptation as an integral component of building performance rather than a secondary concern.

Ultimately, the long-term success of smart environments depends on the establishment of a harmonious relationship between human behavior and technological responsiveness. Adaptive systems must be designed not only to react intelligently to changing conditions but also to support human understanding, confidence, and engagement. When occupants can successfully adapt to dynamic environments, smart buildings become more than efficient technological systems; they evolve into responsive environments that foster comfort, productivity, and meaningful interaction between people and the spaces they inhabit.

13.2.4 Psychological and cultural dimensions

Human responses to smart environments are shaped not only by technical performance but also by psychological and cultural factors. Different user groups may interpret automation differently depending on cultural norms, prior experiences, and technological literacy. In some cultures, automation is associated with efficiency and progress, while in others it may be perceived as intrusive or controlling. Similarly, generational differences influence acceptance levels, with younger users generally showing higher tolerance for digital systems. Key psychological factors influencing user experience include:

- perceived control;
- transparency of system behavior;
- trust in automation;
- cognitive load;
- environmental predictability.

Designing successful smart environments therefore requires an understanding of human cognition and behavioral patterns, not just technical optimization.

As smart buildings increasingly rely on data-driven decision-making, the collection, storage, and processing of occupant and operational data have become critical ethical considerations. The effectiveness of intelligent building systems depends on the continuous acquisition of information generated through sensors, connected devices, and digital platforms that monitor environmental conditions, energy consumption, occupancy patterns, and user interactions. While these technologies enable enhanced operational efficiency, predictive maintenance, energy optimization, and personalized environmental control, they also raise significant concerns regarding privacy, autonomy, transparency, and data governance. One of the most critical ethical challenges in smart architecture is the extensive collection of data related to human behavior. Smart buildings frequently employ sensing technologies capable of monitoring occupancy, movement, environmental interactions, energy-use patterns, and, in some cases, biometric indicators. Although such information is essential for optimizing building performance and improving user comfort, it simultaneously introduces risks associated with surveillance and privacy intrusion.

Ethical concerns emerge from several interconnected issues, including the continuous monitoring of occupants, the lack of informed consent regarding data collection practices, the presence of invisible or poorly understood sensing systems, the secondary use of personal information beyond its original purpose, and the potential for data breaches or unauthorized access. These challenges are further complicated by the often ambiguous distinction between operational building data and personal data. Information collected for building management purposes may inadvertently reveal sensitive details about individual behaviors, daily routines, occupancy schedules, social interactions, work patterns, or lifestyle habits. For example, occupancy data intended to optimize heating, ventilation, and air-conditioning operations may also provide insights into when occupants arrive, leave, or move within a building. Consequently, even seemingly anonymous datasets can generate detailed behavioral profiles when analyzed over extended periods.

Table 13.3 Ethical risks in data collection systems.

Data type	Example	Ethical concern
Occupancy data	Room usage patterns	Behavioral profiling
Environmental interaction	Temperature adjustment behavior	Indirect personal tracking
Biometric data	Access control systems	Identity sensitivity
Energy usage data	Appliance monitoring	Lifestyle inference
Location tracking	Indoor navigation systems	Surveillance risk

The increasing sophistication of artificial intelligence, machine learning algorithms, and predictive analytics further amplifies these concerns by enabling the extraction of meaningful patterns from large datasets. As a result, the ethical management of data in smart buildings extends beyond technical cybersecurity measures and encompasses broader questions of transparency, accountability, user rights, data ownership, and responsible governance. To ensure that technological innovation remains aligned with human values, smart building systems must establish clear policies regarding data collection, consent mechanisms, information access, retention periods, and protection against misuse. In this context, understanding the ethical risks associated with data collection systems is essential for developing intelligent built environments that balance technological performance with the protection of occupant privacy and trust. Therefore, Table 13.3 summarizes the principal ethical risks associated with data collection systems in smart buildings and highlights the implications of these risks for occupants, building operators, and society as a whole.

13.2.5 Data ownership and governance structures

Data ownership represents one of the most complex ethical and managerial challenges in smart architecture. Unlike conventional buildings, where operational information is relatively limited and decentralized, smart buildings generate continuous streams of data through sensors, automation systems, occupancy monitoring devices, and digital management platforms. This data is valuable for improving building performance, optimizing energy consumption, enhancing occupant comfort, and supporting predictive maintenance. However, determining ownership and control of such information is often difficult because multiple stakeholders participate in its generation, collection, management, and use.

Building owners may regard operational data as part of the asset management process, while tenants and occupants may view information related to their activities and behaviors as personal data. Facility managers require access to operational information to maintain building performance, whereas technology providers and service contractors may collect and process data through proprietary software platforms. As a result, ownership rights, access privileges, and responsibilities become distributed across a complex network of actors, each with different interests and objectives (Bakar et al., 2024).

Without clearly defined governance frameworks, significant ethical risks may emerge. Building-generated data can potentially be commercialized without the knowledge of occupants, shared with third parties without explicit consent, stored in insecure digital environments, or repurposed for uses that extend beyond the original intent of data collection. Such practices can undermine trust, create legal liabilities, and raise concerns regarding privacy and individual rights.

To address these challenges, smart buildings require comprehensive data governance structures that clearly define ownership responsibilities, access permissions, and data management procedures. Effective governance frameworks typically include explicit definitions of data ownership, hierarchical access control systems, consent-based data collection mechanisms, anonymization and encryption protocols, and transparency requirements that inform occupants about how information is collected, stored, and utilized. Increasingly, the concept of privacy-by-design has become a fundamental principle in smart building development. Rather than treating privacy as a secondary consideration addressed after system deployment, privacy protections should be integrated into the architecture of digital systems from the earliest design stages. Such an approach helps ensure that ethical safeguards evolve alongside technological capabilities and become embedded within the operational logic of intelligent environments.

13.2.6 Ethics of automation and algorithmic decision-making

As smart buildings become more sophisticated, decision-making processes are increasingly delegated to artificial intelligence systems, machine learning models, and predictive algorithms. These technologies regulate a wide range of building functions, including heating and cooling operations, lighting control, security management, energy distribution, occupancy scheduling, and space utilization. Through continuous analysis of environmental and behavioral data, automated systems can respond more rapidly and efficiently than traditional manual control mechanisms. However, the growing reliance on algorithmic decision-making introduces important ethical questions concerning accountability, transparency, fairness, and human oversight.

One of the primary concerns involves the opacity of artificial intelligence systems. Many advanced algorithms operate as complex computational models whose internal decision-making processes are difficult for occupants, facility managers, and even system developers to fully understand. When users cannot determine why a particular environmental adjustment has occurred, trust in the system may decline. Furthermore, algorithms are inherently dependent on the quality and representativeness of the data used to train them. If datasets contain biases, omissions, or inaccuracies, these shortcomings may be reflected in operational decisions.

Algorithmic bias can produce unequal outcomes across different users or spaces within a building. For example, an energy optimization system may prioritize efficiency targets while inadvertently reducing comfort levels in specific zones. Similarly, predictive models may favor commonly occupied areas while overlooking the needs of less frequently used spaces. Such situations raise concerns regarding fairness and equitable treatment of building occupants.

Another ethical issue relates to the gradual reduction of human involvement in decision-making processes. Excessive dependence on automation may weaken human situational awareness and limit opportunities for intervention when unexpected circumstances arise. In safety-critical environments, including healthcare facilities, transportation hubs, and emergency response centers, maintaining appropriate levels of human oversight is particularly important.

To mitigate these risks, many researchers advocate the adoption of human-in-the-loop approaches, in which automated systems support rather than replace human decision-making. Under this framework, artificial intelligence functions as an intelligent advisor capable of processing large volumes of information and generating recommendations, while final authority remains with human operators. Such arrangements promote accountability, preserve professional judgment, and ensure that critical decisions remain subject to ethical and contextual evaluation.

13.2.7 Responsible innovation in smart architecture

The concept of responsible innovation has emerged as an important framework for guiding the development of intelligent built environments. Responsible innovation emphasizes that technological progress should not be evaluated solely according to technical performance or economic benefits. Instead, innovation must be assessed in relation to its broader social, environmental, and ethical implications (Bastidas et al., 2023). In the context of smart architecture, this requires continuous reflection on how technological systems influence human well-being, social equity, environmental sustainability, and long-term societal development.

Responsible innovation encourages designers and decision-makers to examine fundamental questions regarding the distribution of benefits and risks associated with smart technologies. It asks who gains access to advanced building services, who may be excluded from technological advantages, how potential risks are distributed among different groups, and what unintended consequences may emerge over time. Such considerations are particularly important because intelligent environments increasingly shape daily routines, patterns of interaction, and access to resources.

Importantly, responsible innovation recognizes that technological sophistication is not always synonymous with improved outcomes. In some contexts, highly automated solutions may introduce unnecessary complexity, reduce user autonomy, or create maintenance challenges that outweigh their benefits. Simpler and more transparent systems may sometimes provide more appropriate responses to human needs than advanced technological alternatives. Consequently, the objective of smart architecture should not be to maximize automation but rather to achieve an appropriate balance between technological capability and human value.

Achieving this balance requires collaboration among a diverse range of stakeholders, including architects, engineers, computer scientists, ethicists, policymakers, facility managers, and end users. Through interdisciplinary cooperation, technical innovation can be aligned with social priorities, ensuring that intelligent buildings contribute positively to both individual and collective well-being.

13.2.8 Human–building interaction framework

The relationship between occupants and intelligent environments can be understood through a human–building interaction framework that conceptualizes buildings as active participants in a continuous feedback process. Within this framework, human behavior influences building operations through occupancy patterns, environmental preferences, movement, and interaction with digital interfaces. Building systems interpret these inputs through sensors and analytical algorithms, generating automated responses that modify environmental conditions. These changes, in turn, affect occupant perceptions, decisions, and behaviors, creating an ongoing cycle of mutual influence.

Rather than operating as isolated entities, occupants and buildings become interconnected components of a dynamic adaptive system. The quality of this interaction significantly influences user satisfaction, trust, comfort, and overall building performance. Effective human–building interaction depends on transparent communication, intuitive interfaces, meaningful feedback mechanisms, and opportunities for user participation. When these elements are successfully integrated, intelligent environments can foster productive relationships between technological systems and human users, enabling buildings to respond effectively while remaining aligned with occupant needs and expectations.

13.2.9 Ethical ecosystem model

Ethical considerations in smart architecture extend beyond individual technologies or isolated design decisions. Intelligent buildings operate within a broader ecosystem that encompasses users, technological infrastructures, data flows, institutional governance structures, regulatory frameworks, and environmental systems. Consequently, ethical decision-making cannot be addressed through a single intervention but must be understood as a multidimensional process involving interconnected actors and influences.

Within this ecosystem, technological decisions affect data management practices, regulatory compliance, environmental performance, and human experiences simultaneously. Similarly, legal regulations influence technological design choices, while user expectations shape governance policies and operational procedures. Recognizing these interdependencies highlights the need for holistic ethical frameworks capable of coordinating decisions across multiple domains. The ethical ecosystem perspective emphasizes that responsible smart architecture requires continuous collaboration among stakeholders and an integrated approach to governance, technology, and human well-being.

13.2.10 Case studies of human experience in smart buildings

Empirical studies of smart building environments demonstrate both the opportunities and challenges associated with intelligent architectural systems. In advanced office buildings, automated climate control, adaptive lighting systems, and occupancy-based environmental management have produced measurable improvements in energy efficiency, thermal comfort, and operational performance. Nevertheless, postoccupancy evaluations frequently reveal concerns related to reduced user control and difficulties understanding automated system behavior. Although occupants often appreciate improved environmental quality, some express dissatisfaction when they are unable to influence settings or interpret system decisions.

Smart residential developments provide additional insights into the interaction between technology and everyday life. Home automation systems, energy monitoring platforms, and intelligent security technologies offer significant benefits in terms of convenience, resource conservation, and safety. However, residential occupants often demonstrate heightened sensitivity to privacy concerns because monitoring technologies operate within personal living environments. Continuous data collection, even when intended for beneficial purposes, may be perceived as intrusive if occupants are not fully informed about its scope and purpose.

Large institutional facilities and smart campuses present another set of challenges. Centralized management systems enable efficient control of energy use, space allocation, and security operations across extensive building portfolios. At the same time, these systems concentrate substantial quantities of personal and operational data within integrated digital platforms. This concentration of information raises important questions regarding surveillance, governance accountability, access control, and institutional responsibility. These examples illustrate that the success of smart architecture depends not only on technical performance but also on the ability to maintain trust, transparency, and ethical stewardship throughout the lifecycle of intelligent environments.

13.2.11 Toward ethical and human-centered smart architecture

The future development of smart architecture depends not only on advances in sensing technologies, artificial intelligence, and automation systems but also on the integration of ethical responsibility into every stage of design, implementation, and operation. Intelligent buildings must be understood as lived environments that shape human behavior, cognition, well-being, and social interaction. Consequently, ethical considerations should occupy a central position within the development of smart architectural systems rather than being treated as secondary concerns.

Future intelligent environments should prioritize transparency in system operation, meaningful user participation, responsible data governance, cultural sensitivity, explainable artificial intelligence, and privacy-centered design strategies. These principles support the creation of environments that remain responsive and efficient while preserving human autonomy and dignity. Ultimately, the long-term success of smart architecture will be determined by its ability to balance technological capability with ethical responsibility. Smart environments must evolve as human-centered ecosystems in which technology functions as an enabling resource that enhances human experience, supports social well-being, and contributes to more equitable and sustainable built environments.

13.3 Commercialization and industry adoption

The successful development of smart structures depends not only on technological innovation but also on the ability of these innovations to achieve widespread commercialization and industry adoption. While advances in sensing technologies, intelligent materials, digital twins, artificial intelligence, building information modeling, and adaptive systems have transformed the possibilities of modern architecture and engineering, the long-term impact of these technologies is ultimately determined by their acceptance within the marketplace. Commercialization represents the process through which scientific discoveries, engineering innovations, and research outcomes are transformed into practical products, services, and solutions that generate economic value. Industry adoption, in turn, refers to the degree to which these innovations become integrated into standard design, construction, operation, and maintenance practices across the built environment sector (Alhassan et al., 2024).

The commercialization of smart structures has accelerated significantly over the past two decades due to several converging factors. Increasing urbanization, growing concerns about energy efficiency, advancements in digital technologies, and the emergence of sustainability-focused regulations have created favorable conditions for intelligent building systems. Simultaneously, declining costs of sensors, communication networks, cloud computing platforms, and data analytics tools have made smart technologies more accessible to a broader range of stakeholders. As a result, smart structures have evolved from experimental research projects and demonstration buildings into commercially viable solutions adopted across diverse building typologies, including airports, healthcare facilities, commercial offices, educational campuses, industrial complexes, and mixed-use developments.

The market evolution of smart structures reflects broader transformations occurring throughout the architecture, engineering, construction, and operations sectors. Historically, buildings were designed as static systems with limited capacity to respond to changing environmental conditions or user needs. Mechanical and electrical systems operated independently, data collection was minimal, and facility management relied heavily on manual inspections and reactive maintenance practices. The introduction of building automation systems during the late twentieth century represented an important milestone by enabling centralized control of heating, ventilation, air-conditioning, lighting, and security systems. However, these early systems remained largely isolated and lacked the interoperability required for comprehensive building intelligence.

The emergence of digital communication technologies, IoT platforms, and cloud-based data management systems fundamentally altered this landscape. Buildings gradually evolved from isolated physical assets into interconnected cyber-physical systems capable of collecting, processing, and responding to real-time information.

Smart structures began integrating sensors that monitored structural performance, environmental conditions, occupancy patterns, energy consumption, and equipment operation. Data generated by these systems enabled predictive maintenance, operational optimization, and evidence-based decision-making. Consequently, buildings increasingly became active participants in their own management processes rather than passive containers of human activities.

The commercialization pathway of smart structures typically follows a multistage progression that transforms innovative concepts into market-ready solutions (Nieto Cubero et al., 2021). Initial research and development activities focus on technological feasibility, proof-of-concept demonstrations, and laboratory testing. Successful technologies then move toward pilot projects and prototype installations where performance can be evaluated under real-world conditions. Following validation, commercialization efforts emphasize product development, regulatory compliance, certification, market positioning, and business model creation. Eventually, widespread industry adoption occurs as technologies become integrated into standard professional practices and supported by mature supply chains, regulatory frameworks, and user acceptance. Fig. 13.1 illustrates the typical commercialization pathway associated with smart structure technologies.

As shown in the illustration, commercialization is not a linear process but rather an iterative cycle involving continuous feedback between technology developers, industry stakeholders, regulatory authorities, investors, and end users. Market adoption often requires multiple cycles of refinement before technologies achieve sufficient reliability, affordability, and usability to become commercially successful.

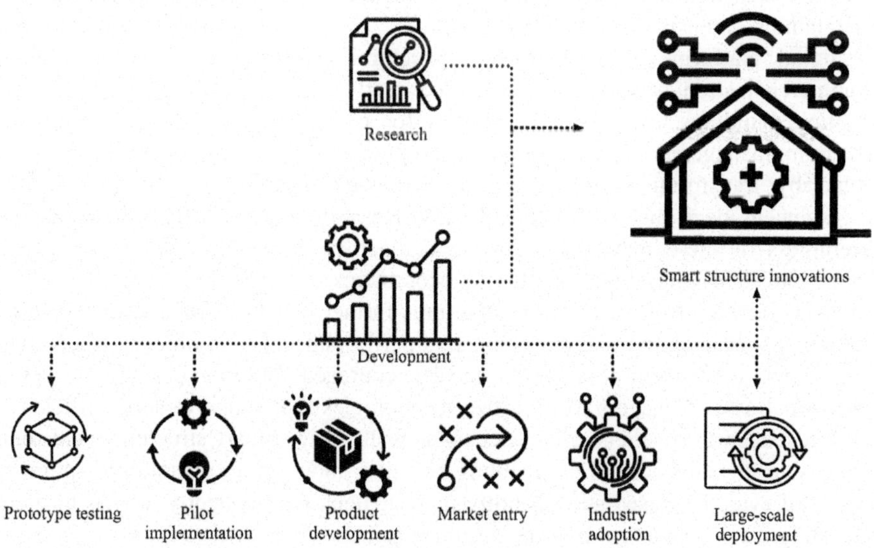

Figure 13.1 Commercialization pathway for smart structure technologies. Commercialization pathway illustrating the progression of smart structure innovations from research and development through prototype testing, pilot implementation, product development, market entry, industry adoption, and large-scale deployment.

The emergence of smart structures has also created new commercial business models within the built environment sector. Traditional construction industries have historically relied on project-based revenue streams centered on design, construction, and asset delivery. Smart structures, however, introduce opportunities for recurring revenue through data-driven services, digital platforms, predictive maintenance programs, and performance-based contracts. These new business models shift the focus from delivering physical infrastructure alone to providing ongoing operational value throughout the lifecycle of the asset.

One increasingly common business model involves performance-based service agreements (Hunhevicz et al., 2022). Under this approach, technology providers are compensated according to measurable improvements in building performance, such as energy savings, reduced maintenance costs, enhanced occupant comfort, or improved operational efficiency. This model aligns incentives between service providers and building owners by emphasizing long-term outcomes rather than short-term equipment sales. As a result, stakeholders share responsibility for achieving performance targets and maintaining system effectiveness throughout the operational phase of the building.

Another important commercial model is the concept of buildings as a service. Rather than purchasing individual technologies outright, clients acquire access to integrated smart building services through subscription-based arrangements. Service providers manage software platforms, data analytics systems, cybersecurity infrastructure, and operational optimization tools while continuously updating technologies to reflect evolving market conditions. This approach reduces upfront investment requirements and enables organizations to benefit from advanced capabilities without maintaining extensive in-house expertise.

Digital platforms have also become central components of smart structure commercialization strategies. Cloud-based platforms aggregate information from multiple buildings, creating opportunities for benchmarking, predictive analytics, portfolio management, and continuous optimization. Platform operators generate value by transforming raw data into actionable insights that support facility management decisions. As data volumes increase, network effects enhance the usefulness of these platforms, creating competitive advantages for organizations capable of managing large-scale digital ecosystems.

The commercialization of smart structures further benefits from growing investor interest in sustainability and environmental performance. Environmental, social, and governance investment frameworks have encouraged financial institutions, developers, and asset managers to prioritize technologies that support energy efficiency, carbon reduction, resilience, and occupant well-being. Smart structures contribute directly to these objectives by enabling real-time monitoring, operational optimization, and evidence-based sustainability management. Consequently, access to green financing mechanisms has become an important driver of market growth.

Public and private sector collaboration plays a critical role in accelerating commercialization efforts. Government agencies frequently support research initiatives, pilot projects, demonstration programs, and innovation clusters that reduce technological uncertainty and encourage market development. Public-sector investments help bridge the gap between laboratory research and commercial deployment by providing opportunities for large-scale testing under realistic operating conditions. Demonstration projects implemented within public buildings, transportation systems, healthcare facilities, and educational campuses often serve as catalysts for broader market acceptance.

At the same time, private sector organizations drive commercialization through product development, investment, manufacturing, marketing, and customer engagement. Technology firms, engineering consultants, construction companies, software developers, and facility management providers each contribute specialized expertise that supports the development of integrated smart structure ecosystems. Successful commercialization therefore depends on collaboration among diverse stakeholders whose interests extend across multiple phases of the building lifecycle.

The complexity of these interactions highlights the importance of viewing smart structures as components of broader innovation ecosystems rather than isolated technologies. Commercial success depends not only on technical performance but also on supply chain coordination, regulatory support, financing mechanisms, workforce development, market awareness, and user acceptance. Fig. 13.2 illustrates the interconnected ecosystem that supports the commercialization and industry adoption of smart structures.

As shown in the illustration, the smart structure ecosystem involves numerous actors whose collaboration influences market success. Research institutions generate innovation, technology providers develop commercial products, investors provide capital, regulatory agencies establish compliance frameworks, and building owners create demand through procurement decisions. The effectiveness of commercialization efforts depends on the alignment of these interconnected participants toward shared objectives that support innovation, sustainability, and long-term value creation.

The commercialization of smart structures is strongly influenced by the complementary roles played by public and private sector organizations. While technological innovation is frequently associated with private industry, many foundational technologies used in intelligent buildings and infrastructure have emerged through publicly funded research programs. Governments, universities, and research institutions often assume the initial risks associated with experimental technologies that may require years of development before achieving commercial viability. Through grants, innovation programs, demonstration projects, tax incentives, and regulatory support mechanisms, public institutions create conditions that encourage technological advancement and facilitate market formation.

Government policies increasingly recognize smart structures as important components of national sustainability, digitalization, and infrastructure modernization strategies. Many countries have established programs that support energy-efficient buildings, smart cities, resilient infrastructure, and digital transformation initiatives. Building performance regulations, carbon reduction targets, and energy-efficiency standards further stimulate demand for intelligent building technologies by encouraging organizations to adopt solutions capable of improving operational performance. Public procurement policies can also influence market development by incorporating

Figure 13.2 Industry ecosystem map for smart structures.

smart technologies into government facilities, transportation hubs, healthcare institutions, and educational campuses. Such projects provide valuable opportunities for technology validation while demonstrating practical benefits to potential adopters.

The private sector, however, remains the primary engine of commercialization. Technology firms transform research outcomes into marketable products, while engineering consultants and construction companies integrate these products into real-world projects. Investors and venture capital organizations provide financial resources that enable innovation to move beyond the laboratory stage. Facility management companies create operational frameworks that sustain smart technologies throughout the building lifecycle, while software developers continuously improve digital platforms that support data collection, analysis, and decision-making. Together, these stakeholders establish the commercial infrastructure required for widespread adoption.

Despite the significant progress achieved during recent decades, industry adoption of smart structures continues to face numerous challenges. One of the most significant barriers is the high initial investment associated with advanced technologies. Smart buildings often require substantial expenditures related to sensors, communication networks, data platforms, software integration, cybersecurity systems, and specialized expertise. Although lifecycle benefits frequently outweigh these costs, decision-makers may hesitate to invest when returns are perceived as uncertain or difficult to quantify. The long operational lifespan of buildings can further complicate investment decisions because benefits may accrue over extended periods that exceed conventional budgeting cycles.

Another major challenge involves interoperability and system integration. Smart structures typically combine technologies originating from multiple vendors and disciplines, including mechanical engineering, electrical engineering, information technology, telecommunications, and facility management. Differences in communication protocols, data standards, and software architectures can create compatibility issues that increase implementation complexity. Organizations may encounter difficulties when attempting to integrate legacy systems with newer technologies, particularly in existing buildings where infrastructure constraints limit upgrade opportunities.

Workforce development represents an additional obstacle to industry adoption. The operation of smart structures requires expertise that extends beyond traditional architectural and engineering knowledge. Professionals must increasingly understand data analytics, cybersecurity, artificial intelligence, digital twins, automation systems, and information management. However, many organizations continue to face shortages of personnel possessing these interdisciplinary skills. Educational institutions and professional training programs therefore play a critical role in preparing future practitioners capable of managing intelligent built environments.

Cybersecurity concerns have also emerged as significant barriers to adoption. The growing connectivity of smart structures increases exposure to digital threats that could compromise operational reliability, data integrity, or occupant safety. Building owners may be reluctant to adopt highly interconnected systems without confidence that adequate security measures are in place. Consequently, cybersecurity has become an essential component of commercialization strategies, requiring continuous investment in monitoring, protection, and risk management practices.

Resistance to organizational change further influences adoption rates. The implementation of smart technologies often requires modifications to established workflows, management practices, procurement procedures, and maintenance strategies. Stakeholders accustomed to conventional approaches may perceive these changes as disruptive or unnecessary. Building trust in new technologies requires clear evidence of performance benefits, transparent communication, and successful demonstration projects that illustrate practical value.

Although these challenges are substantial, numerous factors continue to drive market growth. Increasing urbanization, sustainability requirements, energy costs, technological advancements, and occupant expectations all contribute to rising demand for intelligent buildings and infrastructure. Understanding the relationship between these drivers and barriers is essential for developing effective commercialization strategies. Table 13.4 summarizes the principal factors influencing market

Table 13.4 Market drivers and barriers affecting the commercialization of smart structures.

Market drivers	Market barriers
Rising energy costs	High initial capital investment
Carbon reduction targets	Uncertain return on investment
Smart city initiatives	Integration complexity
Digital transformation strategies	Lack of interoperability
Advances in artificial intelligence	Workforce skill shortages
Internet of Things expansion	Cybersecurity concerns
Improved sensor affordability	Data governance challenges
Growing sustainability awareness	Regulatory fragmentation
Demand for occupant comfort	Organizational resistance to change
Predictive maintenance benefits	Limited awareness among stakeholders
Green financing opportunities	Technology obsolescence concerns
Infrastructure modernization programs	Long procurement processes
Increasing urbanization	Legacy system constraints
Enhanced operational efficiency	Market maturity differences
Real-time performance monitoring	Limited standardization

adoption. The growth of smart structures is influenced by a combination of technological, economic, environmental, regulatory, and social factors. While several drivers encourage investment and adoption, numerous barriers continue to affect implementation across different market sectors.

The above table demonstrates that successful commercialization requires not only technological innovation but also strategic efforts to address financial, organizational, regulatory, and operational barriers. The balance between drivers and barriers varies across regions, sectors, and project types, influencing adoption rates and market maturity.

Emerging markets represent one of the most promising areas for future growth in smart structure technologies. Rapid urbanization across developing economies is creating unprecedented demand for buildings, transportation systems, utilities, and public infrastructure. In many cases, these regions have the opportunity to implement advanced technologies directly rather than following the gradual modernization pathways observed in more mature markets. By incorporating digital infrastructure during initial development phases, emerging economies can potentially avoid some of the challenges associated with retrofitting older assets.

The expansion of smart city programs further accelerates demand for intelligent buildings and infrastructure (Umeike et al., 2025). Urban authorities increasingly seek integrated solutions capable of improving energy efficiency, mobility, environmental quality, public services, and resource management. Smart structures serve as critical components within these broader urban ecosystems by generating data, supporting automation, and enhancing operational performance. As smart city investments continue to increase worldwide, demand for intelligent building technologies is expected to grow correspondingly.

Another emerging market opportunity involves industrial facilities and logistics infrastructure. The growth of e-commerce, advanced manufacturing, and automated supply chains has increased interest in intelligent warehouses, distribution centers, and production facilities. Smart structures enable real-time monitoring of equipment performance, environmental conditions, energy consumption, and operational processes. These capabilities contribute to productivity improvements while supporting sustainability objectives and risk management strategies.

The healthcare sector also represents a rapidly expanding market for smart technologies. Hospitals and healthcare facilities require highly controlled environments that support patient safety, operational efficiency, and resource optimization. Intelligent building systems can monitor air quality, occupancy patterns, equipment performance, and energy use while enhancing comfort for patients, visitors, and staff. As healthcare systems face increasing pressures related to aging populations and resource constraints, smart facilities offer opportunities to improve service delivery while controlling operational costs.

Future economic opportunities associated with smart structures extend far beyond individual buildings. As digital technologies become increasingly integrated into urban systems, entire ecosystems of products and services are emerging around data management, predictive analytics, artificial intelligence, digital twins, and performance optimization. These developments are creating new employment opportunities, investment markets, and business sectors that support the ongoing evolution of intelligent built environments.

One of the most significant economic opportunities involves the growing importance of data as a strategic asset. Buildings generate vast quantities of information related to occupancy, energy consumption, environmental performance, equipment operation, and user behavior. Organizations capable of transforming this information into actionable knowledge can create substantial value through operational improvements, predictive maintenance, asset optimization, and enhanced user experiences. Consequently, data analytics services are becoming important components of the smart structure economy.

Digital twin technologies are expected to play a particularly influential role in future commercialization strategies. By creating continuously updated virtual representations of physical assets, digital twins enable real-time monitoring, simulation, forecasting, and optimization. These capabilities support more efficient facility management while reducing operational risks and maintenance costs. As digital twin platforms become increasingly sophisticated, their adoption is expected to expand across commercial, industrial, healthcare, transportation, and infrastructure sectors.

The transition toward net-zero carbon buildings further strengthens future market opportunities. Governments, investors, and organizations are increasingly committing to ambitious sustainability objectives that require improved building performance. Smart structures provide essential tools for monitoring emissions, optimizing energy consumption, integrating renewable energy systems, and supporting evidence-based sustainability management. Consequently, environmental performance requirements are likely to remain powerful drivers of commercialization throughout the coming decades.

The practical impact of commercialization can be observed through numerous real-world examples. Smart airports represent some of the most advanced applications of intelligent building technologies due to their operational complexity, large user populations, and critical infrastructure requirements. Modern airports increasingly employ integrated systems that monitor passenger flows, energy consumption, security operations, environmental conditions, and equipment performance. Real-time data enables operators to optimize resource allocation, improve passenger experiences, reduce energy consumption, and enhance operational resilience. Predictive maintenance systems help prevent equipment failures, while intelligent wayfinding technologies support efficient navigation throughout complex facilities. These capabilities demonstrate how smart structures create measurable value in large-scale transportation environments.

Healthcare facilities provide another important example of successful commercialization. Hospitals operate continuously and require highly reliable environmental control systems to support patient care. Smart healthcare facilities utilize interconnected sensors, automation platforms, and analytics systems to monitor air quality, temperature, humidity, lighting, energy use, and equipment status. Intelligent systems can automatically adjust environmental conditions according to occupancy and clinical requirements while supporting infection control measures and energy-efficiency objectives. Predictive maintenance technologies help ensure the reliability of critical medical infrastructure, reducing downtime and improving service continuity. These benefits illustrate the economic and operational value that smart structures can deliver within highly demanding environments.

Smart mixed-use developments represent a third significant area of industry adoption. These developments combine residential, commercial, retail, hospitality, and recreational functions within integrated environments that require sophisticated management strategies. Smart technologies facilitate coordination among diverse building systems while supporting sustainability, security, mobility, and occupant comfort objectives. Shared digital platforms can monitor resource consumption across multiple facilities, optimize energy distribution, coordinate maintenance activities, and provide personalized services to occupants. The integration of smart technologies enhances property value while creating more efficient and attractive urban environments.

These case studies demonstrate that commercialization success depends on the ability of smart technologies to address practical challenges while generating measurable benefits for stakeholders. Organizations are more likely to adopt intelligent systems when clear evidence exists regarding improvements in operational efficiency, sustainability performance, occupant satisfaction, risk reduction, or financial returns. Consequently, demonstration projects and documented success stories remain important mechanisms for accelerating market adoption.

The future trajectory of smart structure commercialization will likely be shaped by increasing convergence among digital technologies, sustainability objectives, and urban development strategies. Artificial intelligence, machine learning, advanced sensors, cloud computing, digital twins, robotics, and autonomous systems are expected to become increasingly interconnected. This convergence will enable higher levels of automation, predictive capability, and operational intelligence while creating new opportunities for value creation throughout the building lifecycle.

At the same time, commercialization strategies must continue addressing challenges related to affordability, interoperability, cybersecurity, workforce development, and ethical governance. Successful industry adoption requires balancing technological innovation with practical considerations that influence stakeholder acceptance. Solutions must not only demonstrate technical excellence but also provide economic value, regulatory compliance, operational reliability, and user trust.

Ultimately, the commercialization and industry adoption of smart structures represent a critical stage in the evolution of intelligent built environments. Research and innovation provide the foundation for technological advancement, but widespread societal impact depends on successful market integration. Through effective collaboration among governments, industry organizations, investors, researchers, technology providers, and end users, smart structures are increasingly transitioning from specialized innovations to mainstream components of contemporary architecture and engineering practice. As urbanization, sustainability requirements, and digital transformation continue to reshape the built environment, the commercialization of smart structures will remain a central driver of economic growth, technological progress, and long-term improvements in building and infrastructure performance.

References

Ageli, O., Chassiakos, A., 2026. Balancing indoor environmental quality and energy efficiency in buildings using digital twins technology: scientometric review and multi-layer conceptual framework. Energy and Buildings 357. Available at https://doi.org/10.1016/j.enbuild.2026.117147, https://www.sciencedirect.com/science/journal/03787788.

Alhassan, M., Alkhawaldeh, A., Betoush, N., Sawalha, A., Amaireh, L., Onaizi, A., 2024. Harmonizing smart technologies with building resilience and sustainable built environment systems. Results in Engineering 22, 102158. Available at https://doi.org/10.1016/j.rineng.2024.102158.

Bakar, A.A., Yussof, S., Ghapar, A.A., Sameon, S.S., Jørgensen, B.N., 2024. A review of privacy concerns in energy-efficient smart buildings: risks, rights, and regulations. Energies 17 (5). Available at https://doi.org/10.3390/en17050977, http://www.mdpi.com/journal/energies/.

Banad, Y.M., Sharif, S.S., Rezaei, Z., 2025. Artificial intelligence and machine learning for smart grids: from foundational paradigms to emerging technologies with digital twin and large language model-driven intelligence. Energy Conversion and Management: X 28. Available at https://doi.org/10.1016/j.ecmx.2025.101329, https://www.sciencedirect.com/science/journal/25901745.

Bastidas, V., Oti-Sarpong, K., Nochta, T., Wan, L., Tang, J., Schooling, J., 2023. Leadership for responsible digital innovation in the built environment: a socio-technical review for re-establishing competencies. Journal of Urban Management 12 (1), 57–73. Available at https://doi.org/10.1016/j.jum.2023.01.004.

Bellini, P., Nesi, P., Pantaleo, G., 2022. IoT-enabled smart cities: a review of concepts, frameworks and key technologies. Applied Sciences 12 (3), 1607. Available at https://doi.org/10.3390/app12031607.

Bibri, S.E., Huang, J., Omar, O., Krogstie, J., Yigitcanlar, T., 2026. The next frontier of generative intelligence in smarter eco-cities: a systematic review of deep generative and hybrid modeling applications for environmental sustainability. Sustainable Cities and Society 145, 107408. Available at https://doi.org/10.1016/j.scs.2026.107408.

Bokrantz, J., Skoogh, A., Berlin, C., Wuest, T., Stahre, J., 2020. Smart maintenance: an empirically grounded conceptualization. International Journal of Production Economics 223, 107534. Available at https://doi.org/10.1016/j.ijpe.2019.107534.

Daher-Armache, G., Armache, J., Ismail, H.N., 2025. Leadership development in U.S. Higher education: strategies for lifelong learning and upskilling. Journal of Innovation & Knowledge 10 (4), 100754. Available at https://doi.org/10.1016/j.jik.2025.100754.

Dwaikat, L.N., Ali, K.N., 2018. Green buildings life cycle cost analysis and life cycle budget development: practical applications. Journal of Building Engineering 18, 303–311. Available at https://doi.org/10.1016/j.jobe.2018.03.015, http://www.journals.elsevier.com/journal-of-building-engineering/.

Frighi, V., 2023. Expanding the wave of smartness: smart buildings, another frontier of the digital revolution. Urban Book Series 813, 285–294. Available at https://doi.org/10.1007/978-3-031-29515-7_26, https://www.springer.com/series/14773.

Han, J., Mo, Y., 2026. Human-building interaction in energy management systems: a systematic review with LLM-based topic modeling for user-responsive and adaptive energy systems. Applied Energy 403, 127079. Available at https://doi.org/10.1016/j.apenergy.2025.127079.

Hernández, J.L., de Miguel, I., Vélez, F., Vasallo, A., 2024. Challenges and opportunities in European smart buildings energy management: a critical review. Renewable and Sustainable Energy Reviews 199, 114472. Available at https://doi.org/10.1016/j.rser.2024.114472.

Himeur, Y., Elnour, M., Fadli, F., Meskin, N., Petri, I., Rezgui, Y., et al., 2023. AI-big data analytics for building automation and management systems: a survey, actual challenges and future perspectives. Artificial Intelligence Review 56 (6), 4929–5021. Available at https://doi.org/10.1007/s10462-022-10286-2, https://www.springer.com/journal/10462.

Hunhevicz, J.J., Motie, M., Hall, D.M., 2022. Digital building twins and blockchain for performance-based (smart) contracts. Automation in Construction 133. Available at https://doi.org/10.1016/j.autcon.2021.103981, https://www.journals.elsevier.com/automation-in-construction.

Hwang, B.G., Ngo, J., Teo, J.Z.K., 2022. Challenges and strategies for the adoption of smart technologies in the construction industry: the case of Singapore. Journal of Management in Engineering 38 (1). Available at https://doi.org/10.1061/(ASCE)ME.1943-5479.0000986, https://ascelibrary.org/journal/jmenea.

Jia, M., Komeily, A., Wang, Y., Srinivasan, R.S., 2019. Adopting Internet of Things for the development of smart buildings: a review of enabling technologies and applications. Automation in Construction 101, 111–126. Available at https://doi.org/10.1016/j.autcon.2019.01.023.

Krakowski, S., 2025. Human-AI agency in the age of generative AI. Information and Organization 35 (1), 100560. Available at https://doi.org/10.1016/j.infoandorg.2025.100560.

Labib, R., 2025. Mapping the contributions of smart buildings to ESG goals: a critical narrative review and socio-technical analysis. Sustainable Cities and Society: Advances 1 (1), 100009. Available at https://doi.org/10.1016/j.scsadv.2025.100009.

Larsson, O., Ferm, R., Bokrantz, J., Skoogh, A., 2026. Implementing smart maintenance in the manufacturing industry. Journal of Manufacturing Technology Management 37 (9), 64–85. Available at https://doi.org/10.1108/jmtm-06-2025-0501.

Martínez Ruiz, I., Cano Suñén, E., Marco Marco, Á., Fernández Cuello, Á., 2025. IoB internet of things (IoT) for smart built environment (SBE): understanding the complexity and contributing to energy efficiency; a case study in mediterranean climates. Applied Sciences 15 (4), 1724. Available at https://doi.org/10.3390/app15041724.

Monaco, R., Bergaentzlé, C., Leiva Vilaplana, J.A., Ackom, E., Nielsen, P.S., 2024. Digitalization of power distribution grids: barrier analysis, ranking and policy recommendations. Energy Policy 188, 114083. Available at https://doi.org/10.1016/j.enpol.2024.114083.

Mulero-Palencia, S., Monzon Baeza, V., 2023. Detection of vulnerabilities in smart buildings using the Shodan tool. Electronics 12 (23), 4815. Available at https://doi.org/10.3390/electronics12234815.

Nieto Cubero, J., Gbadegeshin, S.A., Consolación, C., 2021. Commercialization of disruptive innovations: literature review and proposal for a process framework. International Journal of Innovation Studies 5 (3), 127–144. Available at https://doi.org/10.1016/j.ijis.2021.07.001, www.keaipublishing.com/en/journals/international-journal-of-innovation-studies/.

Nour, B., Sharif, K., Li, F., Biswas, S., Moungla, H., Guizani, M., et al., 2019. A survey of internet of things communication using ICN: a use case perspective. Computer Communications 142-143, 95–123. Available at https://doi.org/10.1016/j.comcom.2019.05.010.

Othman, H., Sieves, G., Guimarães, T., Azari, R., 2026. A calibration chamber framework for low-cost indoor air quality sensor validation. Building and Environment 287. Available at https://doi.org/10.1016/j.buildenv.2025.113856, https://www.sciencedirect.com/science/journal/03601323.

Pantelic, J., Nazarian, N., Miller, C., Meggers, F., Lee, J.K.W., Licina, D., 2022. Transformational IoT sensing for air pollution and thermal exposures. Frontiers in Built Environment 8. Available at https://doi.org/10.3389/fbuil.2022.971523, journal.frontiersin.org/journal/built-environment.

Peiris, S., Lai, J.H.K., Kumaraswamy, M.M., Hou, H., 2023. Smart retrofitting for existing buildings: state of the art and future research directions. Journal of Building Engineering 76. Available at https://doi.org/10.1016/j.jobe.2023.107354, http://www.journals.elsevier.com/journal-of-building-engineering/.

Shah, R., Doss, A.S.A., Lakshmaiya, N., 2025. Advancements in AI-enhanced collaborative robotics: towards safer, smarter, and human-centric industrial automation. Results in Engineering 27, 105704. Available at https://doi.org/10.1016/j.rineng.2025.105704.

Tariq, S., Loy-Benitez, J., Yoo, C.K., 2024. Multi-sensor fault detection and correction for automated IAQ monitoring in smart buildings through attention-aware autoencoders with spatial prediction module. Journal of Building Engineering 96, 110573. Available at https://doi.org/10.1016/j.jobe.2024.110573.

Toderean, L., Cioara, T., Anghel, I., Sarmas, E., Michalakopoulos, V., Marinakis, V., 2025. Demand response optimization for smart grid integrated buildings: review of technology enablers landscape and innovation challenges. Energy and Buildings 326, 115067. Available at https://doi.org/10.1016/j.enbuild.2024.115067.

Umeike, R., Dao, T., Crawford, S., 2025. State-of-the-art review of resilient smart cities: progress and challenges. Urban Governance 5 (2), 222–239. Available at https://doi.org/10.1016/j.ugj.2025.05.007.

Xenakis, A., Arsenopoulos, A., Papias, I., Serepas, F., Rizou, S., Psarras, J., 2025. Advancing smart building upgrades through an innovative decision support framework for smart readiness indicator implementation: a greek case study. Building and Environment 281. Available at https://doi.org/10.1016/j.buildenv.2025.113189, https://www.sciencedirect.com/science/journal/03601323.

Conclusions and future perspectives

14

This book has presented a comprehensive and systematic exploration of the convergence between architectural design, structural engineering, and smart systems, positioning smart structures not as an auxiliary technological enhancement but as a fundamental paradigm shift in the conception, design, and operation of the built environment. Across the chapters, architecture is progressively redefined as an adaptive, intelligent, and performative system—one that continuously interacts with environmental conditions, structural demands, and human behaviors.

The central contribution of the book lies in demonstrating that contemporary architecture is no longer constrained by static form or deterministic structural logic. Instead, it is increasingly governed by dynamic interactions between materials, computation, environmental forces, and real-time data feedback systems. This shift is consistently illustrated through interconnected themes such as smart materials, kinetic systems, adaptive envelopes, computational design, and interdisciplinary collaboration. Collectively, the chapters establish a unified conceptual framework in which architecture operates as a cyber-physical ecosystem, integrating sensing, actuation, computation, and human experience into a continuous loop of performance optimization and spatial adaptation.

A key trajectory emerging from the book is the transition from static, object-based architecture toward responsive and adaptive systems. Traditional architectural paradigms prioritize permanence, rigidity, and fixed performance assumptions. In contrast, the systems discussed throughout the chapters—ranging from kinetic structures to smart façades and tensegrity frameworks—demonstrate that adaptability is now a core design principle.

This transformation is not only technological but also conceptual. Buildings are no longer understood as finished artifacts but as evolving systems that respond to external stimuli such as climate variation, occupancy patterns, and energy demands. Adaptive building envelopes, kinetic mechanisms, and sensor-driven control systems collectively redefine architectural performance as a continuous process rather than a static outcome.

The book consistently highlights the convergence of structural engineering principles with material innovation and computational intelligence. Structural systems such as tensegrity frameworks illustrate how efficiency and elegance emerge from optimizing tension and compression rather than mass and rigidity. Similarly, smart materials—including shape memory alloys, electrochromic glazing, and phase change materials—extend structural behavior into the realm of responsiveness. When combined with embedded sensors, actuators, and data-driven control systems, these material and structural innovations form the basis of intelligent architecture.

Smart Civil and Structural Engineering. DOI: https://doi.org/10.1016/B978-0-443-30223-7.00012-7

This integration enables buildings to monitor their own performance, adjust to changing conditions, and optimize energy and environmental behavior in real-time.

Another major conclusion is the critical role of computational design methodologies in enabling the complexity inherent in smart and adaptive structures. Parametric modeling, generative algorithms, building information modeling (BIM) environments, and AI-driven design processes collectively allow architects and engineers to navigate multivariable performance criteria.

Rather than treating geometry, structure, and performance as separate domains, computational design integrates them into unified systems of interdependent parameters. This shift has enabled the realization of highly complex projects discussed in the book, where form emerges directly from performance logic, environmental constraints, and structural optimization. In this sense, computation is not merely a tool but a design medium that mediates between conceptual intent and physical realization. Across all chapters, collaboration emerges as a structural necessity rather than a procedural choice. The increasing complexity of smart structures requires the integration of architects, engineers, material scientists, computational designers, and data specialists from the earliest design stages.

Case studies such as large-scale cultural landmarks and high-performance buildings demonstrate that successful outcomes depend on shared digital environments, interdisciplinary workflows, and iterative feedback loops. The breakdown of disciplinary silos enables the synthesis of structural efficiency, environmental performance, and architectural expression. Thus, collaboration is not an external factor but an internal condition of contemporary architectural production.

Sustainability in this book is positioned beyond conventional energy efficiency metrics. Instead, it is framed as a regenerative and adaptive condition in which buildings actively contribute to environmental balance, resource optimization, and long-term ecological resilience. Smart structures extend sustainability through continuous monitoring, adaptive control, and lifecycle optimization. Buildings become active participants in environmental systems rather than passive consumers of resources. The integration of renewable energy systems, circular material strategies, and adaptive environmental controls demonstrates a shift from reduced impact toward positive impact design strategies. This positions smart architecture as a key agent in addressing global challenges such as climate change, resource scarcity, and urban density. The book also demonstrates that technological integration does not diminish architectural aesthetics but fundamentally redefines it. Form is no longer fixed or purely symbolic; it becomes temporal, responsive, and performance-driven.

Aesthetic value emerges from transformation, movement, and environmental interaction. Kinetic façades, adaptive shading systems, and parametric geometries generate dynamic visual expressions that evolve over time. In this context, beauty is no longer static but operational—embedded in the behavior of systems rather than only in their appearance. This represents a shift from object-based aesthetics to system-based aesthetics. As buildings become data-rich and algorithmically controlled environments, ethical considerations become central. Issues of data ownership, privacy, cybersecurity, transparency, and user autonomy define the governance landscape of smart architecture.

The book highlights that technological advancement must be accompanied by clear ethical frameworks that define who controls building data, how it is used, and how occupants are protected. Without such frameworks, the risks of surveillance, misuse of information, and technological inequality increase significantly. Therefore ethical governance is not an optional layer but a foundational requirement for the deployment of smart structures.

The future of architectural design and structural engineering, as outlined in this book, is shaped by several converging trajectories:

- Smart structures will increasingly evolve into fully autonomous systems supported by artificial intelligence, real-time simulation, and predictive analytics. Digital twins will enable continuous feedback between physical buildings and their virtual counterparts, allowing for real-time optimization of performance and maintenance.
- Robotic fabrication and additive manufacturing will redefine construction processes, enabling greater precision, reduced waste, and increased customization. Materials will become more adaptive, multifunctional, and bio-integrated, further blurring the boundaries between structure, environment, and computation.
- At the urban scale, buildings will no longer function as isolated entities but as interconnected nodes within intelligent infrastructures, sharing data, energy, and environmental responses.
- Ultimately, architecture will transition toward a fully integrated ecological-technological system, where design, operation, and adaptation occur continuously over the building lifecycle.

The overarching vision of this book is the redefinition of architecture as an intelligent, adaptive, and collaborative discipline. Smart structures are not simply technological enhancements but represent a fundamental rethinking of how buildings are conceived, constructed, and experienced. By integrating structural intelligence, computational design, material innovation, sustainability principles, and ethical governance, architecture moves toward a future in which buildings are no longer static objects but active participants in environmental and human systems. This transformation positions architecture at the intersection of technology, ecology, and human experience—offering a pathway toward more resilient, responsive, and sustainable built environments.

Index

Note: Page numbers followed by "*f*" and "*t*" refer to figures and tables, respectively.

Printed and bound by CPI Group (UK) Ltd, Croydon, CR0 4YY
22/09/2026
02220473-0002